To my wife, Annette

ESSENTIALS OF
HUMAN PHYSIOLOGY

Owl

PHILLIP SHEELER
California State University, Northridge

WCB Wm. C. Brown Publishers

Dubuque, IA   Bogota   Boston   Buenos Aires   Caracas   Chicago
Guilford, CT   London   Madrid   Mexico City   Sydney   Toronto

Book Team

Editor *Colin H. Wheatley*
Developmental Editor *Kristine Noel*
Production Editor *Kay Driscoll*

 Wm. C. Brown Publishers

President and Chief Executive Officer *Beverly Kolz*
Vice President, Publisher *Kevin Kane*
Vice President, Director of Sales and Marketing *Virginia S. Moffat*
Vice President, Director of Production *Colleen A. Yonda*
National Sales Manager *Douglas J. DiNardo*
Marketing Manager *Craig S. Marty*
Advertising Manager *Janelle Keeffer*
Production Editorial Manager *Renée Menne*
Publishing Services Manager *Karen J. Slaght*
Royalty/Permissions Manager *Connie Allendorf*

 A Times Mirror Company

Cover design by National Graphics

Cover photo by Becker/Custom Medical Stock Photo

Library of Congress Catalog Card Number: 94–73280

ISBN 0–697–26058–5

Printed in the United States of America by Times Mirror Higher Education Group, Inc.,
2460 Kerper Boulevard, Dubuque, IA 52001

10   9   8   7   6   5   4   3   2   1

ESSENTIALS OF
HUMAN PHYSIOLOGY

TABLE
OF
CONTENTS

CHAPTER 6

CHAPTER 7

CHAPTER 10

CHAPTER 11

CHAPTER 15

APPENDIX I

APPENDIX II

USING THE EHP © FOR WINDOWS SOFTWARE 439

APPENDIX III

ANSWERS TO SELF TEST QUESTIONS 445

INDEX 449

PREFACE

I have been teaching the Human Physiology course at California State University, Northridge for nearly 30 years. At this institution, the course is designed for students majoring in a broad spectrum of fields, including health science, nutrition science, physical education, physical therapy, pre-nursing studies, psychology, and a number of related disciplines. The varied and usually modest scientific backgrounds of the students and the diversity of their career goals have been important considerations in planning and writing this book, especially when deciding on the breadth and depth of the book's coverage and what preliminary foundations ought to be set.

For a majority of students, human physiology is the last undergraduate science course in their curriculum (perhaps in their college careers), and this too has guided me in selecting the topics to be presented and the extent of their detail. I've deliberately focused on topics whose understanding will help young, maturing adults be better informed about their body's functions and make prudent decisions where their personal health and well-being (and society's too) are concerned.

Over the years, I've used many textbooks for my course, most of them quite good. In recent years, however, there has been a trend in human physiology texts toward broader and broader coverage and greater and greater detail. The result is textbooks of human physiology that are needlessly elaborate and in which the coverage is so extensive and detailed that only a fraction of the book's content is covered in a one-semester course. Human physiology texts approaching 1000 pages in length are not uncommon. Probably the most alarming aspect of this trend is that many students are overwhelmed or intimidated by the text's size and do not do the necessary reading. Moreover, some students share the purchase of a single book and some buy no textbook at all. The need for an inexpensive book that can and will be read by students has been the primary motivation in writing this book and reigning in its scope and depth by limiting its size to a little over 400 pages. I'm hopeful the result is a text that realistically interfaces with the lecture coverage of a one-semester human physiology course.

In writing the book, I've tried to simplify concepts as well as present them in a concise manner. By preparing all of the illustrations myself, I've ensured that the figures and the text complement one another.

As a brief look at the table of contents reveals, *Essentials of Human Physiology* takes a "systems" approach to the subject. That is to say, the human body is viewed as an integrated assemblage of different organ systems (such as the digestive system, the circulatory system, the excretory system, and so on). This is a classical approach and one that in my experience presents the subject in units that are easier to relate to, easier to conceptualize, and, most importantly, easier to understand. Throughout the text, however, I've tried to emphasize the interactions of the body's organ systems and their interdependence.

Essentials of Human Physiology has two particularly unusual features: (1) the book was prepared by the author using "desktop publishing" methods and (2) with each copy of the book, the student receives computer diskettes containing HyperCard[1] and Windows[2] software that augments the concepts presented in the text using *interactive* demonstrations and simple animations. Although the book is independent of the software, using the software materially assists the understanding of particular concepts, while at the same time giving the reader the opportunity to test his or her understanding.

To ensure that the software can be used effectively with minimum effort or computer expertise on the part of the student, clicking the computer's mouse at the appropriate time and with the cursor positioned at the appropriate location on the computer screen is the only requirement for using the interactive software. Students are not asked to enter text or data via the keyboard. There is no way to harm the software, even if the computer crashes or suffers a power loss during use. Once the computer is re-started, the software will function normally. The software also includes traditional multiple

choice and true-false tests that challenge the student's understanding of the text. Guides to the installation and use of the software are presented in Appendix I (for HyperCard) and Appendix II (for Windows), at the back of the book.

I should like to express my appreciation to a number of people at Wm. C. Brown Publishers for their help with this project. Thanks are extended to my editor Colin Wheatley for his support and enthusiasm for the project, and to developmental editor Kris Noel, production editor Kay Driscoll, and copyeditors Kennie Harris and Kay J. Brimeyer. My thanks are also due to Hal Peters of Educational Software Products (Iowa City, Iowa) for preparing the Windows version of the interactive software.

My appreciation is also extended to a number of my colleagues at CSUN who reviewed and critiqued selected chapters of the book; my thanks go to Professor Joseph Moore, Professor Mary Lee Sparling, Professor Linda Caren, Professor Anthony Gaudin, and Professor Randy Cohen. My thanks are due also to Professor John McGill (Alpena Community College), Professor C. Thomas Wiltshire (Culver-Stockton College), and Professor John P. Harley (Eastern Kentucky University) for their reviews. My thanks also to Professor George Bloom (University of Minnesota) for comments on the illustrations. I should like also to thank CSUN graduate student Cynthia Lee Hockman who patiently copyedited the manuscript and rigorously tested the HyperCard software. Finally, I extend my thanks in advance to readers who may bring errors to my attention.

Phillip Sheeler
Northridge, California
May, 1995

---

[1] HyperCard is a trademark of Apple Computer, Inc.

[2] Windows is a trademark of Microsoft Corporation.

CHAPTER 1

INTRODUCTION

ORGANIZATION OF THE BODY

Chapter Outline

Levels of Organization of the Body
  Organ Systems
  Tissues and Cells
Body Regions and Spatial Planes
  Major Body Regions
  Spatial Planes and Relative
    Movements of the Body
Homeostasis

Physiology is the branch of biological science that attempts to explain in chemical, physical, and molecular terms the multitude of phenomena that are displayed by living things. Physiology has several branches of its own, including *animal physiology, plant physiology*, and *microbial physiology*. This book is concerned with a specific aspect of animal physiology, namely *human physiology*–the study of how the human body works.

LEVELS OF ORGANIZATION OF THE BODY

Organ Systems

The human body is an extremely complex structure. To simplify the study of the body's physiology, it is helpful to subdivide the body into a number of different functional parts, each of which can then be considered separately. The major functional subdivisions of the human body are the body's so-called **organ systems**, each organ system having a rather specialized function. The body's major organ systems are listed in table 1.1.

Each organ system also has a hierarchy of structural and functional parts. Organ systems are comprised of a number of **organs**. For example, the eyes and ears are organs of the receptor system; the heart, arteries, and veins are organs of the circulatory system; the stomach and small intestine are organs of the digestive system; and the kidneys and urinary bladder are organs of the excretory system.

TABLE 1.1	THE BODY'S MAJOR ORGAN SYSTEMS*
Muscle system	
Nervous system	
Receptor system	
Circulatory system	
Immune system	
Respiratory system	
Digestive system	
Excretory system	
Endocrine system	
Reproductive system	

* Some physiologists consider the skin (or integument) a separate organ system. However, we will consider the functions of the skin in connection with other organ systems.

Tissues and Cells

Each of the organs of an organ system is formed by an assemblage of **tissues**, each tissue contributing in a particular way to the organ's overall function. For example, the stomach contains *muscle tissue* (which is responsible for contractions and other movements), *epithelial tissue* (which produces and releases the stomach's digestive enzymes and other secretions), *connective tissue* (which provides the organ's structural integrity), and *nerve tissue* (which carries information between the stomach and the brain in the form of nerve impulses). The various kinds of tissues that comprise the body's organs are listed in table 1.2.

Each tissue is made up of large numbers of individual **cells**. The cells of each tissue share properties that are common to the cells of other tissues, but they also possess tissue-specific properties (e.g., muscle cells *contract*; nerve cells *conduct* impulses; endocrine cells *secrete* hormones; and so on).

Even cells can be subdivided into distinct structural and functional components called the subcellular **organelles** (e.g., nucleus, mitochondria, and ribosomes). The organization and functions of the cellular organelles are reviewed in chapter 3. Finally, each organelle is comprised of a specific array of **molecular** (and **atomic**) constituents.

Thus, in order of decreasing scope and increasing organizational and functional specificity, the levels of organization of the human body may be described as follows:

Whole Body
    Organ Systems
        Organs
            Tissues
                Cells
                    Organelles
                        Molecules

Most of the remaining chapters in this book deal with the organization and functions of the tissues and organs that make up the body's organ systems.

BODY REGIONS AND SPATIAL PLANES

Major Body Regions

The subdivision of the body into a number of organ systems is based on physiological (i.e., functional) distinctions. However, the body can also be subdivided into regions based on position or anatomical location. Even though we will not be concentrating on anatomy, it is helpful (and important) to be familiar with the body's general anatomical plan and how one region of the body is described in relation to other regions.

TABLE 1.2   THE BODY'S TISSUES

TISSUE	SUBTYPES	FUNCTIONS
Muscle	Striated	Contraction
	Smooth	Contraction
	Cardiac	Contraction
Nervous	Neuronal	Conduction and transmission
	Neuroglial	Support
Epithelial	Mucous membrane	Absorption and/or secretion
	Serous membrane	Secretion; lining of organs
	Endothelium	Lining of vessels
	Glandular epithelium	Secretion
Connective	Adipose	Fat storage
	Cartilage	Support
	Bone	Support
	Blood	Gas transport and immunity

The major body regions are (1) the **head** (**cranial** and **facial** subdivisions), (2) the **neck** (or **cervical** region), (3) the **trunk** (**thoracic**, **abdominal,** and **pelvic** subdivisions), (4) the right and left **upper limbs**, and (5) the right and left **lower limbs**. These regions are depicted in figure 1.1 and are listed in table 1.3.

In addition to the major body regions, one may distinguish the **dorsal** surface of the body (i.e., the rear or **posterior** surface) and the body's **ventral** surface (i.e., front or **anterior**).

| 1-1 |

Spatial Planes and Relative Movements of the Body

Just as there are three planes in space (i.e., three *dimensions)*, the body has three spa-

tial planes; these are (1) the **frontal** (or **coronal**) plane; (2) the **median sagittal** (or **midsagittal)** plane; and (3) the **transverse** (or **horizontal**) plane (fig. 1.2). The frontal plane passes vertically through the body, dividing the body into front (anterior or ventral) and rear (posterior or dorsal) halves. The median sagittal plane also passes vertically through the body, but this plane divides the body into right and left halves. The transverse plane passes horizontally through the body from front to rear and divides the body into upper and lower halves.

| 1-2 |

Specific terms are used when describing spatial relationships among different parts of the body or when describing movement or progression through the body. For example, **anterior** movement implies forward movement through the body (e.g., from the dorsal

3

TABLE 1.3	THE BODY'S MAJOR REGIONS
MAJOR REGION(S)	SUBDIVISIONS
Head	Cranial region
	Facial region
Neck (cervical region)	
Trunk	Thoracic region
	Abdominal region
	Pelvic region
Upper Limbs	
Lower Limbs	

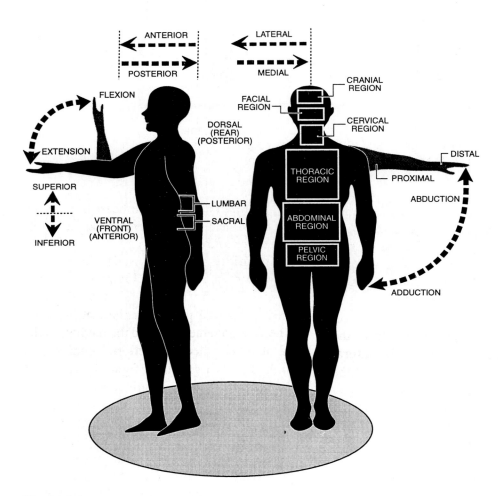

Figure 1.1
The body's major regions and spatial relationships.

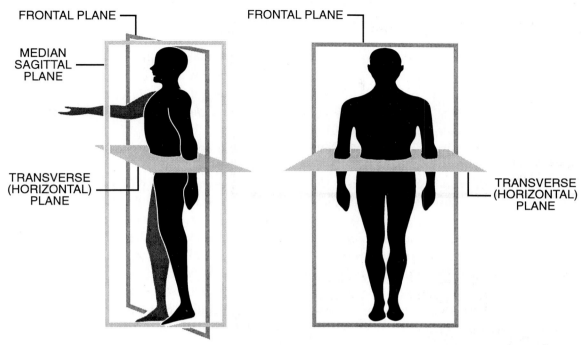

FRONTAL PLANE

MEDIAN SAGITTAL PLANE

TRANSVERSE (HORIZONTAL) PLANE

FRONTAL PLANE

TRANSVERSE (HORIZONTAL) PLANE

Figure 1.2
Like the three dimensions of space, the body has three spatial planes; these are (1) the frontal, (2) the median sagittal, and (3) the transverse plane. The median sagittal plane is not seen in the front view (right half of diagram).

surface to the ventral surface). **Posterior** movement implies backward movement through the body (e.g., from the ventral surface to the dorsal surface). In a similar manner, a part of the body that lies in front of another part is said to lie anterior (or ventral) to that part.

Movement to the right or left from the median sagittal plane is said to be **lateral** movement, whereas movement toward the median sagittal plane from either the right or left sides is said to be **medial** movement. In a like manner, the ears are positioned lateral to the nose (or, conversely, the nose is positioned medial to the ears).

Upward progression in a direction that is perpendicular to the transverse plane is said to be **superior**, whereas downward progres-

sion is **inferior**. Thus, the heart is superior to the stomach (or, conversely, the stomach is inferior to the heart).

Specific terms are also used to describe the movements of the limbs. Movement of a limb (e.g., the left arm in figure 1.1) away from the median sagittal plane is called **abduction**, whereas movement toward the median sagittal plane is called **adduction**. Bending of a limb (e.g., the right arm in figure 1.1) is called **flexion,** whereas the straightening of a limb is called **extension**. The nearness of one part of the body relative to other parts is reflected in the terms **proximal** and **distal** (again see figure 1.1). For example, the upper arm is proximal to the shoulder, whereas the lower arm is distal to the shoulder. These and other terms of spatial and anatomical position will become increasingly familiar to you as the actions of the body's parts are considered in detail in subsequent chapters.

HOMEOSTASIS

The basic structural and functional units of the body are its cells, and it is estimated that the body of the average adult is comprised of *thousands of billions* of cells. Each one of these cells is a living, functioning entity that attends to the metabolic activities that sustain its own life as well as those that contribute to the specific functions of the tissue and organ of which the cell is a part.

In order to play its assigned role, each cell must have access to all of the nutrients and raw materials that it needs for its metabolism and must rid itself of the wastes that are produced by this metabolism. Accordingly, each cell of the body is bathed in a fluid (called **tissue fluid** or **extracellular fluid**) from which it withdraws its material needs and into which it empties its wastes.

If the fluid that bathes the surfaces of the body's cells were fixed, then this fluid would soon be depleted of nutrients and raw materials and would quickly accumulate large amounts of cellular waste. Clearly, there must be a continuous turnover of the composition of this fluid, so that fresh materials replace those that are withdrawn by the cells and wastes are carried away. In this way, each cell's external environment (i.e., the tissue fluid) is able to remain constant.

Each of the body's organ systems contributes in some fashion to maintaining the constancy of the tissue fluid (fig. 1.3). The principal role is played by the **circulatory system**, which, by circulating the blood, brings a continuous supply of fresh materials to the body's tissues and carries away cellular wastes. The ongoing exchanges that take

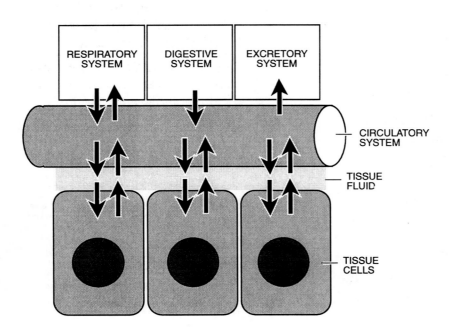

Figure 1.3
During their metabolic activity, the cells of the body acquire nutrients from the tissue fluid that bathes them. At the same time, cellular wastes and carbon dioxide are emptied into the tissue fluid. The circulatory, respiratory, digestive, and excretory systems act to maintain the chemical constancy of the tissue fluid by removing the cellular wastes and replenishing depleted nutrients.

place between the bloodstream and the tissue fluid ensure that the composition of tissue fluid hardly changes at all over time. The constancy of the cellular environment that results from this activity is called **homeostasis**.

Other organ systems maintain homeostasis by contributing to the blood's constancy. For example, the **respiratory system** transfers oxygen from the surrounding atmosphere into the bloodstream and thereby replaces oxygen that was withdrawn from the blood as it circulated through the body's tissues. At the same time, carbon dioxide picked up by the blood from the body's tissues is expelled into the atmosphere.

By digesting food and transferring the products of this digestion to the bloodstream, the **digestive system** replenishes the nutrients and raw materials that are removed from the blood by the body's tissues. Wastes produced during metabolism are emptied into the tissue fluid and pass from there into the bloodstream. By removing these wastes from the blood and conveying them out of the body, the **excretory system** contributes to the maintenance of the blood's constancy.

SELF TEST[*]

True/False Questions

1. The major functional subdivisions of the body are the body's organ systems.

2. Although the cells of all body tissues share some features in common, they also possess tissue-specific properties.

3. The median-sagittal plane divides the body into dorsal and ventral halves.

4. It is estimated that the body of an adult contains about 100 million cells.

5. The body's tissue fluid mediates the exchange of nutrients and wastes between tissue cells and the bloodstream.

Multiple Choice Questions

1. In humans, the dorsal surface of the body may be equated with the (A) anterior surface, (B) posterior surface, (C) ventral surface, (D) lateral surface.

2. Cells may be subdivided into a number of different functional components; these components are called (A) organs, (B) organelles, (C) tissues, (D) molecules.

3. The horizontal plane divides the body into (A) right and left halves, (B) front and rear halves, (C) upper and lower halves.

4. The bending of an arm or leg is known as (A) medial movement, (B) lateral movement, (C) flexion, (D) extension.

5. The immediate source of oxygen for liver and kidney cells is the (A) lungs, (B) circulating blood, (C) tissue fluid, (D) respiratory organelles.

---

[*] *The answers to these test questions are found in Appendix III at the back of the book.*

CHAPTER 2

CHEMISTRY OF THE BODY

Although the cells of one tissue may differ in form and function from the cells of other tissues, all of the body's parts are fundamentally similar in their chemical composition. In this chapter, we will explore the body's chemistry in order to determine the nature, properties, and actions of the body's chemical components.

ELEMENTS AND THE STRUCTURE OF ATOMS

Elements are defined as the *fundamental units of matter*. Although about 106 different elements occur in nature, only a small number of these are found to any appreciable extent in the cells and tissues of the human body. The principal elements found in the body's tissues are listed in table 2.1; among these *hydrogen*, *oxygen*, *carbon*, and *nitrogen* are overwhelmingly the most common. Each element is identified by a one-letter or two-letter symbol as well as a name.

Elements are said to be "fundamental" in that they cannot be changed into other elements as a result of conventional chemical reactions. To be sure, there are instances in which one element is changed into another as the result of *radioactive decay*; however, such changes involve the atomic nucleus and do not fall within the realm of chemical reactions that characterize living things. (The chemical reactions that characterize living things involve changes in the electrons surrounding the nucleus of an atom and do not involve intranuclear rearrangements.)

An **atom** is the basic unit of an element and is, in turn, comprised of *subatomic* (or *elementary*) *particles*. The atoms of all known elements are formed from the same subatomic particles and differ only in the numbers of these particles. Insofar as human physiology is concerned, three kinds of subatomic particles take on importance– **protons**, **neutrons**, and **electrons**. It is the electrons that participate in the chemical reactions that characterize a cell's (and ultimately the body's) activities.

Of the three types of subatomic particles, the protons and neutrons are found in the "core" or **nucleus** at the center of the atom, whereas the electrons are found some distance from the nucleus (for convenience, they may be thought of as orbiting the nucleus, much as the planets orbit the sun; fig. 2.1). The protons possess *positive* electrical charge, whereas the neutrons are electrically *neutral*; thus an atomic nucleus carries positive charge. In contrast, the electrons that surround the nucleus are *negatively*

TABLE 2.1	CHEMICAL COMPOSITION OF THE BODY	
ELEMENT	SYMBOL	PERCENT*
Major Elements		
Hydrogen	H	10
Oxygen	O	65
Carbon	C	19
Nitrogen	N	3
Minor Elements		
Calcium	Ca	<1.0
Phosphorus	P	<1.0
Potassium	K	<0.4
Sulfur	S	<0.2
Sodium	Na	<0.2
Chlorine	Cl	<0.1
Magnesium	Mg	<0.1
Iodine	I	<0.1
Iron	Fe	<0.1
Manganese	Mn	<0.1

* Percent expressed in terms of the body's total weight.

Figure 2.1
Model of the structure of an atom. The nucleus of the atom contains protons and neutrons. The protons carry positive charges, whereas the neutrons are electrically neutral. Around the nucleus orbit a number of electrons, each electron carrying a negative charge.

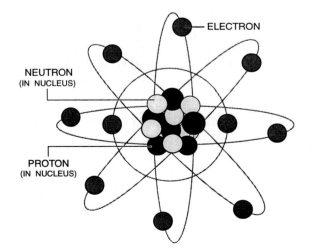

10

charged. Because like charges repel one another and opposite charges attract one another, there is a force attracting the negatively-charged electrons of an atom to the positively-charged nucleus.

In a *complete* atom, the number of electrons around the nucleus is equal to the number of protons in the nucleus. Therefore, the atom as a whole is electrically neutral. To gain some insight into the relative size of the atom and the space occupied by its parts, consider the following analogy. If the nucleus of a carbon atom were magnified to the size of a baseball, then the outermost electrons would be about 100 yards away (the length of a football field). Thus, it is clear that an atom is largely empty space.

The electrons of an atom occur in layers or **shells** at specific distances from the atomic nucleus. Each shell can accommodate a certain maximum number of electrons, and the electrons of each shell are said to possess a certain amount of energy. The energy of an electron depends upon the shell in which it resides, the energy increasing in relation to the shell's distance from the nucleus. The innermost electron shell (called the *first* or **K shell**) can contain up to 2 electrons; K-shell electrons have the lowest energy. The *second* shell (called the **L shell**) can accommodate up to 8 electrons (and these have a higher energy than K-shell electrons). The *third* shell (called the **M shell**) accommodates up to 18 electrons. Succeeding shells accommodate either the same or a greater number of electrons than the shell before (i.e., the order is 2, 8, 18, 32, etc.). Larger and larger atoms contain correspondingly greater numbers of electrons. Figure 2.2 shows the distributions of electrons among the first, second, and third shells of atoms of hydrogen, carbon, nitrogen, oxygen, sodium, and chlorine.

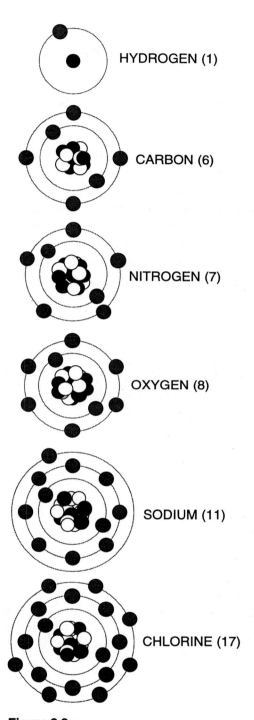

Figure 2.2
The distributions of electrons among the first, second, and third (K, L, and M) shells of atoms of hydrogen, carbon, nitrogen, oxygen, sodium, and chlorine.

The electrons in the shells that surround an atom's nucleus are apportioned among a number of **subshells**. The innermost shell (the K shell) has only one subshell called the *1s* subshell. The second shell (the L shell) has two subshells called *2s* and *2p*. The third shell (the M shell) has three subshells called *3s*, *3p*, and *3d*. The 1s, 2s, and 3s subshells can accommodate up to 2 electrons; the 2p and 3p subshells can accommodate up to 6 electrons; and the 3d subshell accommodates up to 10 electrons. The relationship between K, L, and M shells, their subshells, and electron distributions is summarized in figure 2.3.

$\boxed{\text{2-1}}$

ATOMIC NUMBER AND ATOMIC WEIGHT

What distinguishes the atoms of one particular element from the atoms of another element is the number of protons found in the atom's nucleus. For example, all carbon at-

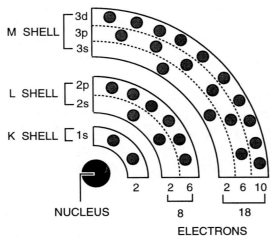

Figure 2.3
The K, L, and M shells, their subshells, and respective numbers of electrons.

oms have 6 protons in their nucleus, whereas all sulfur atoms have 16 protons in their nucleus. Since it was noted earlier that there are about 106 different elements, atoms have from 1 to 106 protons in their nucleus.

The number of nuclear protons in an atom is called the **atomic number**, and this number differs for all of the known elements. For whole atoms, the atomic number is also the number of electrons in the shells that surround the atom's nucleus. Listed in table 2.2 are the atomic numbers and electron distributions of a number of elements found in the human body. As you examine this table, notice that all of the elements listed have partially filled outermost shells. For example, hydrogen atoms have only one electron in their K shell. Since the K shell can accommodate two electrons, there is room for one more electron.

Carbon's K shell is filled and so is the 2s subshell of carbon's L shell; however, the 2p subshell has only two electrons in it, leaving room for four more. An iron atom's K and L shells are filled and so are the 3s and 3p subshells of iron's M shell, but the 3d subshell has only 8 electrons in it. Therefore, two more electrons can be accommodated in the 3d subshell of an iron atom. As you will soon see, when the atoms of an element have partially-filled outer shells, the element is *reactive*. Partially filled shells are nearly always the outermost ones, and an atom never has more than one partially filled shell.

Not listed in table 2.2 are atoms whose outermost shells are filled. However, such atoms do exist and include the gases *helium* (He) and *argon* (Ar). Because their outermost shells are filled, these gases are chemically unreactive and are said to be *inert*.

TABLE 2.2 ATOMIC NUMBERS, ATOMIC WEIGHTS, AND ELECTRON DISTRIBUTIONS OF ELEMENTS FOUND IN THE BODY'S TISSUES

ELEMENT	ATOMIC NUMBER	ATOMIC WEIGHT	K	L		M			N			
			1s (2)	2s (2)	2p (6)	3s (2)	3p (6)	3d (10)	4s (2)	4p (6)	4d (10)	4f (14)
Hydrogen	1	1	1									
Carbon	6	12	2	2	2							
Nitrogen	7	14	2	2	3							
Oxygen	8	16	2	2	4							
Sodium	11	23	2	2	6	1						
Magnesium	12	24	2	2	6	2						
Phosphorous	15	31	2	2	6	2	3					
Sulfur	16	32	2	2	6	2	4					
Chlorine	17	35	2	2	6	2	5					
Potassium	19	39	2	2	6	2	6	1				
Calcium	20	40	2	2	6	2	6	2				
Manganese	25	55	2	2	6	2	6	7				
Iron	26	56	2	2	6	2	6	8				
Iodine	53	127	2	2	6	2	6	10	2	6	10	7

* For each of the subshells (e.g., 2s, 2p, 2d, etc.), the maximum number of allowable electrons is shown below in parentheses.

The weight (or mass) of an atom is referred to as its **atomic weight** (or *atomic mass*) and is the sum of the weights of the atom's protons, neutrons, and electrons. Since an electron weighs only about 1/2000th as much as a proton or neutron, the atomic mass contributed by electrons is exceedingly small. The combined weights of a proton and electron are equal to one **dalton** (or one *atomic mass unit*). A neutron also weighs one dalton. Therefore, the atomic weight of an atom, expressed in daltons, is the sum of the numbers of protons and neutrons in the atom's nucleus.

Isotopes

Nearly all elements possess different **isotopic** forms. That is, while all atoms of a given element have the same number of protons in their nucleus, not all atoms of the same element have the same number of neutrons. For example, all carbon atoms contain 6 nuclear protons, but carbon atoms exist in nature that contain 5, 6, 7, or 8 neutrons in their nuclei. All are said to be **isotopes** of carbon. Each isotope of an element has a different atomic weight because of the different numbers of neutrons in its nucleus.

The isotopes of carbon we just mentioned have atomic weights of 11 (i.e., 6 protons + 5 neutrons), 12 (i.e., 6 protons + 6 neutrons), 13 (i.e., 6 protons + 7 neutrons), and 14 (i.e., 6 protons + 8 neutrons). To specifically identify a particular isotope of an element, the element's symbol is preceded by a superscript indicating the atomic weight. Thus, ^{14}C is the isotope of carbon that contains 8 neutrons in its nucleus, whereas ^{12}C is the isotope of carbon that contains 6 neutrons in its nucleus.

Usually, one isotope of an element occurs on the earth in far greater abundance than any of the element's other isotopes. In the case of carbon, the most abundant isotope is ^{12}C. Some isotopes of an element may be physically unstable, and in their transition to a stable state, they may release energy in the form of *rays*. For example, ^{14}C is a radioactive isotope of carbon that emits *beta rays*. ^{59}Fe is a radioactive isotope of iron that emits both beta rays and *gamma rays*. Radioactive isotopes are especially valuable research tools because they can be detected and traced even when they are mixed with other, stable isotopes of the same element. Consequently, they are often employed as tracers when studying the chemical reactions that take place in cells and tissues.

MOLECULES AND CHEMICAL BONDS

When two or more atoms of the same or different elements are linked together, a **molecule** is formed. The linkages between the atoms of these molecules are referred to as **chemical bonds.** Although molecules can be comprised of two or more atoms of the same element (e.g., hydrogen gas, oxygen gas, and ozone), molecules are usually comprised of more than one kind of element. In the latter instance, the molecules are also called **compounds**. Water, the most abundant substance in the human body, is a compound that is formed from one atom of oxygen and two atoms of hydrogen.

Valence

Each type of atom can combine with a specific *maximum* number of other atoms. For example, a hydrogen atom can combine with only *one* other atom; an oxygen atom can combine with up to *two* atoms; and a carbon atom can combine with as many as *four* atoms. The number of additional atoms with which a single atom can form chemical bonds is known as **valence**; accordingly, hydrogen atoms have a valence of one, oxygen atoms a valence of two, and carbon atoms a valence of four.

An atom's valence is based on the distribution of electrons in its outer shells. Elements whose outermost shells (and subshells) are not filled may be thought of as either being "short of electrons" or as "having electrons to spare." Let's consider a few examples that illustrate this concept. Oxygen atoms (see table 2.2) have 8 electrons; these fill the K shell and the 2s subshell of the L shell, but they occupy only four of the 6 positions in the 2p subshell. Thus, because oxygen is 2 electrons short of filling its 2p subshell, its valence is 2. Now consider nitrogen (again see table 2.2). Nitrogen atoms have 7 electrons; these fill the K shell and the 2s subshell of the L shell, but occupy only three of the 6 positions in the 2p subshell. Thus, because nitrogen is 3 electrons short of filling its 2p subshell, its valence is 3.

In both of the above examples, the valence is based upon the electron deficiency of an outer subshell. However, valence can also be based upon an electron excess. Consider, for example, sodium. Sodium atoms have 11 electrons; 10 of these fill the K and L shells, but the M shell contains only one electron (it is in the 3s subshell). If sodium had only 10 electrons (i.e., if there were no 3s electron), sodium would have no incomplete shells. Thus, the solitary 3s electron of the sodium atom may be thought of as an excess electron. Because there is only one "excess" electron, sodium's valence is 1. Finally, let's consider calcium atoms, which contain 20 electrons. Ten electrons fill the K and L shells and 8 of the remaining 10 electrons fill the 3s and 3p subshells of the M shell. Calcium's 3d subshell, which can accommodate 8 electrons, contains the two remaining electrons. If calcium had only 18 electrons (i.e., if there were no 3d subshell electrons), then there would be no partially filled subshells. Thus, calcium may be thought of as having 2 electrons in excess and, therefore, a valence of 2.

Chemical Bonds

From the preceding discussion it should be clear that the valence of an atom is based upon (1) the number of electrons that are needed to fill the atom's outermost shell or subshell, or (2) the number of electrons that an atom would have to lose in order to leave the atom with a filled outer shell or subshell. Accordingly, hydrogen and sodium have a valence 1, oxygen and calcium a valence of 2, and so on. An atom's valence, in turn, determines the number of bonds that can be formed with other atoms (i.e., hydrogen and sodium can form one bond, oxygen and calcium can form two bonds, etc.).

The formation of **bonds** between atoms reflects the atoms' tendency to avoid partially filled outer shells and subshells. At-

oms can avoid partially filled outer shells and subshells in two ways: (1) by **ionic bonding** or (2) by **covalent bonding**. Ionic bonding involves the *transfer* of electrons from an atom with an outer shell surplus to an atom with an outer shell deficiency. Covalent bonding involves the *sharing* of outer shell electrons between two atoms.

Ionic Bonds and Ions

Ionic bonds are formed by the transfer of one or more outer-shell electrons from one atom to one or more other atoms. Atoms that receive electrons in such a transfer become more negative, while atoms donating electrons become more positive (or less negative). The bond that holds the two atoms together is the attractive force that exists between the two opposite charges.

A common example of an ionic bond is the one that occurs between sodium and chlorine in sodium chloride (NaCl) crystals (i.e., ordinary table salt; figure 2.4). Each sodium atom in a salt crystal gives up its single 3s electron to the 3p subshell of a chlorine atom. By accepting the electron, chlorine's 3p subshell becomes filled. As a

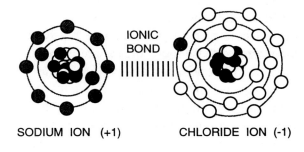

SODIUM ION (+1)   CHLORIDE ION (-1)

Figure 2.4
In salt crystals, ionic bonds are formed between the positively charged sodium ions and the negatively charged chloride ions.

result, neither the sodium nor the chlorine have any partially-filled outer subshells. By giving up an electron to chlorine, the sodium becomes positively charged, and by accepting an electron, the chlorine becomes negatively charged.

Atoms that have either gained electrons or lost electrons are called **ions**. A *positively charged* ion is called a **cation**, and a *negatively charged* ion is called an **anion**. In sodium chloride, the sodium is the cation (represented by the symbol Na^+) and the chloride is the anion (represented by the symbol Cl^-). Ions can also be comprised of more than one atom. In the salt *ammonium chloride* (i.e., NH_4Cl), the ammonium portion, which is composed of one nitrogen and four hydrogens (NH_4^+), becomes a cation when the atoms give up an electron to chlorine (Cl^-).

As another illustration, consider the case of the salt *calcium chloride* ($CaCl_2$) in which *two* electrons are transferred from calcium to the two chlorine atoms. Calcium has only two electrons in its outermost shell; one electron is transferred to each of the chlorines. Thus, calcium forms *two* ionic bonds–one with each of the two chlorines. Because two electrons are transferred, the calcium cation carries two positive charges and is represented by the symbol Ca^{++}.

Covalent Bonds

Another way that an atom can fill its outermost electron shell or subshell is through the formation of **covalent bonds**. Covalent bonds are not formed by the transfer of electrons, but by the *sharing* of electrons. A simple example of covalent bonding occurs in molecular hydrogen or H_2 (fig. 2.5). In molecular hydrogen, two hydrogen atoms share their 1s electrons (i.e., each hydrogen atom contributes its single 1s electron). As a result, the K shell of each atom has two electrons in it and, therefore, is filled. In water (H_2O), two hydrogen atoms share their 1s electrons with oxygen. Simultaneously, oxygen shares two of its 2p electrons with the hydrogen atoms (i.e., one electron for each hydrogen atom). As a result, the 2p subshell of oxygen is filled and so is the 1s subshell of each hydrogen atom (fig. 2.5). Covalent bonds that result from the sharing of one pair of electrons between two atoms (as in the cases of molecular hydrogen and water) are called **single bonds**. When depicting the chemical structure of a molecule, single bonds are usually represented by a single dash mark (i.e., water would be represented by H-O-H).

In a number of compounds, certain of the atoms share *more than one* pair of electrons in order to complete their outermost shells or subshells. Bonds formed by sharing more than one electron pair are called **multiple bonds**. For example, a **double bond** is one in which *two* pairs of electrons are shared; in a **triple bond**, *three* pairs of electrons are shared.

Gaseous oxygen occurs in the form of diatomic molecular oxygen (i.e., O_2). Because an oxygen atom has four 2p electrons, two more are needed in order to complete the 2p shell. In molecular oxygen, this occurs when two oxygen atoms share two pairs of electrons (i.e., one pair contributed by each oxygen atom; see figure 2.5). Thus, the oxygen atoms are linked by a double bond.

The shorthand for a double bond is two parallel dash marks. Therefore, an oxygen molecule may be depicted O=O. Another

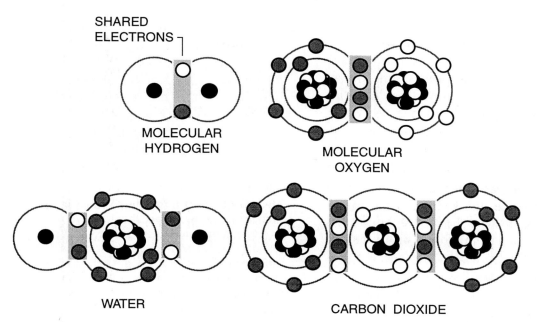

SHARED
ELECTRONS

MOLECULAR
HYDROGEN

MOLECULAR
OXYGEN

WATER

CARBON DIOXIDE

Figure 2.5
Covalent bonds. Covalent bonds are formed whenever two or more atoms share electrons, such that each atom's outermost subshell is filled. In molecular hydrogen, each hydrogen atom shares its single 1s electron. When two electrons are shared (as shown here for molecular hydrogen and water), the covalent bond is called a single bond. When four electrons are shared (as shown here for molecular oxygen and carbon dioxide), the covalent bond is called a double bond.

example of a molecule that contains double bonds is carbon dioxide (CO_2). As we have already seen, oxygen atoms have four 2p electrons. A carbon atom's 2p subshell contains only two electrons. Since the 2p subshell has a capacity of six electrons, if each oxygen atom shares a pair of its 2p electrons with the carbon atom, and the carbon atom simultaneously shares its two 2p electrons with each of two oxygen atoms, then the 2p subshells of all three atoms are completed (fig. 2.5). The carbon dioxide molecule may be depicted as O=C=O.

MOLECULAR WEIGHT

Just as atomic weight is the sum of the weights of the elementary particles of an atom, **molecular weight** (expressed in *daltons*.) is the sum of the weights of all of the atoms that form a molecule. For example, water's molecular weight is 18 daltons, which is the sum of the atomic weights of oxygen (16 daltons) and the two hydrogen atoms (one dalton each). The molecular weight of carbon dioxide (CO_2) is 44 daltons (12 + [2 x 16]), and the molecular weight of ammonia (NH_3) is 17 daltons (14 + [3 x 1]).

POLAR COVALENT BONDS

When two identical atoms form a covalent bond, the electrons are shared equally between the two atomic nuclei. When different atoms form a covalent bond, the electrons are not shared equally. Instead, one of the two atoms attracts the shared electrons more strongly than does the other atom. The attraction of an atom for electrons is called **electrophilia**. In the O-H bond (as occurs, for example, in a water molecule), the oxygen atom is more electrophilic than the hydrogen atom; in the N-H bond, the nitrogen atom is more electrophilic than the hydrogen atom.

The magnitude of the electrophilia that is possessed by an atom is based on the number of protons in the atom's nucleus. Accordingly, oxygen with its six nuclear protons is more electrophilic than hydrogen, which has only one proton. Nitrogen, with its seven nuclear protons, is also more electrophilic than hydrogen. Because shared electrons may be more strongly attracted to the nucleus of one atom than the other, the bond that the shared electrons create is said to be **polar**; that is, at one end of the bond the electrical charge is slightly different than at the other end.

The bonds between the oxygen atom and each of the two hydrogen atoms in water are examples of polar covalent bonds. In a water molecule, the shared electrons are more strongly attracted to the oxygen atom than they are to the hydrogen atoms. As a result, the shared electrons spend more time near the oxygen nucleus than they do near each hydrogen nucleus. Each hydrogen atom is therefore said to possess a *partial positive charge* (symbolized by δ^+), and the oxygen atom is said to possess a *partial negative charge* (that is, δ^-). Because the hydrogen atoms of a water molecule are not symmetrically distributed on either side of the oxygen atom, water molecules behave like **dipoles** (fig. 2.6).

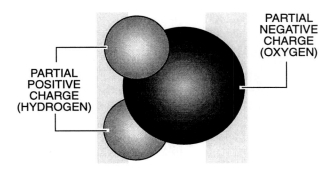

PARTIAL POSITIVE CHARGE (HYDROGEN)

PARTIAL NEGATIVE CHARGE (OXYGEN)

Figure 2.6
The water molecule. Because the oxygen atom is more electrophilic than the two hydrogen atoms, the hydrogen atoms are partially positive and the oxygen atom is partially negative. Also, because the hydrogen atoms are not symmetrically distributed about the oxygen atom, water acts like a dipole.

SOME SPECIAL PROPERTIES OF WATER

Water is the most abundant molecular constituent of the body and accounts for about 60% of the body's weight. Indeed, in most tissues cellular water exceeds 70% of the total cell weight. Many of the physical and chemical events that take place in cells are the result of reactions between compounds that are dissolved in the cellular water. Since many of these reactions depend upon the special chemical properties of the surrounding water molecules, it is apparent that the role of water is an active one.

The essential role played by water can be attributed to water's unique physical and chemical properties. Because of their overall shape and dipole character, water molecules tend to form weak bonds with each other (fig. 2.7). These bonds result from the attraction of a partially positive region of one water molecule (i.e., the region near

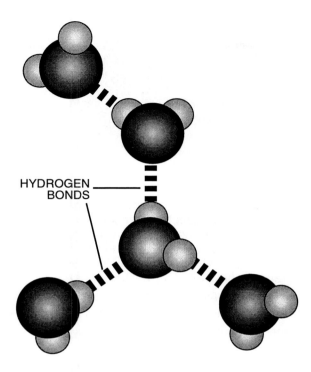

HYGROGEN BONDS

Figure 2.7
Hydrogen bonding among water molecules in liquid water.

each hydrogen atom) with the partially negative region of another water molecule (i.e., the region near the oxygen atom). These weak bonds are called **hydrogen bonds**.

In addition to water, other molecules may be weakly associated through hydrogen bonds. Indeed, hydrogen bonds frequently occur *within* a molecule. Although hydrogen bonds are individually weak, their sheer numbers create important stabilizing forces between and within the very large molecules that are found in the body (e.g., proteins and nucleic acids). Moreover, as you will learn later in the chapter, many enzymes owe their catalytic properties to the precise shapes they are able to assume through intramolecular hydrogen bonding.

As explained earlier, crystals of salts such as sodium chloride consist of ions that are held together by the attractive forces of the ions' opposite charges. When salts are dissolved in water, they **dissociate** or **ionize**; that is, the ions comprising the salt molecules separate from one another. Because the separated salt ions carry charges, they attract water molecules. As a result, the ions become surrounded by **spheres of hydration** (fig. 2.8). The ability to form spheres of hydration around ions is what makes water such a good solvent for salts (and other molecules carrying positive and/or negative charges).

A small percentage of the water molecules in a sample of water behave in a manner similar to salts and undergo ionization. When a water molecule ionizes, one of the two hydrogen nuclei is pulled away from the oxygen nucleus with which electrons were being shared and binds instead to the oxygen atom of a *different* water molecule (i.e., a water molecule with which a hydrogen bond formerly existed). This phenomenon, which is depicted in figure 2.9, produces a **hydroxyl ion** (OH^-) and a **hydronium ion** (H_3O^+). The notation H_3O^+ is rarely used; instead H^+ (called a **hydrogen ion** or proton) is used, even though free protons do not generally exist *per se* but are bound to water to form hydronium ions.

SOLUTIONS AND THE CONCEPT OF CONCENTRATION

When a substance (e.g., salt or sugar) is dissolved in a liquid (e.g., water), a **solution** is formed. The substance that is dissolved is called the **solute** and the liquid in which the dissolution occurred is called the **solvent**. Nearly all solutions of biological importance

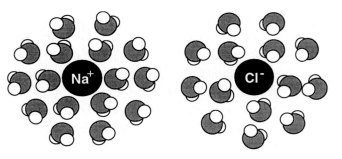

HYDRATED SODIUM ION   HYDRATED CHLORIDE ION

Figure 2.8
Spheres of hydration. Because they act like dipoles, water molecules are attracted to ions and form enclosing spheres of hydration. Shown here are the spheres of hydration formed around sodium and chloride ions.

are *aqueous* solutions; that is, the solvent is water. Although most solutes are solids, gases (e.g., oxygen and carbon dioxide) can also be dissolved in a liquid. When solutes are dissolved in water, it is usually through the formation of spheres of hydration around the solute molecules or ions. Nonpolar substances, such as fats, do not dissolve in wa-

TWO WATER MOLECULES

HYDRONIUM ION   HYDROXYL ION
(H_3O^+)   (OH$^-$)

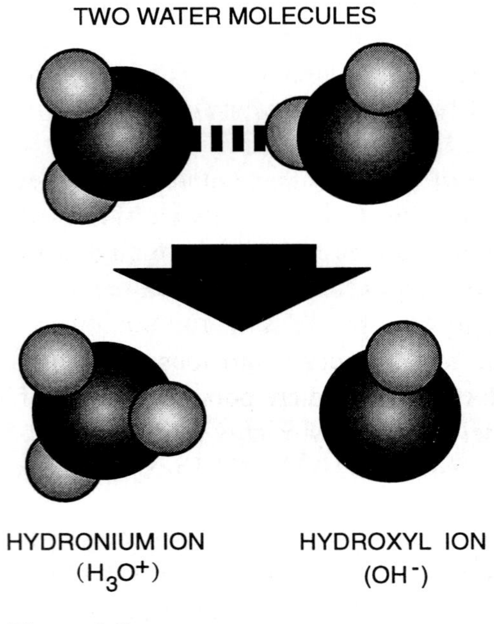

Figure 2.9
Ionization of water. A small percentage of the water molecules in a sample of water undergo ionization, thereby forming hydronium ions and hydroxyl ions.

ter but do dissolve in *organic* solvents such as acetone.

It is important to be able to express quantitatively how much of a particular solute is dissolved in a given volume of solvent. This is what is meant by a solution's **concentration**. Although concentration can be expressed in a number of different ways, each method reflects the *number of solute molecules per unit volume of solution*. Most physiologists use **molarity** to express the concentration of a solution. Recall that atoms of each element have a characteristic atomic weight and that a compound has a characteristic molecular weight. Another important measure of weight is the **gram-atomic weight**, which is defined as the number of grams of an element that is numerically equal to the element's atomic weight. For example, one gram-atomic weight of carbon weighs 12 grams and one gram-atomic weight of magnesium weighs 24 grams (i.e., carbon's atomic weight is 12, and magnesium's atomic weight is 24; see table 2.2). However, although a gram-atomic weight of magnesium weighs twice as much as a gram atomic weight of carbon, both contain *the same number of atoms*. The number of atoms that are present in one gram-atomic weight of *any* element is a *constant* called **Avogadro's number** and is equal to 6.023×10^{23}.

The molecular counterpart of gram-atomic weight is the **gram-formula weight** (or **gram-molecular weight**). One gram-formula weight of a compound is a number of grams numerically equal to the compound's molecular weight. For example, the molecular weight of water is 18 (i.e., 16 + [2 x 1]), and therefore one gram-formula weight of water weighs 18 grams. Just as a gram-atomic weight of an element contains Avogadro's number of atoms, a gram-formula weight of a compound contains Avogadro's number of molecules (e.g., one gram-formula weight of water contains 6.023×10^{23} water molecules).

Another quantitative measure of an element (or compound) is the **mole**. The term *mole* refers to the same quantity of a substance as do the terms *gram-atomic weight* and *gram-formula weight*. However, *mole* is the more frequently used expression of quantity, because it applies to *both* individual elements and compounds.

As noted previously, a solution consists of a liquid in which one or more substances are dissolved; the liquid in which the dissolution occurs is called the **solvent** and the substances that are dissolved in the liquid are called **solutes**. The concentration of a solution is identified on the basis of the number of moles of solute that have been dissolved in a given volume of the solution's solvent. For example, a solution that has a **molarity** of 1.0 contains one mole of solute per *liter* of solution (a liter is a metric unit of volume and is equal to 1.06 quarts). The symbol for molarity is **M**; therefore, a one molar solution of *sucrose* (i.e., table sugar) would be identified as 1.0 **M** sucrose; a one-tenth molar solution would be identified as 0.1 **M**, and so on. (The molecular weight of sucrose is 342 daltons; thus, a 1.0 **M** sucrose solution contains 342 grams of sucrose per

liter of solution; a 0.1 **M** sucrose solution has 34.2 grams of sucrose per liter of solution, and so on.).

In the body's cells and tissues, the concentrations of most solutes are much lower than one molar. Therefore, for solutes present in the body, it is more practical to express concentration in units that are correspondingly smaller, such as *millimolar* (thousandths of a mole per liter), and *micromolar* (millionths of a mole per liter), or even *nanomolar* (billionths of a mole per liter). For example, a one millimolar solution (abbreviated 1.0 mM) contains one-thousandth of a mole of solute per liter. In the same way that a millimole is one-thousandth of a mole, a *milligram* is one-thousandth of a gram. Thus, a 0.5 mM sucrose solution contains 171 milligrams of sucrose per liter (i.e., 0.0005 x 342 grams = 0.171 grams = 171 milligrams).

ACIDS, BASES, AND THE CONCEPT OF pH

As already noted, salts and other substances dissociate into ions when they are dissolved in water. These compounds are called **electrolytes** because they are able to conduct electricity through the solution. Compounds that do not dissociate into ions are called **nonelectrolytes** and are poor conductors of electricity. Electrolytes may be divided into three classes: (1) **acids**, (2) **bases**, and (3) **salts**.

Acids and bases can be defined in terms of their effects on the concentrations of hydrogen ions (H^+) and hydroxyl ions (OH^-) in an aqueous solution. Even pure water contains some H^+ and OH^- because water itself undergoes a small amount of dissociation. The dissociation of water molecules in pure water produces H^+ and OH^- concentrations that have a molarity of only 0.0000001 (i.e., 1.0 x 10^{-7} **M**). An **acid** *increases* the concentration of H^+ when added to a solution (or *decreases* the concentration of OH^-); whereas a **base** *increases* the concentration of OH^- (or *decreases* the concentration of H^+). For example, *hydrochloric acid* (HCl) dissociates in water to form H^+ and Cl^-, thereby raising the concentration of H^+. On the other hand, *sodium hydroxide* (NaOH) is a base because it dissociates in water to form Na^+ and OH^-, thereby raising the hydroxyl ion concentration of the solution (i.e., the added hydroxyl ions make the solution more alkaline; figure 2.10).

When salts such as NaCl and KCl dissociate in water, neither a hydrogen ion nor a hydroxyl ion is produced; therefore, the amounts of hydrogen ions and hydroxyl ions in the water remain unchanged. Mixing an acid with a base in the right proportions produces water and salt ions; this is illustrated in figure 2.11, which shows what hap-

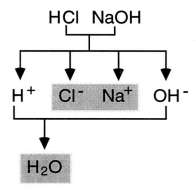

Figure 2.11
When an acid and base are mixed together, water and salt ions are formed.

pens when hydrochloric acid and sodium hydroxide are mixed.

Hydrochloric acid is said to be a "strong" acid because it undergoes a high degree of dissociation when it is dissolved in water. When HCl is dissolved in water, nearly all of the HCl molecules dissociate into ions and only a few HCl molecules remain intact. In contrast, *acetic acid* dissociates far less completely when dissolved in water. When acetic acid is added to water, *acetate* ions and H^+ are produced, but large amounts of undissociated acetic acid molecules remain. Acetic acid is considered a "weak" acid because of its small degree of dissociation. In a like manner, there are also weak bases and weak salts.

As you will learn later, most of the chemical reactions that take place in the body are catalyzed by enzymes that are especially sensitive to the concentration of hydrogen ions. For example, the digestion inside the stomach of proteins in the food that we eat requires a highly acidic environment. The opposite is true in the small intestine, where digestion of the food requires a basic

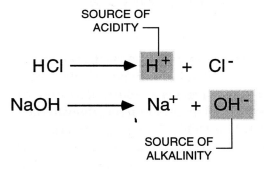

Figure 2.10
Dissociation of HCl in water releases hydrogen ions (H^+), thereby making the solution more acidic. On the other hand, the dissociation of NaOH in water releases hydroxyl ions (OH^-), making the solution more basic (or alkaline).

(alkaline) environment. The internal fluid of most cells is neither acidic nor basic; rather it is said to be *neutral* because the concentrations of H$^+$ and OH$^-$ are about the same as in pure water (namely, 1.0×10^{-7} **M**). Small variations in the H$^+$ or OH$^-$ concentrations of the intracellular fluid prevent a cell from functioning properly.

The **pH** of a solution is a measure of the solution's acidity or alkalinity. The pH value of a solution is defined by the following equation:

$$pH = -\log_{10}[H^+]$$

Thus, the pH of pure water, in which [H$^+$] = 1.0×10^{-7} **M** is $-\log_{10}(10^{-7})$, which equals 7. Water is said to have a *neutral* pH because the concentrations of H$^+$ and OH$^-$ are equal (i.e., the concentration of OH$^-$ is also 1.0×10^{-7} **M**). Thus, on the pH scale, 7.0 represents neutrality. Solutions that have pH values less than seven are said to be "acidic," whereas solutions that have pH values greater than seven are said to be "basic" (or "alkaline"). The pH scale and pH values of some familiar liquids are shown in figure 2.12.

Because pH is a logarithmic function, one unit of pH represents a 10-fold difference in hydrogen ion concentration. A solution having a pH of 6.0 has ten times the hydrogen ion concentration as a solution of pH 7.0. A solution of pH 5.0 has 100 times the hydrogen ion concentration as a solution of pH 7.0, and so on.

Since most intracellular processes operate at a pH that is near 7.0, **buffers** are used in order to maintain this neutral pH. The body's natural buffers are mixtures of either (1) a weak acid and the salt of that acid or (2) a weak base and the salt of that base; these

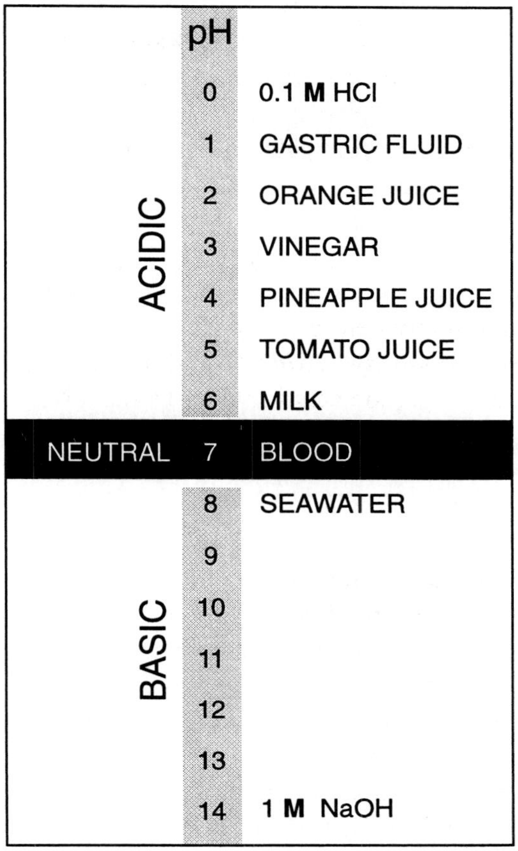

pH	
0	0.1 **M** HCl
1	GASTRIC FLUID
2	ORANGE JUICE
3	VINEGAR
4	PINEAPPLE JUICE
5	TOMATO JUICE
6	MILK
NEUTRAL 7	BLOOD
8	SEAWATER
9	
10	
11	
12	
13	
14	1 **M** NaOH

(ACIDIC for pH 0–6, BASIC for pH 8–14)

Figure 2.12
The pH scale and pH values of some familiar liquids.

combinations are known as *buffer pairs*. Probably the most common of the buffer pairs is the *bicarbonate pair*, consisting of carbonic acid and sodium bicarbonate (i.e., $H_2CO_3/NaHCO_3$). Using the bicarbonate pair as an example, figure 2.13 shows how the buffer pair can act to maintain the pH of a solution to which acid or base has been added. As seen in the figure, the buffer combines with hydrogen (or hydroxyl) ions that are added to the solution, thereby forming additional weak acid (or water), so that the solution's pH is not changed.

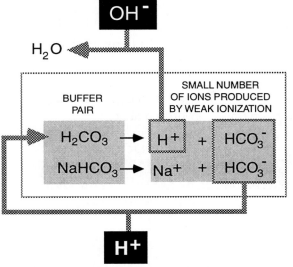

IF STRONG BASE IS ADDED TO SOLUTION
CONTAINING BUFFER PAIR, HYDROXYL IONS
COMBINE WITH HYDROGEN IONS FROM
WEAK ACID TO FORM WATER.

OH^-

H_2O

BUFFER
PAIR

SMALL NUMBER
OF IONS PRODUCED
BY WEAK IONIZATION

$H_2CO_3 \rightarrow$ H^+ $+$ HCO_3^-

$NaHCO_3 \rightarrow$ Na^+ $+$ HCO_3^-

H^+

IF STRONG ACID IS ADDED TO SOLUTION
CONTAINING BUFFER PAIR, HYDROGEN
IONS COMBINE WITH BICARBONATE
TO FORM MORE WEAK ACID.

Figure 2.13
When buffer pairs are present in a solution,
they act to prevent changes in pH. Shown in
this figure is the action of the carbonic acid/
sodium bicarbonate buffer pair. When a strong
base is added to the solution, hydroxyl ions
produced by the base's dissociation combine
with hydrogen ions to form water. When a
strong acid is added to the solution, hydrogen
ions produced by the acid's dissociation
combine with bicarbonate ions, thereby forming
additional undissociated acid.

BIOCHEMISTRY

In the average adult male, water accounts
for more than 60 percent of the body weight;
in the average adult female, water accounts
for about 50 percent of the total body weight.
The remaining molecules of the body can
be divided into two broad categories: **organic** and **inorganic**.

In a general sense, organic substances are
compounds peculiarly associated with living things, whereas inorganic compounds
are found both in living things and in the non-
living world. Biochemists employ a much
stricter definition of an organic substance, limiting this category to compounds that contain
both *carbon* and *hydrogen* atoms. Consequently, there are some substances (e.g.,
carbon dioxide) that do not properly fit the
biochemical definition of "organic" but which
clearly are associated with the activities of
living things. Inorganic substances include such
things as salts and minerals and typically are
compounds of relatively low molecular weight.

In most organic compounds of the body,
carbon forms the backbone of the molecule,
as for example, in the common body sugar
called **glucose** (fig. 2.14). Other atoms (or
groups of atoms) are attached to the carbon
backbone, and usually these give the compound its unique chemical and physiological properties. Because organic compounds
usually contain several carbon atoms, each
of the carbon atoms is assigned a number so
that it (or the group of atoms attached to
that carbon) can be identified specifically.

Many biologically important compounds
form *ringed* structures. For example, the six
carbon atoms of *benzene* (fig. 2.14) form a
closed loop in which the carbon atoms are
linked by alternating single and double
bonds. The remaining carbon bonds are
formed with hydrogen atoms. For convenience and simplicity, ringed compounds
like benzene are usually drawn as geometric figures (i.e., polygons) in which it is
understood that each corner of the polygon
is occupied by a carbon atom or CH-group.
Even in this abbreviated form, the locations

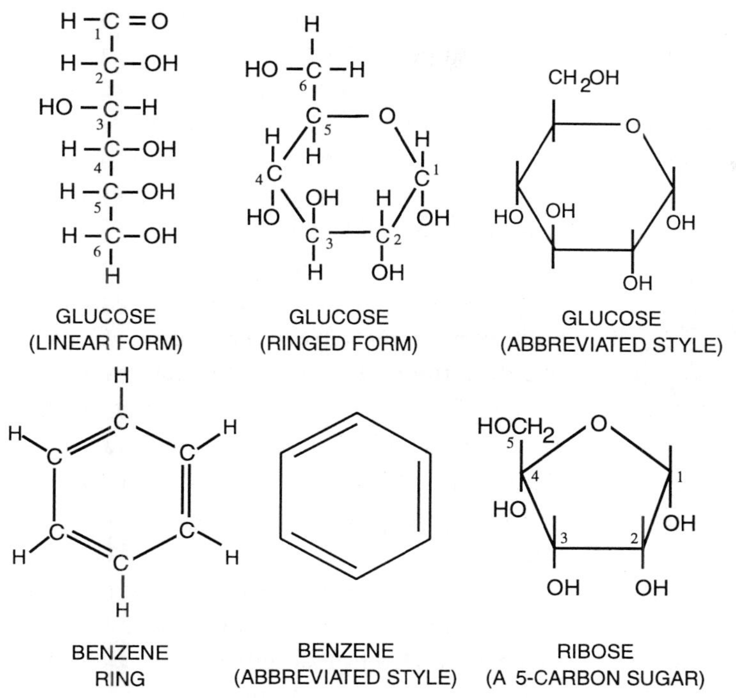

GLUCOSE
(LINEAR FORM)

GLUCOSE
(RINGED FORM)

GLUCOSE
(ABBREVIATED STYLE)

Figure 2.14
Chemical formulas of glucose, benzene, and ribose. Glucose exists in linear and ringed forms. In the abbreviated formulas, the individual carbon atoms of a ring are not shown (they occur at each vertex). When a bond appears to have no attached group, this implies that the attached group is a hydrogen atom.

BENZENE
RING

BENZENE
(ABBREVIATED STYLE)

RIBOSE
(A 5-CARBON SUGAR)

of the double bonds are usually identified. In some ringed structures, nitrogen or oxygen atoms are members of the ring. The ringed forms of the sugars glucose and ribose are common examples and are shown in figure 2.14. Note that the positions of the noncarbon members of a ring (e.g., the oxygen atoms in the rings formed by glucose and ribose) are shown using the element's symbol. Nearly all of the major ringed structures of importance in the body contain 5 or 6 members.

2-2

Major Groups of Organic Compounds

For convenience, we can assign all of the common organic compounds of the body to five major groups: (1) **organic acids**, (2) **amino acids**, (3) **sugars**, (4) **nucleotides**, and (5) **lipids**.

Organic Acids

Organic acids are compounds that contain one or more **carboxyl** (i.e., COOH) groups. One of the simplest and most important of

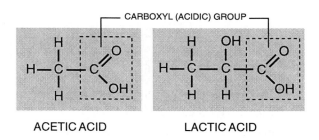

ACETIC ACID

LACTIC ACID

Figure 2.15
The organic acids acetic acid and lactic acid.

25

these compounds is *acetic acid* (fig. 2.15), which is the essential component of vinegar. Another important organic acid is *lactic acid*, a major end product of strenuous muscle activity. Acids that contain a carboxyl group dissociate incompletely, and they are therefore considered weak acids.

Amino Acids and Proteins

Amino acids are especially important organic compounds because they are the building blocks of **proteins**. Amino acids are organic acids that contain a basic **amino** group in addition to a **carboxyl** group. In this regard, amino acids are somewhat unusual since they contain *both* acidic and basic groups. The general formula of an amino acid is shown in figure 2.16. The central (or **alpha**) carbon atom of each amino acid is covalently bonded to four chemical groups: (1) a **hydrogen atom**, (2) an **amino group** (specifically referred to as the "alpha amino group"), (3) an **acid** (or **carboxyl**) group (specifically referred to as the "alpha carboxyl group"), and (4) a side chain called an **R group**. The letter **R** represents one of a number of different chemical groups that vary from a single hydrogen atom to more complex chemical structures. Thus, it is the

different R groups that give rise to the different kinds of amino acids. The structural formulas of the most common amino acids are given in figure 2.17, where they are also subdivided into different categories according to the functional properties or chemical nature of their R groups.

In most tissues of the body, more than 90% of the total mass (excluding water) is represented by large molecules called **macromolecules**. The macromolecules vary in size from several hundred to several hundred million daltons. Three major classes of macromolecules can be identified; these are the **proteins, polysaccharides**, and **nucleic acids**.

Macromolecules consist of long chains of smaller *subunits*. Among the most important macromolecules are the **proteins** formed by combinations of amino acids. The typical protein contains hundreds of amino acids that form one or more long chains called **polypeptides**. The amino acids of a polypeptide are linked together by **peptide bonds** formed between the amino group of one amino acid and the carboxyl group of its neighbor (fig. 2.18). When two amino acids are linked together by a peptide bond, a **dipeptide** is produced; the linking of three amino acids by peptide bonds forms a **tripeptide**, and so on. Polypeptide chains

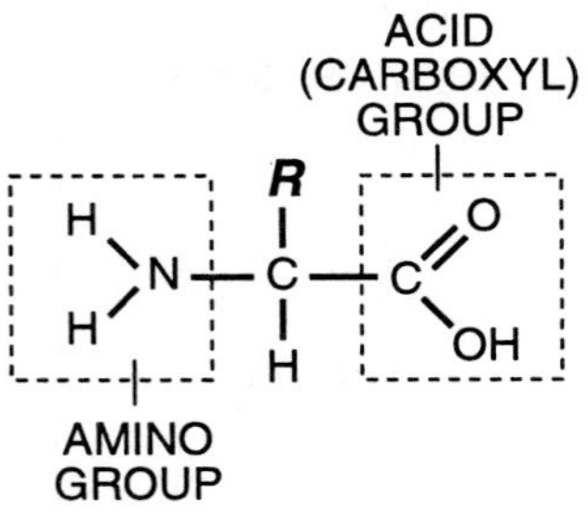

Figure 2.16
General chemical formula of an amino acid.

Figure 2.17 (Opposite page)
The 20 different amino acids that regularly occur in proteins. It is the varying chemical composition of the R-groups or side-chains that distinguishes one amino acid from another. The conventional three-letter and one-letter abbreviations of the amino acids are also shown.

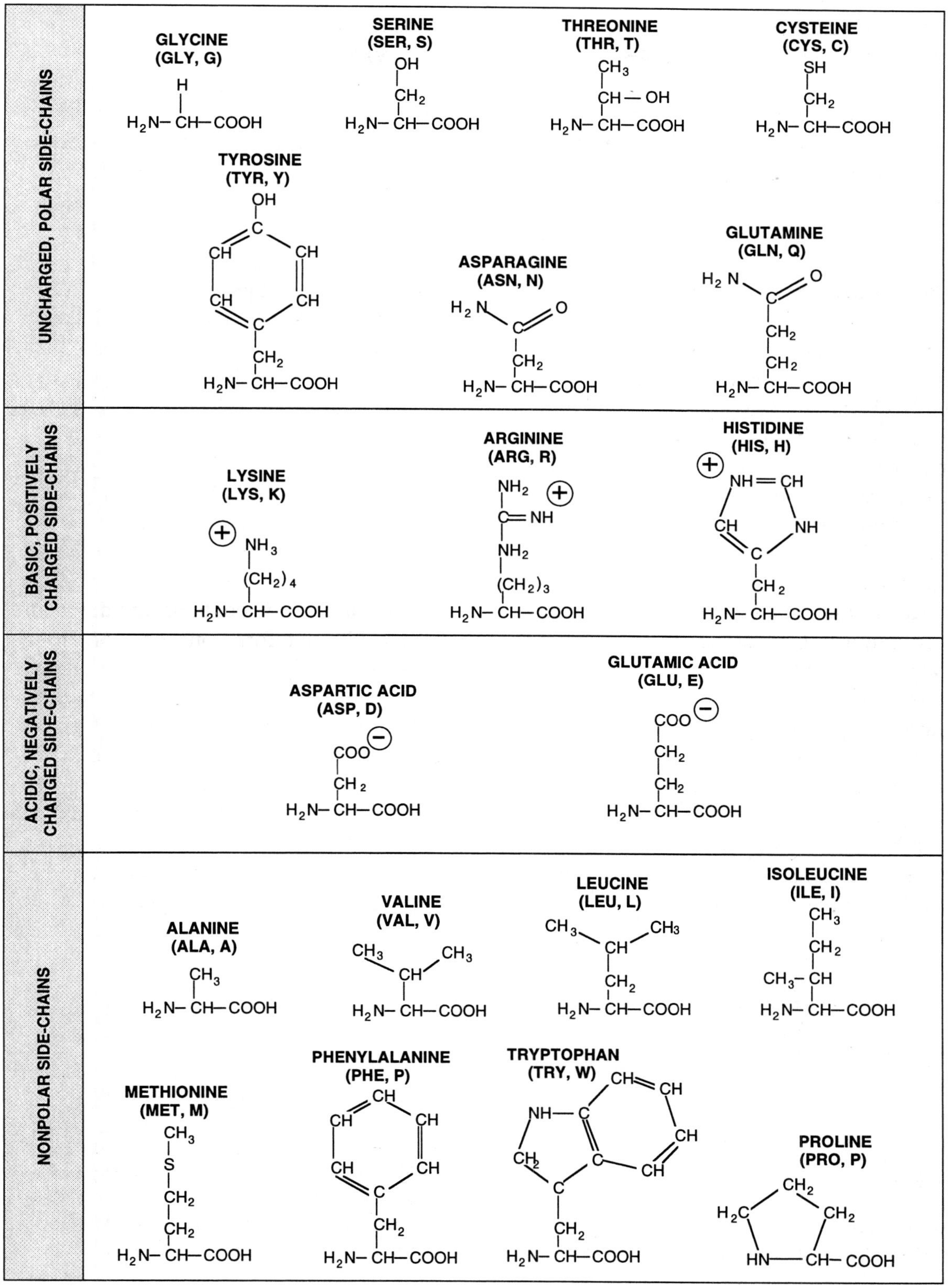

UNCHARGED, POLAR SIDE-CHAINS	

GLYCINE
(GLY, G)

H

$H_2N-CH-COOH$

SERINE
(SER, S)

OH

CH_2

$H_2N-CH-COOH$

THREONINE
(THR, T)

CH_3

$CH-OH$

$H_2N-CH-COOH$

CYSTEINE
(CYS, C)

SH

CH_2

$H_2N-CH-COOH$

TYROSINE
(TYR, Y)

OH

C

CH   CH

CH   CH

C

CH_2

$H_2N-CH-COOH$

ASPARAGINE
(ASN, N)

H_2N O

C

CH_2

$H_2N-CH-COOH$

GLUTAMINE
(GLN, Q)

H_2N O

C

CH_2

CH_2

$H_2N-CH-COOH$

BASIC, POSITIVELY CHARGED SIDE-CHAINS	

LYSINE
(LYS, K)

⊕ NH_3

$(CH_2)_4$

$H_2N-CH-COOH$

ARGININE
(ARG, R)

NH_2 ⊕

C

NH

NH_2

$(CH_2)_3$

$H_2N-CH-COOH$

HISTIDINE
(HIS, H)

⊕

NH   CH

CH   NH

C

CH_2

$H_2N-CH-COOH$

ACIDIC, NEGATIVELY CHARGED SIDE-CHAINS	

ASPARTIC ACID
(ASP, D)

COO ⊖

CH_2

$H_2N-CH-COOH$

GLUTAMIC ACID
(GLU, E)

COO ⊖

CH_2

CH_2

$H_2N-CH-COOH$

NONPOLAR SIDE-CHAINS	

ALANINE
(ALA, A)

CH_3

$H_2N-CH-COOH$

VALINE
(VAL, V)

CH_3 CH_3

CH

$H_2N-CH-COOH$

LEUCINE
(LEU, L)

CH_3 CH_3

CH

CH_2

$H_2N-CH-COOH$

ISOLEUCINE
(ILE, I)

CH_3

CH_2

CH_3-CH

$H_2N-CH-COOH$

METHIONINE
(MET, M)

CH_3

S

CH_2

CH_2

$H_2N-CH-COOH$

PHENYLALANINE
(PHE, P)

CH

CH   CH

CH   CH

C

CH_2

$H_2N-CH-COOH$

TRYPTOPHAN
(TRY, W)

CH   CH

NH   C

CH

CH_2 C

C

CH_2

$H_2N-CH-COOH$

PROLINE
(PRO, P)

CH_2

H_2C CH_2

HN   CH   COOH

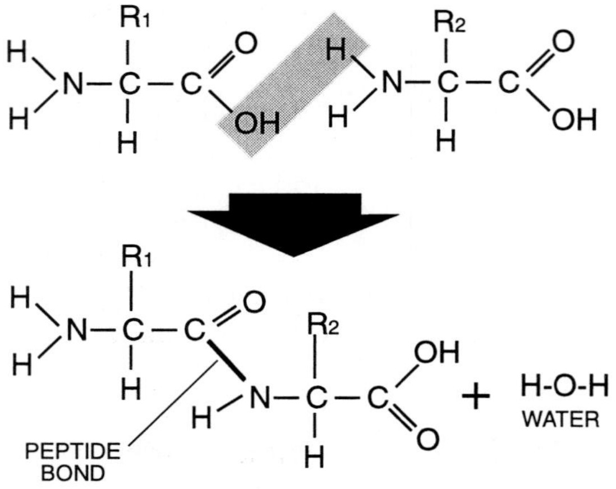

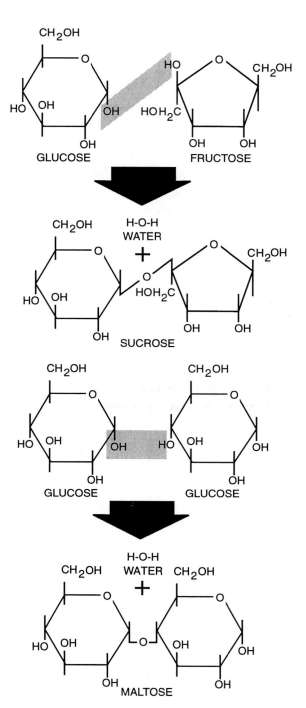

Figure 2.18
Two amino acids combine to form a dipeptide by a reaction involving the alpha amino group of one amino acid and the alpha carboxyl group of the other amino acid. The bond formed is called a peptide bond.

are very flexible and can assume a diversity of three dimensional shapes.

Sugars and Polysaccharides

Sugars are a diverse collection of compounds in which the carbon backbone of the molecule forms bonds with several hydroxyl groups and contains at least one **carbonyl** (i.e., C=O) group. In nearly all common sugars, the numbers of oxygen and hydrogen atoms are such that the sugar may be abbreviated $C_n(H_2O)_n$. That is, for every carbon atom in the molecule there is the equivalent of one molecule of water. For most of the *simple* sugars, the value of n is 5 or 6. When 5 carbons are present, the sugar is a **pentose**; 6 carbon sugars are called **hexoses**. The chemical structures of the hexose *glucose* and the pentose *ribose* are shown in figure 2.14.

Figure 2.19
Formation of the disaccharides sucrose (top) and maltose (bottom). Sucrose is formed from glucose and fructose, whereas maltose is formed from two glucose molecules.

The most important simple sugar is **glucose** (also called **dextrose**). As seen in figure 2.14, glucose (as well as many other simple sugars) can exist in an "open" (i.e., linear chain) form or as a ringed structure; in cells and tissues, it is the ringed forms of the sugars that predominate.

One molecule of glucose (or other simple sugar) can be linked to a second simple sugar molecule, producing another form of sugar called a **disaccharide** (fig. 2.19). Each simple sugar of the disaccharide is called a **monosaccharide** and the two monosaccharides are linked by a **glycosidic** bond. One of the most abundant disaccharides is **sucrose**, which is formed from one molecule of glucose and one molecule of the simple sugar **fructose**. Sucrose is the "table sugar" that we use to sweeten drinks (coffee, tea, etc.). Another common disaccharide is **maltose**, which is formed from two glucose molecules.

Polysaccharides are formed by linking very large numbers of sugar molecules together. The most common polysaccharides found in nature are the plant polysaccharides **starch** and **cellulose**, which are formed from glucose. Humans store glucose in the liver and in the skeletal musculature as a polysaccharide called **glycogen** (also known as "animal starch" because of its chemical resemblance to plant starch). Each glycogen molecule is comprised of many thousands of glucose units arranged to form a large and highly branched complex (fig. 2.20).

Nucleotides and Nucleic Acids

Nucleotides are the subunits of a family of extremely important macromolecules called **nucleic acids**. The nucleic acids are information-carrying molecules, which among other things make up the units of heredity called **genes**.

A nucleotide consists of three different chemical groups: (1) a 5-carbon sugar (called a **pentose**), (2) a **phosphate** group, and (3) one of several different ringed compounds called **nitrogenous bases** (fig. 2.21). In nucleic acids, the nucleotides are strung

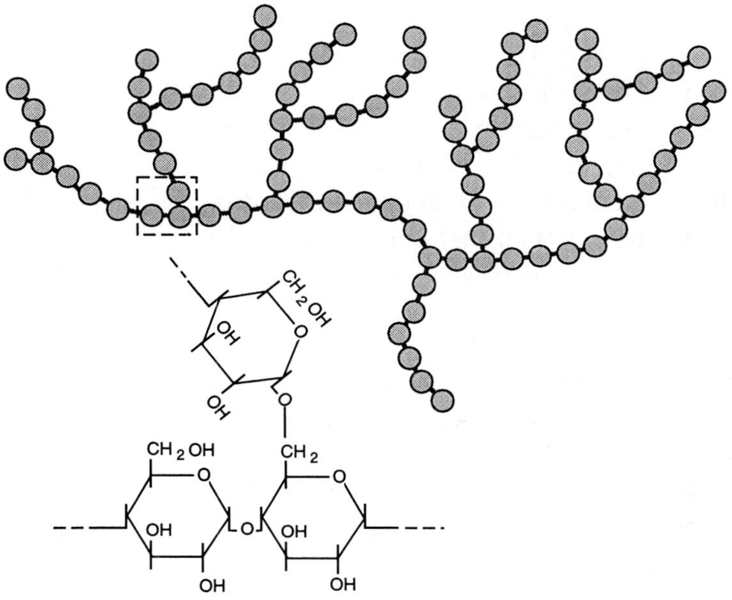

Figure 2.20
Structure of glycogen. Glycogen consists of large numbers of interconnected linear chains of glucose units (i.e., the circles). In each linear chain, neighboring glucoses are connected by bonds between their number 1 and 4 carbon atoms (e.g., see figs. 2.14 and 2.19). Branching occurs via bonds between the number 1 carbon atom of a glucose unit at the end of one linear chain and the number 6 carbon of a unit in another chain. The chemical structure of the branch point enclosed by the dashed square is shown in the enlarged view.

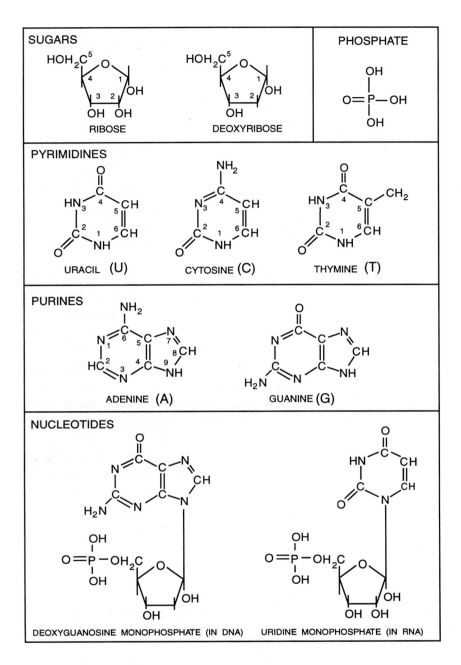

Figure 2.21
Nucleotides–the building blocks of nucleic acids. Each nucleotide is comprised of a pentose sugar (either ribose or deoxyribose), a purine or pyrimidine, and a phosphate group. Two of the eight different nucleotides that regularly occur in DNA and RNA are shown at the bottom of the figure.

together to form long chains or **polynucleotides**. The backbone of the chain consists of alternating pentose and phosphate groups. The nitrogenous bases are bonded to each of the sugars but project away from the molecule's backbone (fig. 2.22).

The pentose of a nucleotide is either **ribose** or **deoxyribose** (in which the hydroxyl group attached to carbon atom number 2 is replaced by a hydrogen atom; see fig. 2.21). Nucleotides containing ribose are known as **ribonucleotides**, whereas nucleotides containing deoxyribose are known as **deoxyribonucleotides**. Nucleic acids in which the nucleotides are ribonucleotides are called **ribonucleic acids** (usually abbreviated **RNA**), whereas nucleic acids in which the nucleotides are deoxyribonucleotides are called **deoxyribonucleic acids** (abbreviated **DNA**.)

The nitrogenous bases of nucleotides are either **pyrimidines** or **purines**. As seen in figure 2.21, the pyrimidines are the simpler chemical groups, consisting of a single ringed structure. Purines, on the other hand, are larger and more complex, consisting of two fused rings. Individual purines and pyrimidines differ from one another according to the chemical groups that are attached to their rings. In *both* DNA and RNA, the purines are **adenine** (abbreviated **A**) and **guanine** (**G**) (fig. 2.21). In DNA, the pyrimidines are **thymine** (**T**) and **cytosine** (**C**), whereas in RNA the pyrimidines are cytosine and **uracil** (**U**).

The genetic material of all living things is comprised of DNA. In viruses (which are not considered to be "alive"), the genetic material may be DNA or RNA. Except in the case of certain DNA viruses, DNA molecules consist of two polynucleotide chains twisted around each other to form a **double helix**. In a DNA double helix, the nitrogenous bases of each chain project toward the center of the double helix. A purine of one strand is always associated with a pyrimidine of the other strand, the two strands held together in part by the formation of two or three hydrogen bonds between the nitrogenous bases (fig. 2.22).

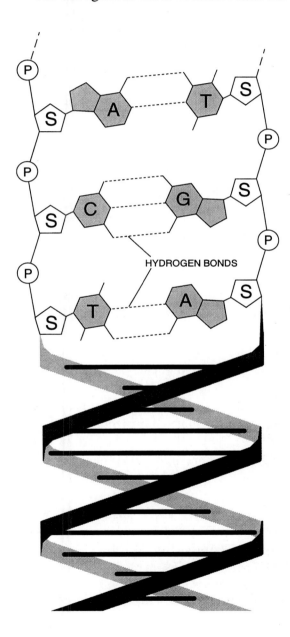

HYDROGEN BONDS

Figure 2.22
DNA molecules consist of two polynucleotide chains twisted around each other to form a double helix. In a DNA double helix, the nitrogenous bases of each chain project toward the center or *axis* of the double helix. A purine of one chain is always associated with a pyrimidine of the other chain, the two chains held together in part by the formation of two or three hydrogen bonds between the nitrogenous bases. P = phosphate group, S = sugar (i.e., deoxyribose), A = adenine, T = thymine, C = cytosine, and G = guanine.

Association of purines and pyrimidines in the center of the double helix is not random. Rather, the purine **adenine** (A) on one chain always faces the pyrimidine **thymine** (T) on the other chain. Similarly, **guanine** (G) is always matched with **cytosine** (C). Implicit in this organization is the fact that the sequence of nitrogenous bases along one chain of a DNA double helix automatically determines the nitrogenous base sequence of the other chain. For example, if the sequence is . . . *A-C-T-G-T-C-T-G-A* . . . on one chain of the double helix, then this implies that on the other chain of the double helix the base sequence is . . . *T-G-A-C-A-G-A-C-T*. . . . Thus, by knowing the base sequence of only one of the two polynucleotide chains of a DNA double helix, the base sequence of the matching chain can readily be determined. Moreover, if the two strands of DNA were separated, each strand's nitrogenous base sequence could be used as a guide for creating a matching chain and, thereby, two new (and identical) double helixes. Therein lies the capacity of DNA for replication–an essential property of the body's genetic material.

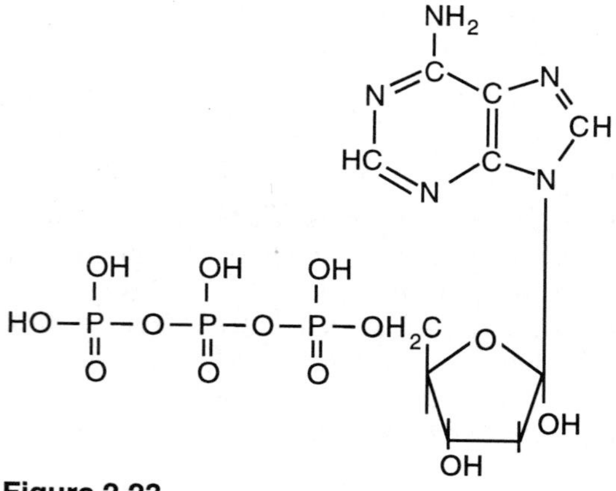

Unlike DNA, RNA is comprised of a single polynucleotide chain. In certain viruses, RNA serves as the genetic material; however, in animals and plants RNA plays an intermediary role in converting the genetic message of DNA into a cell's proteins.

In addition to being the building blocks of nucleic acids, nucleotides are important in other ways. For example, **adenosine triphosphate** (abbreviated **ATP**; figure 2.23), a compound of major importance in energy transfer during the body's chemical reactions (see chapter 4), consists of a purine (i.e., *adenine*), ribose, and three phosphate groups. When we say that ATP is important in energy transfer, we mean that it often serves as a link between chemical reactions in the body that release energy and reactions that consume energy.

Lipids (Fats)

Lipids are a heterogeneous collection of compounds that are grouped together because they share a common physical property; namely, they are insoluble in water but are soluble in a variety of organic solvents (such as acetone and chloroform). The simplest lipids are the **fatty acids**; in these molecules, a carboxyl group is attached to a long **hydrocarbon chain** (fig. 2.24). In the *saturated* fatty acids, all of the carbon atoms of the chain are linked together by *single* bonds, and the molecules conform to the general formula $CH_3\text{-}(CH_2)_n\text{-}COOH$. In *unsaturated* fatty acids, two or more of the carbon atoms of the hydrocarbon chain are linked together by *double* bonds. The chemical structures of some of the most common fatty acids are given in figure 2.24.

Figure 2.23
Chemical structure of adenosine triphosphate (i.e., ATP).

SATURATED FATTY ACIDS		UNSATURATED FATTY ACIDS	
$CH_3-(CH_2)_n-COOH$	GENERAL FORMULA		
$CH_3-(CH_2)_{10}-COOH$	LAURIC ACID (n = 10)	$CH_3-(CH_2)_7-CH=CH-(CH_2)_7-COOH$	OLEIC ACID
$CH_3-(CH_2)_{12}-COOH$	MYRISTIC ACID (n = 12)		
$CH_3-(CH_2)_{14}-COOH$	PALMITIC ACID (n = 14)	$CH_3-(CH_2)_4-CH=CH-CH_2-CH=CH-(CH_2)_7-COOH$	LINOLEIC ACID
$CH_3-(CH_2)_{16}-COOH$	STEARIC ACID (n = 16)		

Figure 2.24
The most common saturated and unsaturated fatty acids (see text for discussion).

Also common in cells and tissues are the **triglycerides** (also known as **neutral fats**) and **phospholipids**. Triglycerides are formed by the combination of three fatty acids and one molecule of **glycerol** (fig. 2.25). In phospholipids, one of the fatty acids of a triglyceride is replaced by a phosphate-containing chemical group (fig. 2.26).

A variety of other chemical compounds are also classified as lipids; included among these are **sterols** such as **cholesterol** (fig. 2.27), certain **hormones** (i.e., the *steroid hormones*) and water-insoluble **vitamins** (e.g., *vitamin D*, found in milk).

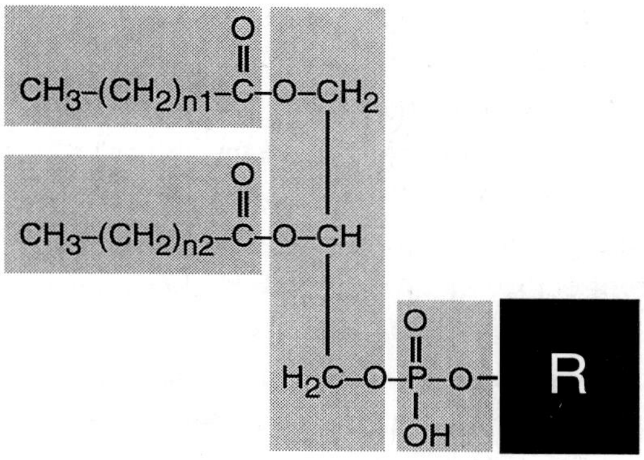

Figure 2.26
Chemical structure of a phospholipid. In a typical phospholipid, one of the fatty acid groups of the triglyceride is replaced either by phosphate or a phosphate-containing chemical group (represented by "R" in the figure).

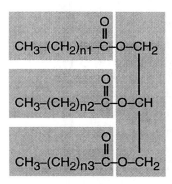

Figure 2.25
Chemical structure of a triglyceride. Triglycerides are formed by a reaction between three fatty acids and a molecule of glycerol.

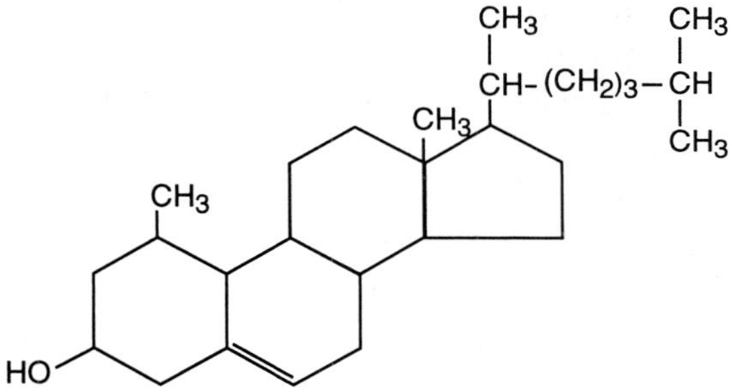

Figure 2.27
The multi-ringed structure of cholesterol is typical of the body's sterols.

METABOLISM AND ENZYMES

During a **chemical reaction**, one or more compounds (called **reactants**) are converted into one or more *different* compounds (called **products**). The changes in the reactants generally take the form of *rearrangements* of atoms, *addition* of atoms, or *removal* of atoms. In the living body, the passage of but a single second of time is accompanied by thousands of different chemical reactions, which collectively are called the body's **metabolism**.

Most of the chemical reactions of metabolism can be reproduced in a laboratory setting; that is, under suitable conditions, they can be duplicated in a beaker (or some other container) of water. Some of the reactions of metabolism are **exergonic** (or *exothermic*); this means that energy is released as the reaction proceeds. When exergonic reactions are duplicated in the laboratory, the energy that is released often takes the form of heat (i.e., the temperature of the solution in the beaker rises).

Other reactions of metabolism are **endergonic** (or *endothermic*); this means that energy is consumed as the reaction proceeds. When endergonic reactions are duplicated in the laboratory, the beaker containing the reactants may have to be heated in order for the reaction to proceed.

In the tissues of the body, the endergonic and exergonic reactions of metabolism are **coupled**; that is, the energy released by exergonic reactions is used to drive the endergonic reactions. As a result, very little of the energy released by exergonic reactions is lost as heat. What heat is produced is reflected in the body's temperature.

Nearly all of the chemical reactions that characterize the body's metabolism require special catalysts that increase the rates at which the reactions proceed. (In the absence of these catalysts, many of the reactions would proceed at an imperceptibly slow pace.) All but a very small number of the catalysts are proteins called **enzymes** and are usually very much larger than the reactants (or **substrates**) on which they work. With specific exceptions, such as the enzymes of the digestive system, most of the enzymes function intracellularly and are immobilized in cells by their attachment to cellular membranes. When substrates come in contact with an enzyme, they become bound to the enzyme, thereby forming a transient complex called the **enzyme-substrate complex** (fig. 2.28). The substrate molecules are bound to the enzyme at a particular site on the enzyme's surface called the **active site**. Soon after the substrate is bound, enzyme catalysis converts the substrate to product(s) (fig. 2.28), so that a transient **enzyme-product** complex is formed. In a final step, the products are released from the enzyme.

During catalysis, the enzyme itself is not changed at all, and at the reaction's completion, the enzyme is again available to catalyze another reaction. It is not unusual for an enzyme to be able to catalyze many thousands of successive (but identical) reactions in one second.

It is the shape and chemical nature of the enzyme's active site that determines the properties of the enzyme. The active site may not only bind the substrate but may also bind a **cofactor** that assists the enzyme in its catalysis. The active site of an enzyme usually occupies only a small portion of the enzyme's surface.

Enzymes range in molecular weight from about 5,000 to many millions of daltons. The biological activity of an enzyme (i.e.,

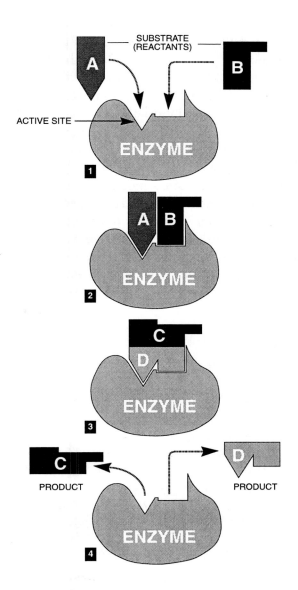

SUBSTRATE
(REACTANTS)

A **B**

ACTIVE SITE

ENZYME

1

A B

ENZYME

2

C
D

ENZYME

3

C
PRODUCT

D
PRODUCT

ENZYME

4

Figure 2.28
Catalysis of a chemical reaction by an enzyme. During catalysis, substrate molecules (or "reactants") are bound to the enzyme's active site (step 1), thereby transiently forming an *enzyme-substrate complex* (step 2). The enzyme then converts the substrate to one or more products, temporarily creating an *enzyme-product complex* (step 3). The products are then released from the active site (step 4), and the enzyme is once again available for another round of catalysis.

its ability to perform its catalysis) is usually lost at temperatures much above body temperature. This loss of activity is known as **denaturation**. The heat sensitivity of an enzyme is related to the molecule's large size and the relatively weak forces (e.g., hydrogen bonds) that maintain its complex and intricate shape. Enzymes are also denatured by small changes in the pH of their surroundings. This is why the digestive enzyme *pepsin* functions properly only in the acidic environment of the stomach, whereas the *salivary enzyme* functions properly only in the neutral saliva of the mouth.

As noted above, the catalytic actions of enzymes often require the participation of a smaller molecule called a cofactor. There are three kinds of enzyme cofactors. These are (1) **coenzymes**, (2) **prosthetic groups**, and (3) **metal ions**.

Coenzymes are organic compounds that are bound to the enzyme only during the course of catalysis. Once catalysis is completed, the coenzyme is released from the enzyme along with the product(s). The same coenzymes serve as cofactors for a variety of different enzyme catalyzed reactions and are used over and over again during metabolism. Coenzymes are usually comprised (at least in part) of **vitamins** (hence the importance of vitamins to the body's metabolism).

Prosthetic groups are also organic compounds but, unlike coenzymes, prosthetic groups are permanently bound (usually by covalent bonds) to the enzyme molecule.

A variety of metal ions can serve as enzyme cofactors, including Ca^{++}, Mg^{++}, and Zn^{++}. Like cofactors, these metal ions are bound to the enzyme only during the course of catalysis.

The catalysis displayed by most enzymes is highly "specific" and most enzymes catalyze a single reaction of metabolism. For example, an enzyme that catalyzes a particular chemical change in a molecule of glucose will not bring about a similar change in some other sugar. The high degree of specificity displayed by enzymes is due to the rather constrained nature of the interaction between the enzyme's active site (which has a particular shape and specific physical and chemical properties) and the substrate of that enzyme (which also has a particular shape and specific properties). Thus, the geometry of the active site determines what substrate molecule(s) will be bound.

Many enzymes catalyze reactions in *both* directions. That is, an enzyme that converts compounds "A" and "B" into compounds "C" and "D" may also be able to convert compounds "C" and "D" into compounds "A" and "B." The direction in which the catalysis takes place is determined by the relative concentrations of "A," "B," "C," and "D." (When there is an excess of "A" and "B," the enzyme converts some "A" and "B" to "C" and "D" [and vice versa]).

Most of the body's metabolism is intracellular, and therefore most of the enzymes of the body are localized within (or on the surfaces of) different cellular components. Consequently, let us now turn to a review of the organization of cells and their functional components.

SELF TEST[*]

True/False Questions

1. There are two broad categories of molecules in the body: organic and inorganic.

2. The body's proteins are formed from one or more chains of nucleotides.

3. Each sugar unit of a polysaccharide is called a disaccharide.

4. The body's major storage form of sugar is glycogen.

5. The reactants upon which an enzyme acts are also referred to as the enzyme's substrate.

6. For some of the body's enzyme-catalyzed chemical reactions, metal ions serve as cofactors.

7. Prosthetic groups are permanently attached to the enzymes they assist.

Multiple Choice Questions

1. In the average adult male, water accounts for about (A) 10%, (B) 25%, (C) 60%, (D) 90% of the body's weight.

2. Biochemists classify a molecular substance as organic if the substance contains (A) water, (B) properties of life, (C) carbon, (D) carbon and hydrogen.

3. Which one of the following is not considered a major group of the body's organic compounds? (A) amino acids, (B) sugars,

---

[*] *The answers to these test questions are found in Appendix III at the back of the book.*

(C) vitamins, (D) nucleotides, (E) lipids.

4. The body's proteins consist of various combinations of (A) 4, (B) 10, (C) 12, (D) 20, (E) 46 different amino acids.

5. Neighboring sugars in a polysaccharide are linked together by (A) peptide bonds, (B) ionic bonds, (C) neutral bonds, (D) glycosidic bonds.

6. Nucleotides are the building blocks of (A) DNA and RNA, (B) proteins, (C) polysaccharides, (D) lipids.

7. Triglycerides are also known as (A) fatty acids, (B) neutral fats, (C) sterols.

8. The body contains chemical catalysts that cause the body's chemical reactions to proceed more rapidly. These catalysts are called (A) sterols, (B) hormones, (C) enzymes, (D) DNA and RNA.

CELLS

The body is comprised of billions and billions of structural and functional units called **cells**. Despite the small size of cells, much has been learned about cell structure and organization using various forms of *light* and *electron* microscopy. Although all cells appear to have certain features in common, all cells are not exactly alike. Rather, cells occur in a variety of shapes and sizes and exhibit a broad range of physiological properties. This range of cellular attributes is directly related to the variety of tissues in which cells are found and to the diversity of functions that the various cells and tissues of the body must perform. For example, not only do human liver cells and nerve cells look quite different, they also exhibit many differences in their physiology.

MICROSCOPY

The existence of cells and the realization that all animal (and plant) tissues are comprised of these fundamental biological units was entirely unknown until the invention of the first microscopes. Quickly following the appearance of the first crude microscopes in the 17th century, it became universally recognized that the tissues of the body (indeed, the tissues of all living things) are comprised of cells. Until the 1940s, most of our knowledge concerning the structure and organization of human cells was obtained from studies carried out with *light microscopes*. Indeed, using light microscopy, nearly all of the major cell structures and *organelles* (see below) were identified and characterized.

In microscopy, the upper limits of useful *magnification* depends on the *wavelength* of the radiation used to illuminate the specimen. Because light microscopes use visible light as the radiation source, useful magnification is limited to about 2000 X (i.e., an object can be magnified to about two thousand times its original size). The practical limits of magnification were significantly advanced with the appearance of the first commercial *electron microscopes* in the 1950s. These instruments use a beam of electrons to "illuminate" the specimen. Ongoing developments and improvements in electron microscope technology have now pushed the limits of magnification to about 500,000 X .

CELL STRUCTURE AND THE CELLULAR ORGANELLES

The internal components of a cell play specific roles in the cell's overall function and behavior. This is not to say that each part of a cell behaves independently of the other parts; rather, the activity of each cell component is influenced by (and at the same time has an influence on) other cell parts. The discrete functioning parts that comprise a cell are referred to as the cell's **organelles**. The organization of a generalized human cell is shown in figure 3.1.

3-1

The Plasma Membrane

At its surface, a cell is bordered by a membrane called the **plasma membrane** (fig. 3.1). This membrane is composed for the most part of thousands of protein and lipid molecules. Most substances that enter or leave the cell must pass through this membrane. Since (1) most human cells obtain their nutrients from and dispatch their wastes into the external cell surroundings, and (2) the greater the ratio of surface area of the plasma membrane to cell volume, the greater can be the rate of exchange between a cell and its surroundings, a cell's shape is often modified in order to maximize the *surface area:volume ratio*. In the case of some cells (e.g., red blood cells), this is achieved by changing the overall shape of the cell to that of a flattened disk. However, in many cells (e.g., the absorptive cells of the digestive tract), the surface area:volume ratio is increased by the presence of numerous tiny projections of the plasma membrane; these projections are called *microvilli*.

The plasma membrane actively regulates the passage of materials between the cell and its surroundings. In some tissues, the plasma membrane is involved in intercellular communication (e.g., nerve tissue), in which materials and information are sent from one cell to another. In most of the body's tissues, it is not unusual to observe special junctions

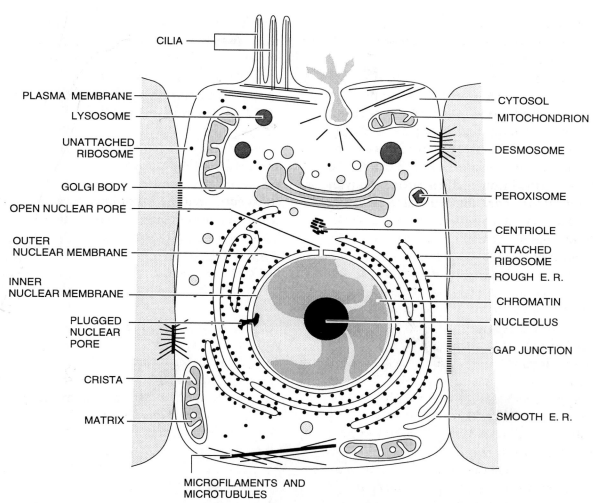

CILIA

PLASMA MEMBRANE

LYSOSOME

UNATTACHED RIBOSOME

GOLGI BODY

OPEN NUCLEAR PORE

OUTER NUCLEAR MEMBRANE

INNER NUCLEAR MEMBRANE

PLUGGED NUCLEAR PORE

CRISTA

MATRIX

MICROFILAMENTS AND MICROTUBULES

CYTOSOL

MITOCHONDRION

DESMOSOME

PEROXISOME

CENTRIOLE

ATTACHED RIBOSOME

ROUGH E. R.

CHROMATIN

NUCLEOLUS

GAP JUNCTION

SMOOTH E. R.

Figure 3.1
A generalized human cell.

between the plasma membranes of neighboring cells (e.g., *desmosomes* and *gap junctions*; see figure 3.1).

Chemically, the plasma membrane consists of protein, lipid, and carbohydrate molecules. The lipid is primarily phospholipid, although small quantities of neutral fats and cholesterol (fig. 2.26) are usually present. According to the **fluid mosaic model** of membrane structure, the phospholipid is organized into

3-2

two layers (fig. 3.2). Protein molecules are associated with the membrane in either of two ways; the proteins may penetrate one or both lipid layers or they may be loosely attached to the outer or inner lipid surfaces. Proteins that are loosely attached to the outer surfaces of the lipid bilayer are called **peripheral** or **extrinsic** proteins. Proteins that penetrate the lipid bilayer or span the membrane entirely are called **integral** or **intrinsic** proteins. Many of the membrane's proteins are enzymes involved in the transfer of substances into or out of the cell. **Network** proteins lying just under the plasma membrane (fig. 3.2) help to anchor many membrane proteins in position.

Peripheral proteins and those parts of integral proteins that occur on the outer membrane surface frequently contain chains of sugar molecules. The sugar chains are believed to be involved in a variety of physiological phenomena, including the adhesion of cells to their neighbors in the tissue. Some membrane carbohydrate may also be chemically bonded to membrane lipid.

The rather simple appearance of the plasma membrane when examined by microscopy belies the heterogeneity of its chemical organization. The kinds of proteins, lipids, and carbohydrates that make up the membrane vary in different regions of the membrane. This heterogeneity may be illustrated using the plasma membrane of a liver cell as an example (fig. 3.3). Some areas of a liver cell's plasma membrane face the plasma membranes of neighboring liver cells; other areas face the bile channels into which bile and other substances produced in the liver cell are secreted. Still other portions of the plasma membrane face the surfaces of capillaries from which substances

Figure 3.2
The fluid-mosaic model of membrane structure (see the text for explanation).

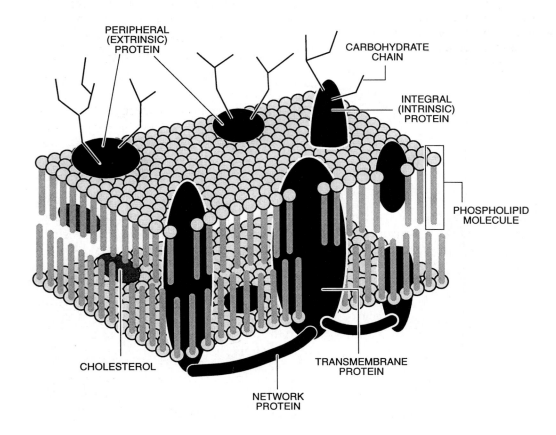

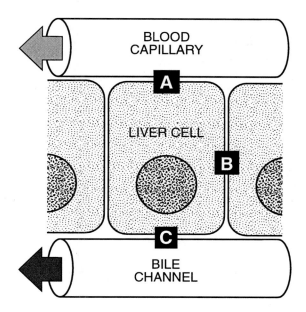

Figure 3.3
The various faces (i.e., "domains") of a liver cell's plasma membrane. Domain *A* faces the bloodstream, from which materials are removed by the cell and into which certain secretions are passed. Domain *B* faces neighboring liver cells, with which cell-cell junctions are formed. Domain *C* faces the bile channels into which bile is secreted. In each of these domains, the membrane's protein and lipid compositions differ.

are absorbed. Each of these regions (called "functional domains") of the plasma membrane is differently composed and differently organized and, in fact, is continually undergoing change and reorganization.

Membrane Transport. The plasma membrane acts as a barrier that separates the cell from its surrounding environment. Therefore, substances that enter or leave the cell must traverse the plasma membrane. The movement of molecules across the plasma membrane can take various forms, including simple **diffusion**, **active transport**, and **bulk transport**. During diffusion, a substance passes through the plasma membrane along a *concentration gradient* (i.e., the substance passes through the membrane from the side where it is more concentrated to the side where it is less concentrated; see figure 3.4, left). Sometimes diffusion through the plasma membrane is facilitated by integral proteins in the membrane that serve as enzymes. One of the special characteristics of transmembrane diffusion is that the passage of molecules (or ions) through the membrane does *not* consume any cellular (metabolic) energy.

During active transport, substances are transferred through the membrane *against* a concentration gradient (i.e., the substance is passed through the membrane from the side where it is less concentrated to the side where it is more concentrated; see figure 3.4, right). As you might suspect, active transport through the plasma membrane consumes cellular energy.

Bulk transport is an energy-driven process that serves to transport large quantities of materials into the cell or out of the cell. Bulk transport *into* the cell is also known as **endocytosis**, and bulk transport *out of* the cell is called **exocytosis**. During endocytosis, portions of the plasma membrane infold to trap extracellular materials in a small vesicle; the vesicle then pinches off, entering the cytosol (figure 3.5, left). During exocytosis, intracellular vesicles containing cellular secretions or wastes fuse with the plasma membrane and empty their contents into the cell's surroundings (figure 3.5, right). Clearly, the plasma membrane is an active and dynamic structure and its depiction as a simple line in drawings such as figure 3.1 belies its great complexity.

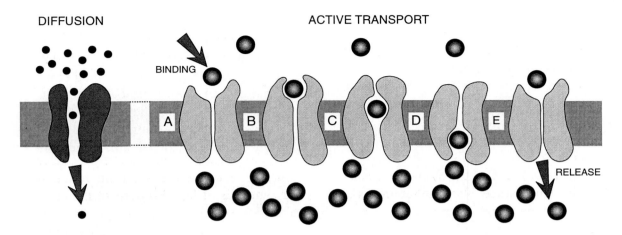

Figure 3.4
Diffusion (left) and active transport (right). During diffusion, molecules and/or ions pass through the plasma membrane in the direction of the concentration gradient. In the diagram above, integral membrane protein molecules form a channel through which the diffusion of molecules (the small black circles) takes place. Diffusion may be facilitated by integral proteins possessing enzymatic activity. No cellular metabolic energy is consumed during diffusion. As is depicted in the right half of the diagram, during active transport, integral membrane proteins transfer molecules through the membrane *against* the concentration gradient. The molecules being transferred through the membrane (the larger spheres) form a complex with the membrane protein serving in the transfer. Molecular rearrangement of the protein acts to transport the molecule through the membrane. Cellular metabolic energy is consumed during active transport.

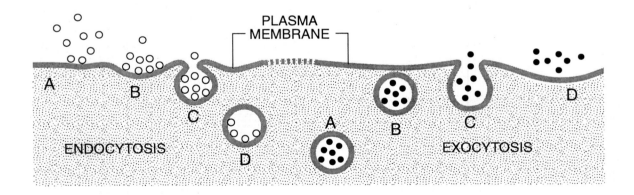

Figure 3.5
Endocytosis (left) and exocytosis (right). During endocytosis, the plasma membrane forms pockets that trap extracellular materials. These pockets ultimately pinch off from the membrane to form intracellular vesicles. During exocytosis, vesicles containing wastes or secretory products fuse with the plasma membrane and release their contents extracellularly.

The Cytoskeleton

Radiating through many cells of the body are the *microfilaments* and *microtubules* of the **cytoskeleton** (fig. 3.1). These structures give shape and form to the cell and are also involved in cell movement. The cytoskeletal elements appear to be interconnected by a network of finer threadlike structures comprising what is called the *microtrabecular lattice*. This lattice also interconnects a number of membranous organelles and ribosomes.

The Endoplasmic Reticulum and Ribosomes

Within the cytoplasm of most cells is an extensive network of branching and fusing membrane-limited channels (or *cisternae*) collectively called the **endoplasmic reticulum** (usually abbreviated **ER**). The membranes of the ER effectively divide the interior of a cell into two phases: the *intracisternal phase* (which includes the space and materials *within* the channels) and the cytosol (the background substance through which the channels radiate). The ER membranes (as all other cellular membranes) are comprised of protein and lipid molecules that are organized in much the same manner as in the plasma membrane.

In the cytosol are large numbers of small particles called **ribosomes**. These particles, which are comprised of RNA and protein molecules, are distributed along the cytosolic surface of the endoplasmic reticulum ("attached" ribosomes) and are also free in the cytosol ("free" ribosomes; see figure 3.1). Free ribosomes may be interconnected by fine threads of the microtrabecular lattice.

Ribosomes synthesize the cell's proteins. Endoplasmic reticulum with attached ribosomes is called **rough ER** (**RER**), the membranes of the RER typically being sheet-like. Endoplasmic reticulum without attached ribosomes is called **smooth ER** (**SER**), the membranes usually forming a network of branching and fusing tubes. Smooth ER appears to be the site for cellular synthesis of fats. Portions of the endoplasmic reticulum may merge with the plasma membrane and the nuclear envelope (see below).

Nucleus

All cells (human or otherwise) may be assigned to one of two major categories: **eukaryotic** (i.e., having a "true nucleus") and **prokaryotic** (i.e., "before nucleus"). In eukaryotic cells, the genetic material is separated from the rest of the cell by a membranous envelope that forms a structure called the cell **nucleus**. In prokaryotic cells, the genetic material is *not* separated from the remainder of the cell by membranes. Essentially all animal and plant cells are eukaryotic, whereas prokaryotic cells include bacteria and the so-called pleuropneumonia-like organisms.

In human cells (which are eukaryotic), the nucleus is a relatively large structure frequently located near the center of the cell. The contents of the nucleus are separated from the remainder of the cell by two membranes that together form the **nuclear envelope** (see figure 3.1). The space between the two membranes of the nuclear envelope is called the **perinuclear space**. At various positions, the outer membrane of the envelope fuses with the inner membrane to form the **nuclear pores**. These pores provide a direct path between the cytosol and the contents of the nucleus. Some of the nuclear pores may be plugged by a granular material. The outer nuclear membrane may have ribosomes attached to its cytosol side and may also merge with the membranes of the endoplasmic reticulum.

The nucleus contains most of the cell's DNA and therefore its genetic apparatus. The DNA is intimately associated with protein to form a fibrous complex called **chromatin**. Just prior to nuclear division, the chromatin of a cell condenses to form a discrete number of visible bodies called **chromosomes,** which are apportioned between the daughter cells produced by cell division (see chapter 15). The nucleus often contains one or more dense, granular structures called **nucleoli**. Nucleoli, which are not bounded by a membrane, contain concentrations of particles that ultimately are incorporated into the cell's ribosomes.

Mitochondria

Distributed through the cytosol are large numbers of organelles called **mitochondria** (figs. 3.1 and 3.6). The number of mitochondria per cell is quite variable; for example, sperm cells have fewer than 100 mitochondria, kidney cells generally contain less than 1000, and liver cells may contain several thousand. Mitochondria are oval bodies enclosed by two distinct membranes called the *outer* and *inner* mitochondrial membranes. The inner membrane separates the organelle's volume into two phases: the **matrix**, which is a gel-like fluid enclosed by the inner membrane, and the fluid-filled **intermem-**

Figure 3.6
Mitochondrion cut open to reveal the complex membranous organization of the organelle's interior (see text for details).

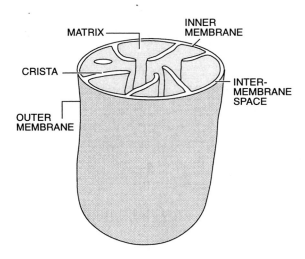

brane space between the inner and outer membranes. The inner membrane has a much greater surface area than the outer membrane because it possesses folds that extend into the matrix. These projections are called **cristae**.

Mitochondria play several different roles in a cell. Probably the most important is the mitochondrion's capacity to break down certain organic acids into carbon dioxide and water, and in so doing produce chemical energy for the cell in the form of *adenosine triphosphate* (ATP; see chapter 4).

Golgi Bodies

Golgi bodies are organelles that consist of sets of smooth, flattened membranous cisternae stacked together in parallel rows and surrounded by vesicles of various sizes (figs. 3.1 and 3.7). The vesicles that surround Golgi bodies represent small

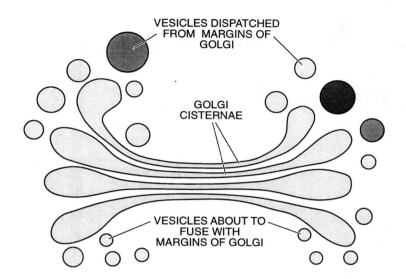

VESICLES DISPATCHED
FROM MARGINS OF
GOLGI

GOLGI
CISTERNAE

VESICLES ABOUT TO
FUSE WITH
MARGINS OF GOLGI

Figure 3.7
Golgi body. A Golgi body consists of a stack of flattened, membranous cisternae. Vesicles fuse with and are also dispatched from the margins of the stack.

membranous chambers that are either about to fuse with the Golgi cisternae or have just been dispatched from the Golgi cisternae. These organelles play a variety of functions including (1) the packaging of chemical substances that are to be secreted from the cell, (2) the chemical modification of proteins that have been synthesized by the cell's ribosomes, (3) the synthesis of certain of the cell's polysaccharides, and (4) the production of membrane components for the cell's plasma membrane and other membranous organelles.

Lysosomes and Microbodies

Most human cells contain small vesicular structures called **lysosomes** (see figure 3.1). Lysosomes contain quantities of various *hydrolytic* enzymes capable of digesting proteins, nucleic acids, polysaccharides, and other materials. Under normal conditions, the activity of these enzymes is confined to the interior of the organelles and is therefore isolated from the surrounding cytosol. However, if the lysosomal membrane is ruptured, the released enzymes can quickly degrade the cell. Among their various roles, lysosomes take part in the intracellular digestion of particles that are ingested by the cell during *endocytosis* and the intracellular scavenging of worn and poorly functioning organelles.

Lysosomes are related to another family of organelles called **microbodies**. Among human cell microbodies, the most common are **peroxisomes**. These small organelles contain a number of enzymes whose functions are related to the breakdown of potentially harmful peroxides that are produced during a cell's metabolism.

Cilia

The surfaces of many human cells possess rows of hairlike extensions called **cilia** (see figure 3.1). These organelles serve to move a substrate across the cell surface (such as the movement of mucus in the respiratory tract, or the movement of an egg cell during its passage through the oviduct from the ovary to the uterus). Each cilium is covered by an extension of the plasma membrane. Internally, cilia contain an array of microtubules that run from the base of the organelle (just below the plasma membrane) toward the cilium's tip.

BACTERIA

Bacteria are small, prokaryotic microorganisms having cell sizes that are rarely greater than that of a mitochondrion of a human cell (fig. 3.8). Bacteria are the causes of a number of human diseases and infections, including *pneumonia*, *strep throat*, *syphilis*, and *tuberculosis* (as you will learn in chapter 11, not all bacteria present in the body are harmful). Bacterial cells are generally enclosed within a protective wall formed from protein, lipid, and polysaccharide. The wall lies outside of the cell's plasma membrane and its content of a particular protein-carbohydrate complex is the basis of the classification of bacteria as "gram-positive" or "gram-negative." Internally, bacteria have a simplified structure, with few organelles. Although some bacteria have membranes within the cytosol, there are no structures comparable to the endoplasmic reticulum of eukaryotic cells. Bacteria contain large numbers of ribosomes, nearly all of which are free in the bacterial cytosol; some ribosomes may be attached to the interior surface of the plasma membrane.

In bacteria, the nuclear material is not separated from the cytosol by membranes, as it is in human (and other eukaryotic) cells. However, the nuclear material of a bacterial cell is usually concentrated in a specific region of the cell referred to as a *nucleoid*. Most of the hereditary material of the cell is carried by a single, circular chromosome; however, a small amount of genetic material is also present in small circular bodies called *plasmids*.

VIRUSES

Viruses are not cells, and although they are smaller even than most prokaryotic cells, they are diverse in size and organization. Viruses are the sources of many infections and diseases; even bacteria are subject to infection by viruses. Among the viruses that

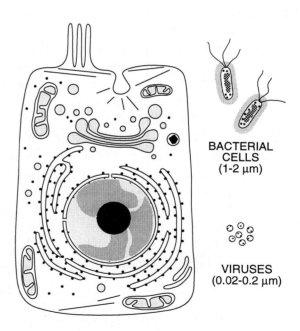

BACTERIAL CELLS (1-2 μm)

VIRUSES (0.02-0.2 μm)

HUMAN CELL (10-20 μm)

Figure 3.8
Relative sizes of a typical human cell, two bacteria, and several viruses. The dimensions shown are micrometers (μm). One micrometer is one-millionth of a meter.

attack human cells, the most notorious are those that cause *smallpox, chicken pox, rabies, poliomyelitis, mumps, measles, influenza, hepatitis, AIDS* (acquired immune deficiency syndrome), and the "common cold." Even certain leukemias and cancers are of viral origin.

Most viruses are either rod-shaped or quasi-spherical and contain a nucleic acid **core** surrounded by a specific geometric array of protein molecules that form a coat or **capsid** (fig. 3.9). In many viruses, a lipoprotein envelope surrounds the capsid (e.g., influenza virus, herpes virus, and smallpox virus).

The Cycle of Infection of a Virus

Solitary viruses do not carry out metabolism and are incapable of reproducing. The re-

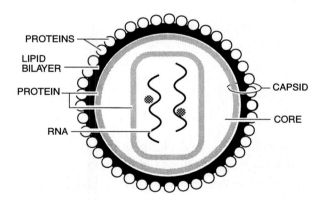

Figure 3.9
Structure of the AIDS virus. Although the organization of viruses is quite variable, in most viruses the viral genetic information (consisting either of DNA or RNA, depending on the virus), is found in the viral core, which is enclosed within a protein and lipid-rich capsid. As seen here, the genetic material of the AIDS virus is RNA.

production of viruses requires a host cell and usually takes the pattern illustrated in the stages of figure 3.10. One or more viruses attach to specific sites on the plasma membrane of the host cell (fig. 3.10A). Following attachment, the viral genetic material (nucleic acid) enclosed within the viral core is inserted through the plasma membrane of the host and into the host's cytosol (fig. 3.10B). With infection of the host cell completed, one of two alternative sequences of events takes place. The viral nucleic acid may immediately take control of the metabolism of the host, with the result that new viral proteins, nucleic acids, and other components are formed (fig. 3.10D).

On some occasions, however, and only for certain viruses, the injected nucleic acid does not cause the immediate proliferation of new viruses inside the host. Instead, the injected nucleic acid is incorporated into the host's own genetic material, and the host cell continues to function in its normal manner (fig. 3.10C). However, the duplication of the infected cell's genetic material prior to cell division is accompanied by duplication of the incorporated viral nucleic acid. As a result, several generations of infected human cells may be produced, each containing a copy of the viral nucleic acid. The dormant viral nucleic acid within the host is referred to as a **provirus**. Sooner or later, in one of the newly produced generations of host cells, the provirus nucleic acid will begin to direct the production of new viral components (fig. 3.10D). These viral components combine in the host to form large numbers of new viruses (fig. 3.10E). The new viruses then exit the cell by disruption of the cell's plasma membrane (fig. 3.10F). (Some viruses exit the host by budding off, enclosed in a small piece of the host cell's

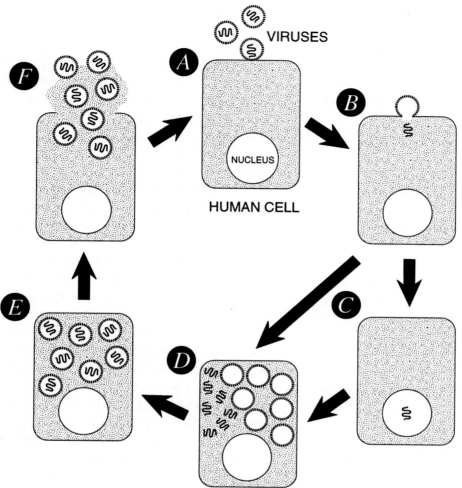

Figure 3.10
The cycle of infection of a virus.

plasma membrane.) The cycle of infection then repeats itself.

Viral Nucleic Acids

Viruses may be assigned to one of two major groups: DNA viruses (viruses in which the hereditary material is DNA) and RNA viruses (viruses in which the hereditary material is RNA). The DNA viruses include those that cause chicken pox, herpes blisters, infectious mononucleosis, and shingles. Among the RNA viruses are those that cause polio, mumps, measles, influenza, AIDS, and colds. Some RNA viruses are also believed to be the causes of certain cancers.

FROM CELLS TO ORGAN SYSTEMS

As noted in chapter 1, in the human body groups of similar cells performing the same function are grouped together to form **tissues**. There are four major types of tissues in the body; these are (1) **epithelial** tissues, (2) **muscle** tissues, (3) **nerve** tissues, and (4) **connective** tissues.

Epithelial tissues line the external and internal body surfaces, such as the skin and the walls of the digestive and respiratory tracts. Their roles are principally *protective*, but much *absorption* of chemicals into the body or the *secretion* of materials from the body involve the actions of epithelium. The principal function of muscle tissue is that of *contraction*, and this tissue is responsible for all body movements (internal movement as well as external movement). Nerve tissue comprises such structures as the brain and spinal cord; this tissue includes cells that are specialized for the *conduction* and *transmission* of signals (e.g., nerve impulses) from one part of the body to another. Connective tissue serves diverse functions. Some forms of connective tissue join other tissues together. For example, tendons connect muscle tissue to bone. Other forms, such as cartilage, give *support* to the body. Fat (or adipose) tissue serves as storage sites.

A group of different tissues held together and working as a unit in order to provide a basic body function is called an **organ**. For example, the heart is an organ comprised of muscle tissue, connective tissue, nerve tissue, and epithelial tissue. All of the tissues of this organ contribute in some manner to the organ's principal function, namely to pump blood through the body.

Groups of organs that serve an overall function—even though the organs may be located in diverse regions of the body—comprise an **organ system**. For example, the heart is just one organ of the organ system known as the circulatory system; the other organs of the circulatory system include arteries and veins. In a like manner, the brain is just one organ of the organ system known as the nervous system; the stomach is just one organ of the organ system known as the digestive system, and so on.

In succeeding chapters of this book, we will consider the physiology of the body by surveying the organization and functions of each of the body's organ systems.

SELF TEST[*]

True/False Questions

1. Cells are the basic functional and structural units of the body.

2. Animal cells are eukaryotic cells, whereas bacterial cells are prokaryotic.

3. In bacteria, the nuclear material is not separated from the cytosol by membranes.

4. In all viruses, as in all animal and plant cells, the genetic material consists of DNA.

5. The bulk transport of materials into a cell is known as endocytosis

_____

[*] *The answers to these test questions are found in Appendix III at the back of the book.*

Multiple Choice Questions

1. The discrete, functional parts of a cell are called (A) molecules, (B) ions, (C) organs, (D) organelles, (E) membranes.

2. Cells are bordered at their surface by (A) a layer of proteins, (B) a layer of lipids, (C) the plasma membrane.

3. The network of branching and fusing channels that radiates through the interior of many cells is called the (A) cytoskeleton, (B) microtrabecular lattice, (C) endoplasmic reticulum, (D) intracellular network.

4. Ribosomes synthesize a cell's (A) nucleic acids, (B) proteins, (C) amino acids, (D) sugars.

5. Diseases such as mumps, measles, chicken pox, and influenza are caused by (A) bacteria, (B) viruses, (C) mycoplasmas, (D) viroids.

6. Groups of similar cells performing the same function comprise a (A) tissue, (B) organ, (C) organ system.

CHAPTER 4

THE PHYSIOLOGY OF MUSCLE

The body's muscles contain a specialized form of tissue whose cells have the capacity to contract, that is, to physically shorten. It is the muscle cells' capacity for **contraction** that is responsible for essentially all movements associated with the body, whether the movements are overt and external (e.g., movements of the arms and legs) or internal (e.g., movements of the digestive organs and the beating of the heart).

TYPES OF MUSCLE TISSUE

There are three different types of muscle tissue found in the body; these are called (1) **striated** (or **skeletal**) muscle tissue, (2)

smooth muscle tissue, and (3) **cardiac** muscle tissue. Among the three, much more is known about the chemistry, organization, and physiology of striated tissue than is known about the others, primarily because striated tissue is so much more accessible to study. Consequently, much of this chapter's discussion of muscle will center around the organization and physiology of striated tissue. Let's begin our consideration of muscle by identifying the principal roles and dispositions of these tissues in the body.

Striated Muscle

The word "striated" means "striped." Striated muscle is given this name because when the tissue is examined with either a light or electron microscope, the cells that comprise the tissue appear to have stripes (striations) running across their widths (i.e., the stripes are oriented perpendicular to the long axes of the cells). Because most of the body's striated muscles are responsible for the movements of the skeleton, striated muscle is also known as **skeletal** muscle. Sometimes striated muscle is also referred to as *voluntary* muscle, because the actions of this type of muscle are generally under conscious control. Skeletal muscles are attached to bones by **tendons** that are located at each end of the muscle. Tendons are bundles of connective tissue that pull on the bones when a muscle contracts, thereby causing one of the bones to pivot at its joint with another bone (fig. 4.1).

Smooth Muscle

Smooth muscle tissue is called "smooth" because when examined with a microscope, this type of muscle lacks the stripes characteristic of striated cells. Smooth muscle tissue is associated with the body's internal organs. For example, it is abundant in the walls of the digestive organs (e.g., the stomach and intestine), the urinary bladder, reproductive tract, and the walls of the major arteries and veins. Like striated muscle, smooth muscle is associated with movement, but the movement is internal and more subtle. Smooth muscle is also found in the skin where it is responsible for the appearance of the "goose bumps" that one experiences when the skin is exposed to cold temperatures. Smooth muscle is also known as *involuntary* muscle because much of the movement of smooth muscles goes on without conscious awareness.

Cardiac Muscle

As its name implies, "cardiac" muscle is found principally in the heart, although the tissue is also found in the walls of the large arteries that are immediately adjacent to the heart. The major role of cardiac muscle is to provide the continuous and rhythmic contractions that propel the blood through the circulatory system. When examined with a microscope, cardiac muscle tissue, like skeletal muscle, has a striated appearance, although the striations are not as pronounced. Generally speaking, the rhythmic contractions of heart muscle are not under voluntary control.

ORGANIZATION OF STRIATED MUSCLE

Altogether, there are about 600 striated (or skeletal) muscles in the human body. The

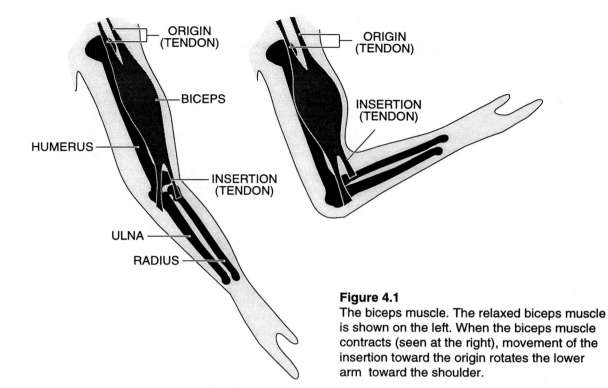

Figure 4.1
The biceps muscle. The relaxed biceps muscle is shown on the left. When the biceps muscle contracts (seen at the right), movement of the insertion toward the origin rotates the lower arm toward the shoulder.

contractions of these muscles control operations that are as delicate as the blinking of an eyelid or as complex as tying your shoelaces. The organization of a typical striated muscle can be illustrated using the biceps muscle as an example (fig. 4.1). The biceps muscle is one of several muscles of the upper arm; its principal function is to draw the lower arm (i.e., the "forearm") upward, toward the shoulder. The action of this muscle demonstrates a characteristic that is typical of the body's striated muscles, namely that during contraction one end of the muscle undergoes very little change in position, whereas the other end moves a considerable distance. In the case of the biceps muscle, the end that undergoes little positional change (called the muscle's **origin**) is the end attached to the shoulder region, whereas the end of the biceps that attaches to the forearm (called the muscle's **insertion**) moves sev-

eral inches. The origins and insertions of the muscles that move the arms, fingers, legs, toes, head, jaw, and eyes are readily identified. However, for many of the body's muscles (such as those of the chest and the upper and lower back), the origin and insertion are not so obvious.

4-1

The movements of the biceps also illustrate another important aspect of muscle action; that is, muscles *pull*, they do not push. The action of a muscle can be likened to that of a tow-rope in that the rope can be used to pull an object along, but it can't be used to push an object along. Even familiar pushing actions such as pushing yourself away from the dinner table or pressing down on a car's accelerator pedal are the results of muscle shortening (i.e., contracting). In the case of pushing yourself away from a table, the contraction of another muscle of the upper arm, called the **triceps** muscle, rotates the forearm away from the shoulder (the elbow acting as the pivot), thereby extending the arm. This action is opposite to that of the biceps. Whenever two muscles have opposite effects or opposite functions they are said to be *antagonistic*. Thus, the biceps and triceps are antagonistic muscles.

The gross actions of a muscle are founded on the individual actions of the thousands (millions in the larger muscles) of contractile cells that make up the muscle. In striated muscles, the muscle cells are long, threadlike structures oriented along the axis that leads from the muscle's origin to its insertion. Because they are threadlike, striated muscle cells are also called striated muscle **fibers**; that is, the term "fiber" is equated with the term "cell." In the longer muscles (e.g., the muscles of the arms and legs), individual fibers may be several inches long and are among the largest cells in the body. The muscle fibers that comprise a striated muscle do not extend the entire length of the muscle. Instead, as noted earlier, the ends of a muscle are formed by a tissue called *tendon* that serves to connect the contractile portion of the muscle to the bone on which the muscle pulls.

The internal organization of a striated muscle is shown in figure 4.2, which depicts a cross section through the muscle's "belly." Each skeletal muscle is comprised of several different tissues that collectively form the muscle trunk. In addition to muscle tissue, a skeletal muscle contains connective tissue, blood vessels, blood, and nerve tissue. The muscle is surrounded by a layer of connective tissue called the **epimysium**. Internally, the muscle is organized into subunits called **fascicles**, each of which is surrounded by a sheath of connective tissue called **perimy-**

sium. Each fascicle contains dozens of individual muscle fibers that are separated from one another by a connective tissue called **endomysium**. The endomysium, perimysium, and epimysium provide the framework that holds the muscle together. Blood vessels pass through these layers of connective tissue, delivering oxygen and nutrients to the organ and carrying away the organ's metabolic wastes. Likewise, nerves pass through the connective tissue bringing the signals that trigger the muscle's contractions.

4-2

Structure of a Striated Muscle Fiber

The fibers seen in figure 4.2 are depicted in cross section; to appreciate their remarkably complex internal organization, we need to withdraw one of these fibers from the muscle and turn it sideways so that we can view its length. This arrangement is shown in figure 4.3. Because most striated fibers are exceedingly long, figure 4.3 depicts only a small portion of the entire cell. The unusual length of these cells stems from the fact that during embryonic development, the muscle fibers are formed by the end-to-end fusion of many smaller cells, thereby forming a continuous tubelike structure. This also explains why striated muscle fibers are multinucleate (i.e., they have many nuclei).

The plasma membrane that encloses each cell is called the **sarcolemma**; in addition to its exceptional length, the sarcolemma is characterized by numerous porelike invaginations that extend into the interior of the cell (i.e., into the cytoplasm or **sarcoplasm**) at right angles to the cell's long axis. Within the cell, these membranous invaginations form a highly branched network of tubules

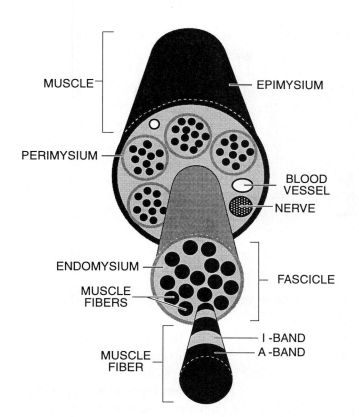

MUSCLE

PERIMYSIUM

EPIMYSIUM

BLOOD VESSEL

NERVE

ENDOMYSIUM

MUSCLE FIBERS

FASCICLE

I -BAND

A -BAND

MUSCLE FIBER

Figure 4.2
Internal organization of a striated muscle. The muscle is comprised of several fascicles. In turn, each fascicle contains a number of muscle fibers.

called T-tubules or the **T-system** (i.e., "T" = "transverse"). Internally, the sarcoplasm is further subdivided by intracellular membranes called the **sarcoplasmic reticulum**. Extending through the cell are many long, cylindrical units called **myofibrils**. These are the cell's contractile units; that is, the myofibrils are the structures that directly bring about the cell's physical shortening. Mitochondria are abundant in striated muscle cells and are usually arranged in rows, sandwiched between groups of myofibrils.

4-3

Each myofibril is divided into units of equal length and content called **sarcomeres**; these are considered to be the *functional units* of muscle contraction. As seen in figures 4.3 and 4.4, each myofibril reveals areas of different density or darkness. The alternating light and dark areas (the stripes or striations

referred to earlier) are respectively called the **I** (i.e., **isotropic** [which means transparent]) and **A** (i.e., **anisotropic** [which means opaque]) bands. The alternating A and I bands are depicted in greater detail in figure 4.4.

At the center of each I-band is a dark line called a **Z-line**. Since myofibrils are long, cylindrical structures, the Z-line is actually the edge of a coin-shaped, circular disk; thus, a Z-line has also been referred to as a *Z-disk*. Within each A-band, there is a region that appears somewhat less dense than the remainder of the band; this is called an **H-zone**. Finally, at the center of an H-zone is the **M-line**. Those parts of each myofibril that extend from one Z-line to the next are termed **sarcomeres**. Accordingly, a myofibril may be thought of as a succession of many sarcomeres.

4-4

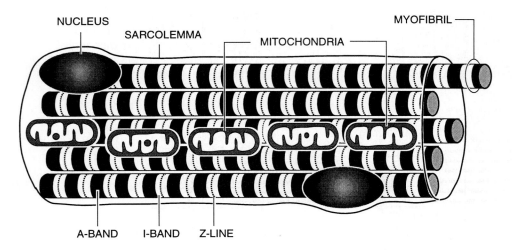

Figure 4.3
Organization of a striated muscle cell (fiber), showing several myofibrils. Only a small portion of the total length of the fiber is depicted here. The T-system and sarcoplasmic reticulum are not shown.

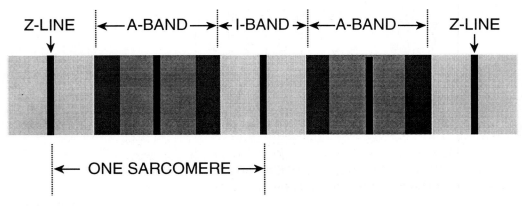

Figure 4.4
Pattern of alternating A-bands and I-bands in a single myofibril. Each I-band is bisected by a Z-line.

Arrangement of Myofilaments Within a Myofibril

The myofibrils of a muscle fiber are comprised of even finer, threadlike structures called **myofilaments** (figure 4.5). There are two kinds of myofilaments in a myofibril: (1) **thin myofilaments**, composed primarily (but not solely) of a protein called **actin**, and (2) **thick myofilaments**, composed of a protein called **myosin**. One end of each thin myofilament is anchored to a Z-line, while the other end projects toward the center of the sarcomere; the thin myofilaments are *isotropic* (light passes through them). The thick myofilaments are *anisotropic* (i.e., they absorb light) and are sandwiched between the thin filaments, overlapping their ends.

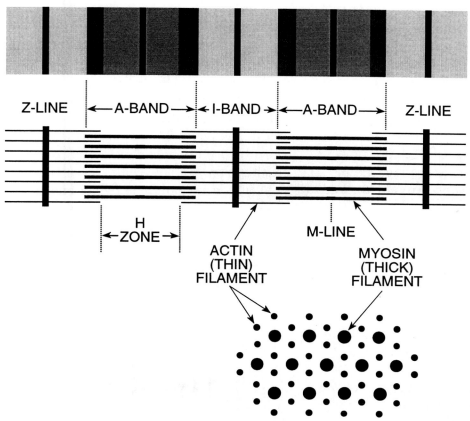

Figure 4.5
One end of each actin (thin) filament is anchored in a Z-line, while the other end
extends toward the center of the sarcomere. Myosin (thick) filaments are sandwiched
between the thin filaments but are not attached to the Z-lines. As seen in the cross-
section (lower right), six thin filaments are equally spaced around one thick filament
and three thick filaments are equally spaced around each thin filament.

Thick myofilaments are not attached to the Z-lines. As seen at the bottom of figure 4.5, which shows the filaments in cross section, six thin filaments are equally spaced about each thick filament in an hexagonal array, and three thick filaments are equally spaced around each thin filament. Dozens of thick and thin myofilaments are bundled together to form each myofibril. It is the light-absorbing properties and specific positions of the thick and thin myofilaments that serve as the basis for the alternating dark (i.e., A) and light (i.e., I) banding pattern that characterizes each myofibril. Within a sarcomere, an A-band extends from one end of a stack of thick filaments to the other end of the stack (fig. 4.5). Note that in addition to including all of the myosin filaments, an A-band includes a small portion of each of the thin filaments (i.e., the segments that overlap the thick filaments). That portion of each A-band that is devoid of actin filaments appears less dense than the remainder of the A-band and creates the H-zone. Finally, at the center of the myosin filaments is the dark M-line. Neighboring myofibrils within a striated muscle fiber are all oriented in register, thereby giving the entire fiber its striped appearance.

During muscle contraction, the thick and thin myofilaments of the myofibrils slide past each other; as a result, the following changes in the appearances of each sarcomere occur: (1) the H-zones almost disappear, (2) the I-bands become very narrow, and (3) the Z-lines are drawn closer together. These changes are depicted in figure 4.6.

Molecular Organization of Myofilaments

Thin filaments contain three kinds of proteins called **actin**, **tropomyosin**, and **troponin**. Among these, actin is by far the most abundant. Individual actin molecules are globular (i.e., sphere-shaped), but in thin myofilaments, the actin molecules are strung together to form long chains (much like a beaded necklace). Two long chains of globular actin molecules are coiled about each other to form a double helical structure (fig. 4.7). Each helical chain contains about 400 globular actin molecules. Tropomyosin molecules lie in the two grooves formed by the double helical chains of actin molecules. Finally, one troponin molecule is bound to each tropomyosin molecule (fig. 4.7).

Unlike actin molecules (which are globular), individual myosin molecules are filamentous; that is, they are long, slender proteins (fig. 4.7). At one end of each myosin molecule there is a globular "head" that projects away from the molecule's main axis or "tail." A complete thick filament contains dozens of myosin molecules that are arranged side-by-side and in such a manner that their globular heads occur only near the two ends of the filament (figures 4.7 and 4.8). The tails of the myosin molecules lie next to one another and create the thick, central segment of the filament. (The middle

Figure 4.6
During contraction, thin and thick filaments slide over each other pulling the Z-lines of neighboring sarcomeres closer together.

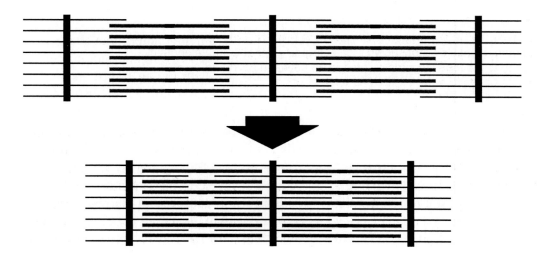

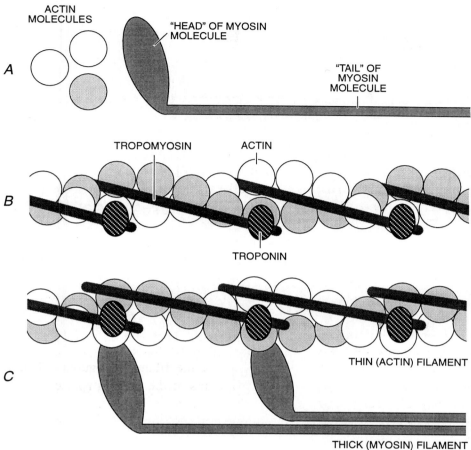

Figure 4.7
Molecular organization of thin and thick myofilaments. *A* shows individual molecules of actin (a globular protein) and myosin (a filamentous protein). Actin molecules are strung together to form two interwoven helical chains. As seen in *B*, seated in the grooves of the actin double helix are molecules of tropomyosin. Associated with each tropomyosin is one molecule of troponin. Together, the actin helixes, tropomyosin, and troponin comprise the thin filaments. Ca^{++} binding by troponin molecules (not shown in the diagram) brings about a configurational change that exposes the sites of interaction between actin and myosin. Binding of myosin heads to actin is depicted in *C*.

region of the filament's backbone lacks the globular myosin heads, since these are found only at the ends of the thick filament.)

Of special importance in the cyclic contraction and relaxation of muscle is the role played by calcium ions (i.e., Ca^{++}). In a noncontracting fiber, these ions are concentrated in small extensions of the sarcoplasmic reticulum. It is the release of Ca^{++} from the sarcoplasmic reticulum that initiates filament sliding (see later). Before Ca^{++} is released, troponin and tropomyosin block the sites of interaction between the myosin heads and actin. When Ca^{++} is released from the

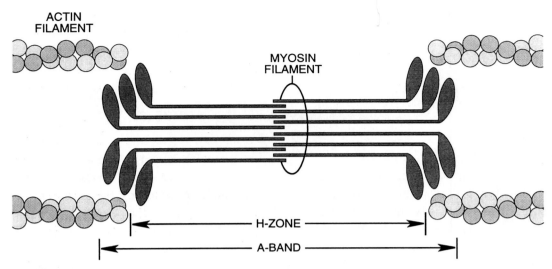

Figure 4.8
Organization of myosin molecules in a thick filament (see also figure 4.7).

sarcoplasmic reticulum onto the filaments, the calcium ions bind to troponin, and this causes tropomyosin to pull away from (and expose) the actin-myosin interaction sites.

In the interaction between actin and myosin that follows, the myosin heads pivot back and forth, pulling the actin filament toward the center of the myosin filament (fig. 4.9).

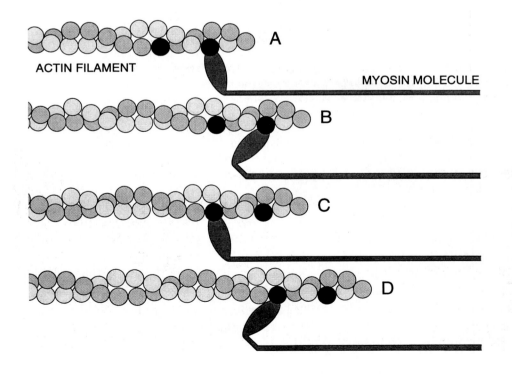

Figure 4.9
The oarlike action of the heads of the myosin filaments pulls the actin filaments toward the center of the myosin filament. (Two of the actin molecules with which the myosin head interacts are shown in black.)

The Contraction/Relaxation Cycle

Contraction of a striated muscle cell is initiated by a nerve impulse travelling along a nerve cell toward its junction with the muscle fiber. In these junctions, called **myoneural junctions** (fig. 4.10), the sarcolemma of the muscle cell forms a number of infoldings or pockets into which branches of the nerve fiber extend. The surface of the nerve fiber does not touch the sarcolemma; rather, a small fluid-filled gap called the **myoneural cleft** separates the two surfaces. The arrival of the nerve impulse at the terminals of the nerve fiber causes the release of a **neurohumor** or **neurotransmitter** into the myoneural cleft. For most myoneural junctions in the body, the neurohumor that is released is a substance called **acetylcholine**. Acetylcholine is manufactured in the terminals of the nerve fiber and is stored there as **synaptic vesicles**. When a nerve impulse reaches these terminals, the synaptic vesicles migrate to and fuse with the nerve fiber's plasma membrane, emptying the acetylcholine into the cleft (fig. 4.10).

4-5

After its release from the terminal branches of the nerve fiber, the acetylcholine diffuses across the fluid-filled gap that separates the nerve cell's surface from the sarcolemma. On reaching the sarcolemma, the acetylcholine attaches to specific receptor molecules in the sarcolemma. Prior to binding the neurotransmitter, the sarcolemma is *polarized*, its outside surface being electrically positive with regard to its inside surface. Binding of acetylcholine to the sarcolemma initiates *depolarization* of the membrane; that is, the charge difference between the outside and inside membrane surfaces diminishes. Indeed, the polarity of

Figure 4.10
The myoneural junction. As seen on the left, the myoneural cleft separates the muscle cell's sarcolemma from the surface of the nerve fiber terminals. When an impulse reaches the nerve fiber terminals, synaptic vesicles rich in neurotransmitter fuse with the nerve fiber's plasma membrane and empty the neurotransmitter into the cleft (shown in the magnified view on the right). Binding of neurotransmitter to the sarcolemma initiates a sequence of actions that culminates in contraction of the muscle fiber.

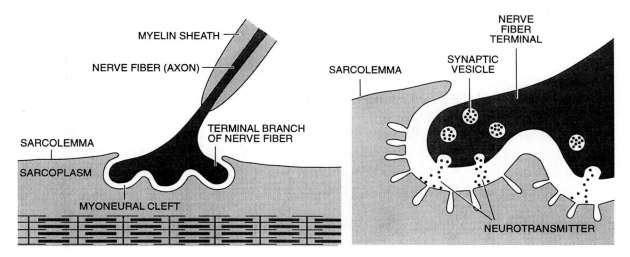

the membrane temporarily reverses as the inside surface becomes electrically positive with regard to the outside surface.

Depolarization and reverse polarization of the sarcolemma begin at the myoneural junction but quickly spread from there in all directions across the surface of the muscle cell. The wave of electrical change spreads into the complex system of invaginations forming the sarcoplasmic reticulum and T-system. As noted earlier, the sarcoplasmic reticulum of a noncontracting (i.e., "resting") muscle cell is rich in Ca^{++}; that is to say, the concentration of Ca^{++} in the sarcoplasmic reticulum is much greater than in the sarcoplasm. This concentration difference exists despite the fact that the membranes of the sarcoplasmic reticulum contain pores that are permeable to Ca^{++} (the pores are called **Ca^{++} channels**). The calcium ion concentration difference is sustained by the actions of enzymes in the sarcoplasmic reticulum that act as "pumps" for Ca^{++}. These calcium pumps act to return to the sarcoplasmic reticulum any Ca^{++} diffusing into the sarcoplasm. As a result, the sarcoplasmic reticulum remains rich in Ca^{++} whereas the sarcoplasm is Ca^{++} poor.

The depolarization of the sarcoplasmic reticulum brings about an increase in the Ca^{++} permeability of the reticulum membranes by widening the Ca^{++} channels. The result is that more calcium diffuses into the sarcoplasm than can be returned by the action of the calcium pumps. The sudden rise in the concentration of Ca^{++} in the sarcoplasm surrounding the myofibrils acts as the trigger for filament sliding.

As seen in figure 4.8, thick filaments are comprised of many individual myosin molecules, and at the two ends of the filament there is an annular array of myosin heads

projecting away from the filament's long axis at an angle of about 90°. Although the myosin heads can interact with actin units in neighboring thin filaments, in resting muscle the actin-myosin interaction sites on the actin units are masked by tropomyosin and troponin.

Also playing a major role in the contraction/relaxation cycle of muscle is **adenosine triphosphate** (abbreviated ATP). Large amounts of ATP are associated with the myofilaments. The heads of myosin molecules contain an enzyme capable of breaking down ATP, converting it to **adenosine diphosphate** (ADP) and **inorganic phosphate** (P). When this reaction occurs, the ADP and inorganic phosphate that are produced remain bound to the myosin head.

When, as described above, the depolarization of the membranes of the sarcoplasmic reticulum widens the Ca^{++} channels, large amounts of Ca^{++} rush from the sarcoplasmic reticulum onto the thin and thick filaments. The Ca^{++} reacts with troponin molecules of the thin filaments, causing the troponin and tropomyosin to move in such a manner as to expose the actin-myosin interaction sites on the actin molecules (see figure 4.7). The configurational change is immediately followed by the release of ADP and P from the myosin heads and the attachment of the myosin heads to the newly-exposed actin sites (figure 4.11, steps A and B). Interaction of the myosin heads with the actin sites is followed by a conformational change in which the head angle (which was about 90°) is changed to about 45° (figure 4.11, step C). This serves to pull the thin filament, causing it to slide along the thick filament in the direction of the thick filament's center. The sliding action, which occurs at the same time for thousands of

ACTIN FILAMENT

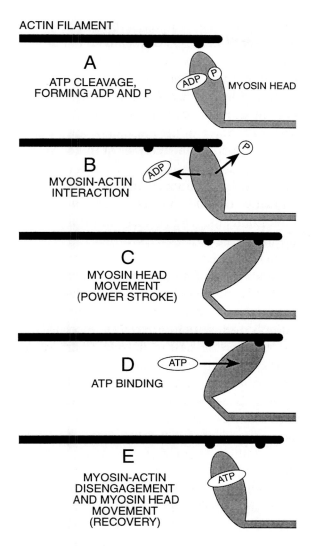

A

ATP CLEAVAGE,
FORMING ADP AND P

MYOSIN HEAD

B

MYOSIN-ACTIN
INTERACTION

C

MYOSIN HEAD
MOVEMENT
(POWER STROKE)

D

ATP BINDING

E

MYOSIN-ACTIN
DISENGAGEMENT
AND MYOSIN HEAD
MOVEMENT
(RECOVERY)

Figure 4.11
Interactions between actin and myosin during
filament sliding. In step A, ATP bound to the
myosin head has been enzymatically cleaved
forming ADP and P (inorganic phosphate).
ADP and P temporarily remain attached to the
myosin head. Release of ADP and P is followed by the formation of a cross-bridge
between the myosin head and specific sites on
the thin filament (step B). Cross-bridge formation is immediately followed by bending of the
myosin head, which pulls the actin filament
toward the center of the myosin filament (step
C). ATP then binds to the myosin head (step
D) and the myosin head detaches from the thin
filament. The cycle is then repeated.

actin and myosin molecules, is referred to
as the **power stroke**. The myosin heads,
which now are free of ADP and P and are
bound to actin sites, react with additional
ATP (figure 4.11, step D). This causes the
myosin heads to detach from the actin site
and reassume their 90° orientation, thereby
completing one full cycle of activity (figure
4.11, step E). So long as the Ca^{++} level remains elevated and the actin-myosin interaction sites of the thin filaments remain exposed, a new cycle is initiated.

Back at the myoneural junction, a muscle
fiber enzyme called *acetylcholinesterase*
(which is localized on the outside surface of
the sarcolemma at the myoneural junction)
begins to degrade the acetylcholine released
by the nerve ending. Degradation of all of
the acetylcholine at the myoneural junction
permits the sarcolemma at the junction to
repolarize (i.e., to restore the former electrical charge distribution across the membrane). The repolarization spreads like a
wave over the sarcolemma from the myoneural junction and passes into the T-system.
Repolarization of the sarcoplasmic reticulum
is accompanied by a narrowing of the Ca^{++}
channels, thereby allowing the calcium
pumps to return Ca^{++} from the sarcoplasm
to the sarcoplasmic reticulum. As Ca^{++} is
withdrawn into the sarcoplasmic reticulum,
troponin and tropomyosin molecules return
to their original positions on the actin filaments and once again mask the actin-
myosin interaction sites. Myosin heads are
no longer able to form cross-bridges with
the thin filament, thereby allowing relaxation of the fiber to occur as thin filaments
slide back to their original positions. The
filament movement that accompanies relaxation usually results from gravity, the contraction of an antagonistic muscle, or the
elasticity of the muscle tissue.

When acetylcholine is enzymatically degraded at the myoneural junction by *acetylcholinesterase*, two products are formed: acetate and choline. These substances are reabsorbed by the nerve ending and are used to produce new molecules of acetylcholine. Therefore, in a sense, acetylcholine is recycled at the myoneural junction.

Motor Units

The contractions of every striated muscle fiber are regulated through their myoneural junctions. However, a single nerve fiber gives rise to branches that form myoneural junctions with a number of *different* striated muscle fibers. The combination of one nerve fiber and all of the striated muscle fibers that the nerve fiber regulates constitutes a **motor unit** (fig. 4.12). The ability to exercise precise control of a striated muscle is related to the sizes of the muscle's motor units. Therefore, you should not be surprised to learn that the motor units of the muscles that move the eyes and the fingers have smaller number of muscle cells than the motor units of the arm and leg muscles or the back muscles. For example, in the muscles that move the eyes, each nerve fiber's terminal branches form junctions with

about 20 muscle cells; whereas in the muscles of the back, several hundred muscle cells are innervated by the terminal branches of a single nerve fiber.

An entire muscle (biceps, triceps, etc.) may be thought of as a collection of motor units. The force of the muscle's contraction is determined by the number of motor units that become active. To lift a lead weight off the table by contracting the biceps muscle requires the participation of many more motor units than are required to lift a pencil off the table.

In response to a *single* nerve impulse, the muscle fibers of a motor unit go through a cycle of contraction/relaxation that lasts less than a second. If a series of nerve impulses quickly follow one another in succession, the contraction of the muscle fibers of a motor unit can be sustained for a longer period of time. To raise an object off of the table and keep it suspended for several minutes requires a rapid, *asynchronous* flow of nerve impulses to different groups of motor units within the contracting muscle.

The "All-Or-None" Response

It is possible to elicit contractile responses from individual muscle fibers *in vitro* by

Figure 4.12
A motor unit. Each motor unit consists of a single nerve fiber plus all of the striated muscle fibers that the nerve fiber stimulates. The hypothetical motor unit depicted here contains three striated muscle fibers. Usually, motor units contain between 20 and 400 striated muscle fibers.

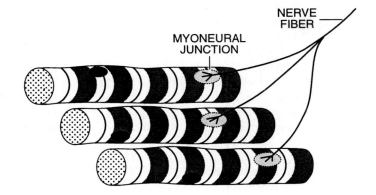

direct electrical stimulation of the fibers. In this way, it has been shown that the electrical stimulus must reach a **threshold** level in order to yield a contraction. Stimuli below the threshold value yield no response. Increasing the stimulus to levels further and further above threshold does not result in more forceful contractions of the muscle fiber. In other words, individual fibers respond to a stimulus by contracting maximally *or they do not contract at all*. This relationship is known as the law of **all-or-none**. It is important not to confuse the all-or-none response of individual fibers (or individual motor units) with graded contractions achieved by a whole muscle (such as the biceps). The graded response of whole muscles is due to the variable numbers of motor units that can be activated in order to move a particular weight. Each motor unit either responds fully or not at all.

Isotonic vs. Isometric Contractions

The contraction of a striated muscle can take either of two forms: (1) the contraction may be **isotonic** (iso = equal and tonic = strength), or (2) the contraction may be **isometric** (iso = equal and metric = length). During an isotonic contraction, the muscle shortens, thereby bringing the insertion closer to the origin (fig. 4.13). An example of an isotonic contraction is the lifting of a book or some other object of similar weight off the surface of a table. In contrast, during an isometric contraction, there is no change in the distance between the insertion and origin. An example of an isometric contraction would be the attempt to lift a 1000-pound weight off the table. The weight could not be lifted, and so the muscle's insertion

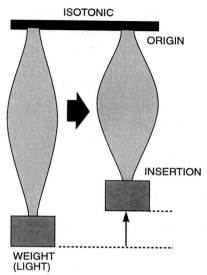

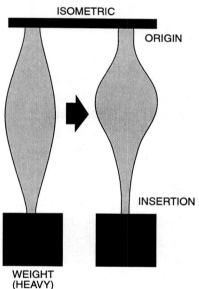

Figure 4.13

Isotonic and isometric muscle contractions. During an isotonic contraction (top), the muscle's insertion is pulled closer to its origin and the weight is moved (lifted). During an isometric contraction, the weight is too great to be moved. The contractile portion of the muscle shortens, but an equivalent amount of stretching occurs in the elastic portions of the muscle (e.g., the tendons). As a result, the distance between the origin and the insertion remains the same (hence "isometric" ["equal length"]).

would remain the same distance from the origin (fig. 4.13). This is not to imply that no contraction is occurring. Indeed, during an isometric contraction, muscle fibers are shortening; however, the shortening of the muscle fibers is accompanied by an equivalent amount of stretching in the muscle's elastic tissues (e.g., in the tendons).

4-6

EXPERIMENTAL STUDIES WITH WHOLE MUSCLES

Experimental studies can be made with a whole muscle that has been surgically removed from an animal. Using a stimulating **electrode**, an electric shock can be applied directly to the muscle, causing it to contract, and the nature of the contraction can be studied. Studies of this sort can be done using an apparatus like that shown in figure 4.14. One end of the muscle (usually the *origin*) is held in position by a stationary clamp, while the other end (the *insertion*) is attached to a pen and a small weight.

When the muscle contracts and relaxes, the pen rises and falls, inscribing a path on a **kymograph** drum that rotates a sheet of chart paper. The tracing or recording produced in this way is called a **myogram**.

Muscle Twitches and Summation

When a muscle is stimulated with an electric shock, either by application of the shock directly to the muscle itself or to the nerve that controls the muscle, the muscle contracts. When a single, brief electrical shock is applied to a muscle, the response is called a **twitch** (fig. 4.15) and is the simplest type of muscle response. Notice that after the stimulus is applied, there is a short delay, called the **latent period**, before contraction occurs. The latent period is due to the time it takes for the wave of depolarization to travel across the surface of the muscle fibers, pass into the T-system, and cause the release of Ca^{++} that triggers the sliding of the muscle filaments. Once the contraction is completed, the muscle returns to its

Figure 4.14
Classical apparatus that can be used to record muscle contractions. When the electrode electrically stimulates the muscle, the muscle contracts, raising the pen and tracing a path on the revolving kymograph drum. The tracing is called a myogram. The myogram shown here is a simple muscle twitch.

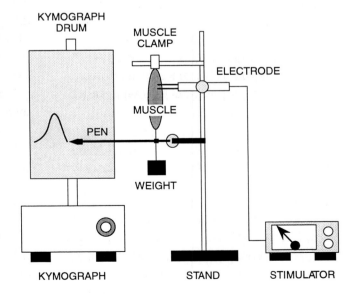

original length (i.e., the muscle relaxes). The duration of the twitch is about 0.1 seconds and is equal to the sum of the latent period (typically 0.01 seconds), contraction (about 0.04 seconds), and relaxation (about 0.05) seconds. Twitches are not common muscle responses, although you've probably experienced an occasional twitch (they are common in eyelid muscles). The duration of a muscle twitch varies from one muscle to another. Quick muscles, like those responsible for blinking, exhibit quick twitches whereas slower muscles, like those of the back, exhibit slower twitches.

If a muscle is stimulated twice in succession, with the second stimulus applied *after* the relaxation phase of the first twitch is completed, the second twitch produces a myogram more or less identical to the first. However, if the second stimulus is applied *before* the relaxation phase of the first twitch is completed, the second contraction raises the pen to a higher position than the first (fig. 4.15). This effect is called **summation**.

As long as two stimuli are not applied too quickly in succession, summation is observed. However, if the second stimulus quickly follows the first, only one twitch is observed. It is as though the second stimulus was ignored. This occurs because after responding to one stimulus, a short amount of time is required before the muscle can respond again; this amount of time is called the **refractory period**.

Tetanus and Treppe

If successive stimuli applied to a muscle are given at intervals greater than the refractory period (but less than the twitch interval), the muscle responds with contractions of greater and greater force, and a myogram is

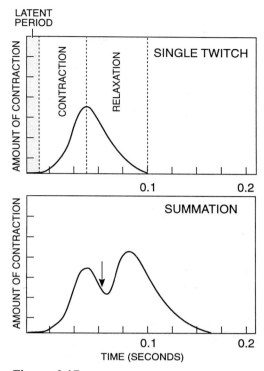

Figure 4.15
Myograms of a single twitch (top) and summation of twitches (bottom). Arrow indicates time of application of the second stimulus.

produced like that shown in figure 4.16, top. This is known as **incomplete tetanus**. It is possible to increase the stimulus frequency to a point at which the individual responses of the muscle are no longer discernible; the resulting myogram, called **complete tetanus**, is seen in figure 4.16, middle.

If stimuli of increasing intensity (not frequency) are applied to a muscle, the muscle responds with increasingly forcible contractions, as greater and greater numbers of motor units are recruited. In this way, the stimulus intensity needed to yield a maximal response on the part of the muscle can be determined. If these "maximal" stimuli are now applied in succession to a muscle, the muscle responds with contractions that continue to increase in force up to some

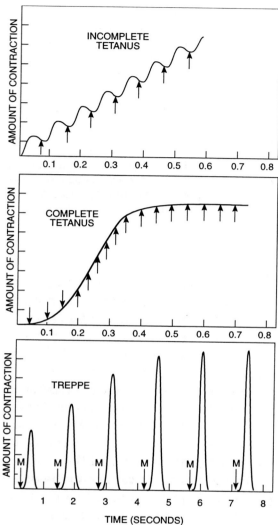

Figure 4.16
Myograms of incomplete tetanus (top), complete tetanus (middle), and treppe (bottom). Arrows indicate stimuli (M is maximal stimulus).

limit. This phenomenon is called **treppe** (or **staircase effect**) and its myogram is shown in figure 4.16, bottom.

SMOOTH MUSCLE

Smooth muscle tissue is found in the walls of many of the body's internal organs. For example, in the walls of the stomach and intestine, contractions of smooth musculature enable these organs to mechanically break down the food and propel it through the digestive tract; contractions of smooth muscle in the walls of the urinary bladder void the body's urine; and contractions (and relaxations) of smooth muscle tissue in the walls of blood vessels regulate the amount of blood circulating to various body parts while also regulating blood pressure.

Smooth muscle tissue is called "smooth" because the cells fail to exhibit the regular dark and light banding pattern that characterizes striated tissue. This is because the thick (myosin) and thin (actin) filaments of smooth muscle cells are not laid out in a highly ordered array (as they are in striated muscle cells) and there are no sarcomeres. The ratio of thin filaments to thick filaments is much higher in smooth muscle than it is in striated muscle. Some smooth muscle thick filaments are particularly long and may bridge more than one cell. Like striated muscle, filament interaction and sliding is triggered by a rise in the Ca^{++} concentration of the sarcoplasm. However, in smooth muscle the Ca^{++} is not derived from the sarcoplasmic reticulum (which is much reduced in smooth muscle) but enters the sarcoplasm from outside the cell. Smooth muscle cells contain tropomyosin but lack troponin. The actions of troponin are supplanted by two other proteins called **calmodulin** and *myosin light-chain kinase*.

There are two types of smooth muscle: **multiunit** smooth muscle and **single unit** (or **visceral**) smooth muscle. Both types of tissue are made up of individual spindle-shaped fibers containing a single nucleus (fig. 4.17). Whereas striated fibers may be several inches long, smooth fibers are much smaller and rarely exceed about 0.02 inches

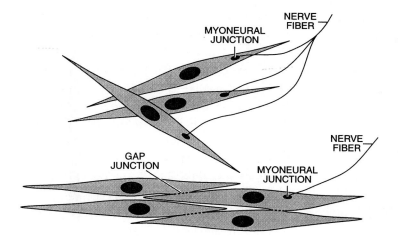

Figure 4.17
Types of smooth muscle. In multiunit smooth muscle (top), each muscle cell forms a myoneural junction with the terminal branches of a nerve fiber. In visceral (or single unit) smooth muscle (bottom), muscle cells are organized into sheets and there are few myoneural junctions; neighboring cells communicate through connections between their plasma membranes called gap junctions.

in length. Multiunit smooth muscle consists of fibers that are individually controlled by the terminal branches of nerve fibers. In this regard, they are similar to the motor units of striated tissue described earlier; an example of such a motor unit is seen in figure 4.17 (top). Multiunit smooth muscle is found in such places as the *ciliary muscles* of the eyes (which alter the shape of the lens), the *iris* muscle of the eyes (which control the size of the pupil), the *arrector pili muscles* that cause "goose bumps" on the skin and which cause hairs to stand erect, and in the walls of large arteries.

Visceral or single unit smooth muscle is more common than multiunit smooth muscle and forms sheets or layers of tissue in the walls of chambered organs such as the stomach, small intestine, large intestine, urinary bladder and uterus (figure 4.17, bottom). As depicted in figure 4.17, some cells form myoneural junctions with the terminal branches of a nerve fiber; most, however, have no myoneural junctions at all. Where myoneural junctions exist, these may be formed between the smooth muscle cell and the terminal branches of nerve fibers that are part of the *sympathetic* division of the **autonomic nervous system** (chapter 5) or with the *parasympathetic* division. During innervation of visceral smooth muscle, polarity changes in the sarcolemma of the innervated cell spread to neighboring cells through special junctions that link the plasma membrane (and sarcoplasm) of one cell to its neighbors; these junctions are called **gap junctions** (fig. 4.17).

Although smooth muscle contractions are elicited by nerve stimulation, smooth muscle also contracts in response to certain chemical and physical stimuli. For example, certain blood-borne hormones can bring about smooth muscle contraction; even simple stretching of smooth muscle tissue may be followed by the tissue's contraction.

As already noted, nerve fibers innervating smooth muscle tissue belong to the sympathetic and parasympathetic divisions of the autonomic nervous system. Sympathetic fibers release the neurohumor **norepinephrine** at their junctions with smooth muscle cells, whereas parasympathetic fibers release acetylcholine. At certain myoneural junctions these neurohumors are stimulatory, whereas at other junctions they are inhibitory (see chapter 5).

In striated muscle tissue, contraction can be very rapid, whereas smooth muscle tissue is characterized by slow contractions that may be sustained for many seconds.

CARDIAC MUSCLE

Cardiac muscle is found in the heart and also in the walls of the large arteries near the heart. The cells that comprise this muscle tissue (depicted in figure 4.18) share some physical properties common to both smooth muscle cells and striated muscle cells.

Like smooth muscle cells, individual cardiac muscle cells are quite small and have only one nucleus. However, unlike smooth muscle, cardiac cells are tubular in shape and display an alternating dark and light striping pattern similar to that of striated cells, although the striations may not be as pronounced and regular. Especially notable in heart tissue is the absence of myoneural junctions. This is because cardiac tissue does not rely on stimulatory impulses

from the nervous system in order to initiate contraction. Instead, cardiac tissue is **myogenic**–the stimulus for contraction arises within the tissue itself. This is not meant to imply that cardiac tissue behaves independently of the nervous system. As you will learn in a later chapter, the rate at which the heart's cells contract is influenced by the nervous system. However, the stimulus for contraction resides within the tissue itself.

Another distinct feature of cardiac tissue is the presence of **intercalated disks**. These are regions where finger-like projections at the ends of one cell interdigitate with the similar projections from one or more neighboring cells (fig. 4.18). Gap junctions (like those present in smooth muscle tissue) are found on the longitudinal surfaces of neighboring cells. The two kinds of junctions provide points through which signals can pass directly from one cell to another.

Unlike striated cells, cardiac muscle cells have a long refractory period. This serves to prevent contraction from occurring until after the muscle has already relaxed following a previous contraction. Therefore, summation and tetanus are not observed in normally-functioning cardiac muscle tissue.

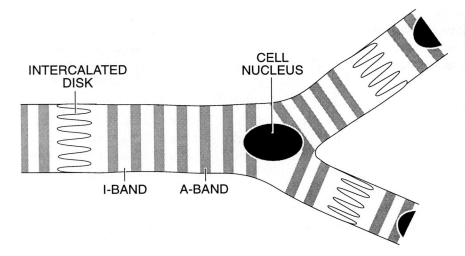

Figure 4.18
Cardiac muscle cells. Individual cells are small, striated, and sometimes branched. The edges of neighboring cells form interdigitations called intercalated disks.

THE METABOLISM OF MUSCLE

Sugar and Glycogen

The actions of muscle rely heavily on the availability of carbohydrate as a fuel. Therefore, before we begin to look closely at the metabolism of muscle, it is of value to consider the body's total picture with regard to its processing of carbohydrate.

Carbohydrate present in food is broken down during its passage through the digestive tract, yielding simple sugar molecules. Among these simple sugars, the most common is **glucose** (also known as **dextrose**), which is then absorbed through the walls of the digestive organs and into the bloodstream (fig. 4.19). Once absorbed into the bloodstream, the glucose is carried through the *hepatic portal system* (see chapter 11) and into the liver. As blood circulates through the liver, much of the glucose is removed by the liver cells. Inside the liver tissue, the glucose is converted

from its simple sugar state into its storage form, called **glycogen**. About 20% of all the glycogen in the body is stored in the liver; most of the remainder is stored in the body's muscles (where it also is produced from glucose derived from the blood). Blood glucose that is not removed by the liver or the musculature circulates to other tissues of the body, satisfying the tissues' fuel needs (all body tissues require a continuous supply of glucose).

The conversion of glucose to glycogen is a reversible process; that is, under appropriate circumstances, glycogen can be degraded to yield individual molecules of glucose. Although the interconversion of glucose and glycogen occurs both in liver and muscle tissue, only the liver can release the sugar back into the bloodstream (fig. 4.19). That is to say, when additional sugar is needed elsewhere in the body, liver glycogen is converted back into glucose, and the glucose is then released into the bloodstream in the needed quantities. During muscle activity, glucose is degraded (in the presence

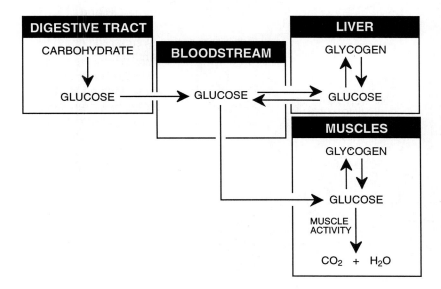

Figure 4.19
Glucose produced by carbohydrate digestion enters the bloodstream and is carried to the liver, where it is converted to glycogen and stored (about 20% of the body stores). As needed, glucose is released from the liver and carried to other tissues. Muscle also stores glucose as glycogen (about 80% of the body stores). During muscle activity, glucose is consumed, being converted to carbon dioxide and water.

of oxygen, O_2) to form carbon dioxide (CO_2) and water (H_2O, figure 4.19).

Roles of Adenosine Triphosphate and Creatine Phosphate

As noted earlier in the chapter, sliding of muscle filaments only occurs in the presence of filament-bound ATP, which is converted to ADP and inorganic phosphate in the process. The breakdown of ATP during muscle activity is a specific example of the rather general rule that energy-requiring processes in the body's tissues consume ATP. As seen in figure 4.20, when ATP is degraded, it is the bond that links the last of the molecule's three phosphate groups to the remainder of the molecule that is broken, thereby releasing free phosphate. The breakdown of sugar in muscle tissue (and

elsewhere) is the source of the energy needed to replace the lost ATP. The overall relationship is summarized in figure 4.20.

The breakdown of sugar to form CO_2 and H_2O takes place in two sequential metabolic phases in muscle cells (and other cells of the body). The first phase, called **glycolysis**, takes place in the cytosol and involves a series of reactions that convert a molecule of glucose into two molecules of **pyruvic acid** (fig. 4.21). The energy derived from these reactions is coupled to the production of a small amount of ATP from ADP and phosphate (namely, two molecules of ATP for every glucose molecule converted into two pyruvic acid molecules). In the second phase, called **respiration**, the products of glycolysis are converted inside the cell's mitochondria into six molecules of CO_2 and six molecules of H_2O. This phase, which consumes six molecules of

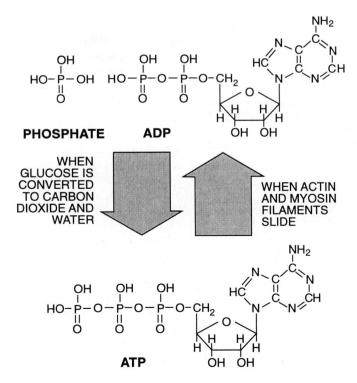

Figure 4.20
Sliding of actin and myosin filaments during contraction (and relaxation) is accompanied by the conversion of ATP to ADP and phosphate. ATP is resynthesized from ADP and phosphate using the energy derived from the conversion of sugar (glucose) to carbon dioxide and water.

O_2, yields much more energy and this energy is used to produce more ATP from ADP and phosphate (36 molecules of ATP per pair of pyruvic acid molecules). Therefore, altogether, the complete breakdown of a molecule of glucose consumes six molecules of O_2 and produce six molecules each of CO_2 and H_2O. The energy released in these reactions is sufficient to produce 38 molecules of ATP from 38 molecules of ADP and phosphate (fig. 4.21). (Glycolysis and respiration are considered further in chapter 12, which deals with the body's metabolism and nutrition.)

Unlike respiration, which requires the presence of oxygen and is therefore said to be **aerobic**, glycolysis occurs in the cytosol even in the absence of oxygen; thus, glycolysis is said to be **anaerobic**.

In the case of muscle cells, although ATP is produced from ADP in mitochondria, it is ATP associated with the cell's actin and myosin filaments that is converted to ADP during filament sliding. In other words, production of ATP in muscle cell mitochondria is not directly coupled to consumption of ATP by filament sliding. Instead, another chemical compound present in the sarcoplasm and called **creatine phosphate** (abbreviated CP) acts as an "intermediary" between mitochondrial ATP and filament ADP. That is, CP donates its phosphate to filament-associated ADP (thereby forming filament-associated ATP) and is subsequently replenished by accepting phosphate from mitochondrial ATP (thereby converting the mitochondrial ATP back into ADP).

The overall mechanism is summarized in figure 4.23 and may be explained as follows. Actin and myosin filaments undergo cyclic transitions between existing in a

Figure 4.21
During glycolysis, a metabolic process that occurs in the cytosol, a series of reactions converts glucose into two molecules of pyruvic acid. The reactions of glycolysis do not require oxygen (glycolysis is anaerobic). During respiration, which occurs inside the cell's mitochondria, a series of reactions convert the two pyruvic acid molecules to carbon dioxide and water. The completion of this series of reactions requires oxygen (respiration is aerobic).

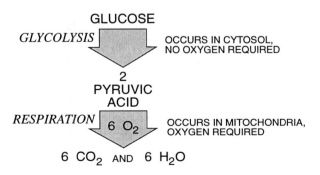

Figure 4.22
In muscle tissue, creatine phosphate (CP) serves as an "intermediary" between filament-associated ADP and mitochondrial ATP. CP donates its phosphate to filament-associated ADP but later reacquires phosphate from ATP produced inside mitochondria.

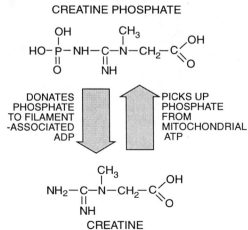

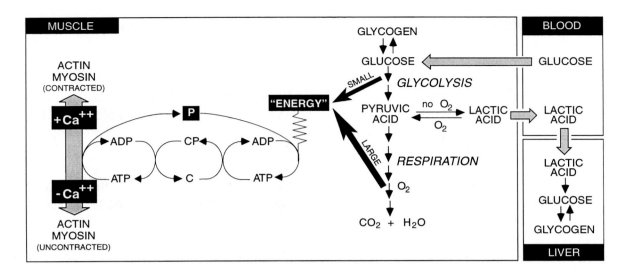

Figure 4.23
Summary of the reactions that characterize the metabolism of
glucose and ATP in muscle tissue.

contracted state (i.e., maximum filament overlap) and being in a relaxed state (i.e., minimum filament overlap). The changes between these two states are accompanied by the consumption of ATP, yielding ADP and phosphate as products (the left side of figure 4.23). Glucose entering muscle cells from the bloodstream or produced by the breakdown of glycogen within the muscle tissue (see the right side of the figure) is converted to pyruvic acid, thereby yielding a small amount of energy that is used to form additional ATP from ADP and phosphate. This metabolic process is called glycolysis and takes place in the muscle cell sarcoplasm; the reactions of glycolysis do not require oxygen. If oxygen is available, pyruvic acid is degraded inside the cell's mitochondria to form carbon dioxide and water; the latter series of metabolic reactions, called respiration, yields large amounts of energy and is coupled to the production of large quantities of ATP from ADP and phosphate.

Creatine acts to mediate the transfer of phosphate from ATP molecules that have been produced inside mitochondria to ADP molecules that are produced during filament sliding. In effect, creatine acquires its phosphate at the mitochondria (thereby forming creatine phosphate) and transfers this phosphate to filament-associated ADP. Phosphate released from filament-associated ATP during muscle activity makes its way into the cell's mitochondria where it is added to ADP at the expense of energy made available during the breakdown of glucose.

The amounts of filament-associated ATP in a muscle cell can support only a few seconds of contraction. Filament-associated ATP that is restored using the cell's creatine phosphate pool supports an additional few seconds of contraction. Therefore, contractions lasting many seconds (or even minutes) draw on ATP and CP that are synthesized *de novo* using the energy that is provided by the breakdown of glucose. Once muscle contraction ends, the cell's creatine phosphate and ATP pools are restored as the breakdown of glucose continues.

Oxygen Debt

So long as ample quantities of oxygen reach the muscle tissue via the bloodstream, muscle cell glucose not only undergoes glycolysis in the sarcoplasm, but the resulting pyruvic acid is converted inside mitochondria to carbon dioxide and water. However, if muscle activity is vigorous and/or prolonged, the oxygen supply to the muscle may fall short of that needed to convert *all* of the glucose to CO_2 and H_2O. Under such conditions, some of the pyruvic acid that cannot be converted to CO_2 and H_2O is converted instead to **lactic acid** (fig. 4.23). Modest amounts of the lactic acid produced may be retained in the exercising muscle, but most of it leaves the muscle and enters the bloodstream. The lactic acid is carried in the blood to the liver, which then removes the lactic acid and converts it back into glucose (an energy-demanding metabolic process called **gluconeogenesis**). Glucose produced in the liver by gluconeogenesis may return to the exercising muscle via the bloodstream, thereby completing a cycle called the **Cori cycle** (fig. 4.24).

When strenuous muscle activity ends and the oxygen supply to the muscle tissue again becomes plentiful, the lactic acid that was retained in the muscle is converted back into pyruvic acid, and the pyruvic acid is then degraded to CO_2 and H_2O (fig. 4.23). The additional oxygen consumed by muscle in order to convert lactic acid to CO_2 and H_2O, together with the oxygen that is needed by the liver to replace the ATP consumed during gluconeogenesis, is called the **oxygen**

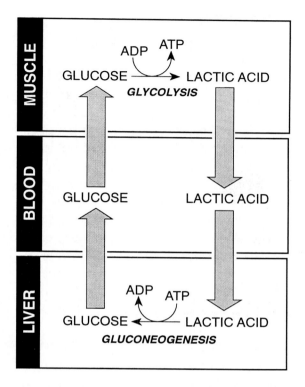

Figure 4.24
The Cori cycle. Lactic acid produced from glucose by exercising muscles is carried in the bloodstream to the liver. In the liver, the lactic acid is converted back into glucose, some of which is carried back to the muscles.

debt. In other words, during vigorous activity a muscle builds an oxygen debt, which it later repays when the activity ceases (or sufficiently diminishes). This is why a person's breathing remains rapid and deep for some time after vigorous exercising has been halted.

Slow- and Fast-Twitch Muscle Fibers

There are two major types of striated muscle fibers: *slow-twitch* fibers and *fast-twitch* fibers. Nearly all striated muscles contain a mixture of the two fiber types, but their ratio varies from one muscle to another. Muscles that are engaged in long, sustained, or rhythmic contractions (such as the muscles of the legs and back) are generally rich in slow-twitch fibers, whereas muscles that provide rapid, precise, short-duration contractions (such as the muscles of the fingers and the eye muscles) are rich in fast-twitch fibers.

The relative amounts of ATP produced in each of these fiber types during glycolysis and respiration vary. Slow-twitch fibers are rich in mitochondria and the reactions of respiration serve as their major source of ATP. A slow-twitch fiber's demand for oxygen is satisfied in part by the presence in these cells of large quantities of an oxygen-storing protein called **myoglobin**. The presence of myoglobin imparts a reddish color to the tissue, so that these fibers are also known as *red fibers*.

Fast-twitch fibers contain few mitochondria and little myoglobin. In these fibers, glycolysis is the primary source of ATP. Because they lack myoglobin, fast-twitch fibers are also known as *white fibers*. In addition to red, slow-twitch and white, fast-twitch fibers, an intermediate form exists that displays the rapidity of fast-twitch fibers but contains myoglobin and relies on respiration as its major source of ATP. The proportion of slow-twitch to fast-twitch fibers in the leg muscles of humans appears to be determined genetically and may account for individual differences between long distance and sprint runners.

SELF TEST*

True/False Questions

1. During an isometric contraction of the biceps muscle, the lower arm is rotated upward toward the shoulder.
2. Some smooth muscle cells have no myoneural junctions.
3. Lactic acid entering the bloodstream from muscle tissue is carried in the bloodstream to the liver, where the lactic acid is then converted to glucose.
4. Myoglobin is found in muscle tissue that contracts slowly and rhythmically.
5. The triceps and biceps muscles are examples of antagonistic muscles.

Multiple Choice Questions

1. Muscles (A) only pull, (B) only push, (C) pull and push.

2. Within a muscle (such as the biceps), fibers are bundled together to form one or more (A) myo-elements, (B) tendons, (C) fascicles, (D) ligaments, (E) sarcomeres.
3. The end of a striated muscle that undergoes the greatest amount of movement when the muscle contracts is called the (A) insertion, (B) origin, (C) fulcrum, (D) momentum, (E) Z-line.
4. The liver can (A) convert lactic acid into sugar, (B) convert sugar into glycogen, (C) convert glycogen into sugar, (D) only choices B and C are correct, (E) choices A, B, and C are correct.

---

* *The answers to these test questions are found in Appendix III at the back of the book.*

5. Most of the ATP that is produced in red muscle is produced in (A) the myofilaments, (B) the mitochondria, (C) the cytosol (i.e., cytoplasm), (D) the cell's nuclei.

6. The calcium pumps of a striated muscle cell act to pump calcium ions into (A) the sarcolemma, (B) the sarcoplasmic reticulum, (C) the cell's sarcoplasm, (D) the cell's mitochondria.

7. Most of the sugar that is stored in the body (A) is stored as glucose in muscle tissue, (B) is stored as glycogen in muscle tissue, (C) is stored as glucose in liver tissue, (D) is stored as glycogen in liver tissue, (E) is stored in nerve tissue, especially the brain.

8. Which one of the following metabolic processes can take place even when oxygen is absent? (A) respiration, (B) glycolysis, (C) conversion of lactic acid to pyruvic acid.

9. A-bands contain (A) actin only, (B) myosin only, (C) both actin and myosin.

THE NERVOUS SYSTEM

Chapter Outline

The nervous system consists of all of the nerve and nerve-associated cells and tissues of the body. The main functions of this complex system are (1) to provide communication between one body part and another, (2) to interpret physical and chemical changes occurring internally, as well as external to the body, (3) to coordinate and regulate the body's activities, and (4) to store information.

The tissues of the nervous system contain two types of cells called **neurons** and **neuroglia** (or **glial cells**). The neuroglia (about 80% of all nerve cells) play a supportive role, whereas the neurons are the basic functional units of the nervous system. Because many neurons are fine, thread-like cells, they are also referred to as nerve **fibers**. The primary and specialized functions of a neuron are the **conduction** and **transmission** of **impulses** from one part of the body to another. In some instances, the impulses travel a distance of several feet, and a single neuron may bridge the entire path.

GENERALIZED ORGANIZATION OF A NEURON

The generalized structure of a neuron is shown in figure 5.1. In certain respects, neurons are similar to most other cells. For example, the cell body contains the typical distribution of subcellular organelles (nucleus, mitochondria, endoplasmic reticulum, ribosomes, etc.). It is the processes (i.e., extensions from the cell body) that make neurons readily distinguishable from other types of cells. These processes are the portions of the cell that are responsible for the conduction of impulses over great (and small) distances. There are two different kinds of nerve cell processes. Processes that conduct impulses *toward the cell body* are

Figure 5.1
Organization of a neuron. In the neuron shown here, the direction of conduction is downward (i.e., from the top of the figure to the bottom of the figure). Therefore, the processes arising from the top of the cell body are dendrites and the process arising from the bottom of the cell body is an axon. Long processes (whether they are dendrites or axons) are usually enveloped in a layer of myelin-rich neuroglial cells called Schwann cells. The surface of the neuron is exposed at gaps between successive Schwann cells (i.e., at the nodes of Ranvier). The break in the axon (shown by the zigzag line) refers to the long portion of the axon that is not shown in the figure.

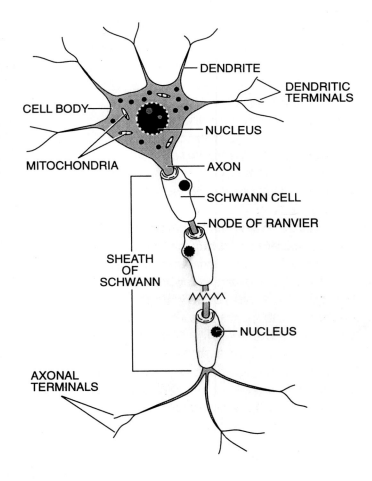

called **dendrites**, whereas processes that conduct impulses *away from the cell body* are called **axons**. Dendrites and axons may give rise to one or more large branches or **collaterals**, each branch eventually ending in a number of fine, finger-like extensions called **terminals** or **terminal branches**. The terminal branches of axons are responsible either for transmitting an impulse to the next nerve cell or for innervating an "effector cell" (see later). The terminal branches of dendrites act either to receive an impulse transmitted by a nerve cell or to receive a stimulus from a "receptor cell" (again, see later). Since axons and dendrites are identified not on the basis of their lengths but on the basis of their physiological roles, the processes labeled "axon" and "dendrite" in figure 5.1 imply that the direction of conduction in that neuron is downward (i.e., down the page). $\boxed{5\text{-}1}$

Myelination of Long Processes

The long processes of many neurons of the so-called *peripheral nervous system* (see later) are associated with another kind of cell called a **Schwann cell**. Schwann cells are small, sheet-like cells rich in a fatty substance called **myelin**. These cells form a spiral envelope around the long processes of neurons, thereby insulating the process from neighboring neurons and other tissue (fig. 5.2). Since many neuronal processes are several inches (or feet) in length, the myelin sheath created by the Schwann cells may consist of several dozen or even several hundred cells. The sheath formed by the Schwann cells is called the **sheath of Schwann**. Neurons with myelin sheaths are called "myelinated" neurons.

A tiny portion of the surface of a long neuronal process is exposed between succes-

sive Schwann cells. These positions are known as **nodes of Ranvier**. Myelinated nerve cells conduct impulses more rapidly than unmyelinated cells. This is because the impulses quickly jump from one node of Ranvier to the next (this rapid form of conduction is called **saltatory conduction**). Unmyelinated nerve fibers conduct impulses more slowly, the impulse passing smoothly and in an uninterrupted manner along the nerve cell surface. The myelination of processes that extend into the arms and legs and the resulting rapid saltatory conduction of impulses reduces the time taken for an impulse to travel these long distances.

Oligodendrocytes form myelin sheaths around the long processes of neurons of the so-called *central nervous system* (i.e., the brain and spinal cord). Unlike Schwann cells, which form an encapsulating sheath around a single nerve fiber, the branches of an oligodendrocyte may encapsulate several *different* nerve fibers.

Other Neuroglial Cells

Schwann cells and oligodendrocytes are two types of neuroglial cells; there are four other

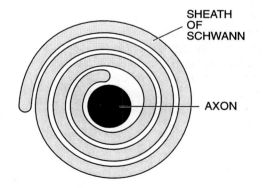

Figure 5.2
Cross section through the myelinated process (axon) of a nerve fiber. Schwann cells form a spiral-shaped sheath around the long process.

types: **microglia**, **astrocytes**, **ependymal cells**, and **satellite cells**. Microglia are small, phagocytic cells that have left the bloodstream to wander through the nervous system removing foreign debris. Astrocytes regulate the transfer of molecules from the bloodstream to the brain. Ependymal cells line the walls of small, fluid-filled chambers in the brain (called *ventricles*) and the central canal of the spinal cord. Finally, satellite cells (also called *amphicytes)* play a supportive role for the cell bodies of neurons that are in the *ganglia* of the peripheral nervous system (see later).

For convenience as we continue our discussion of the nervous system, we will use a simple, abbreviated representation of a neuron. In this form, the neuron's cell body is depicted as a small circle (containing a central, dark spot representing the nucleus) and the axonal and dendritic processes are shown as thin lines extending from the cell body (fig. 5.3). Using this abbreviated style, three common types of neurons may be distinguished: (1) *unipolar* neurons, in which the dendrite and axon form one continuous length, with the cell body attached via a stalk-like extension; usually unipolar neu-

rons conduct *sensory* impulses; (2) *bipolar* neurons, in which a single dendrite and axon exit from opposite sides of the cell body; bipolar neurons are uncommon in the body, but are abundant in the eye's retina; and (3) *multipolar* neurons, in which a number of dendrites and a single axon exit the cell body at multiple points on the cell body's surface; multipolar neurons conduct *motor* and *association* impulses (fig. 5.3). The physiological significance of the terms "motor," "sensory," and "association" will become apparent later.

Conduction vs. Transmission

The terms "conduction" and "transmission" have specific meanings with respect to the actions of nerve cells. Conduction refers to the movement of an impulse from one end of a neuron to the other end of the *same* neuron (e.g., from the dendritic endings to the axonal endings). Transmission, on the other hand, refers to the passage of an impulse from the axonal endings of one neuron to the dendritic endings (or cell body) of another neuron. Accordingly, impulses are conducted by a single neuron but are transmitted *between*

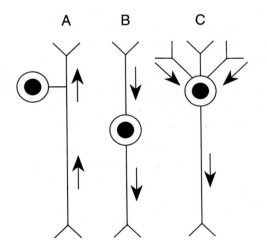

Figure 5.3
A simplified scheme showing the various structural forms that a neuron can take. In unipolar fibers (A), the axonal and dendritic processes form a continuous length, and the cell body is attached to the long axis of the processes by a short, stalk-like extension; typically, unipolar neurons are responsible for the conduction of sensory impulses. In bipolar neurons (B), the dendritic and axonal processes emerge from opposite points on the cell body. Bipolar fibers are rare but are found, for example, in the retina of the eye where they conduct sensory impulses. Finally, in multipolar fibers (C), several dendritic processes and a single axonal process emerge from the cell body; most motor and association fibers take this form.

neurons. (It would *not* be correct to say that an impulse is "transmitted" from one end of a neuron to the other, or that an impulse is "conducted" from one neuron to another.) As you will see later, the conduction and transmission of impulses are the results of quite different physical and biochemical phenomena.

CONDUCTION OF NERVE IMPULSES

Neurons conduct impulses from one part of the body to another. Under normal circumstances, each impulse begins at the terminal branches of the cells' dendrites (occasionally at the cell body) and spreads across the surface of the cell to the terminal branches of the cells' axons. Experimentally, an impulse can be initiated anywhere on the cell surface and can be elicited by applying a variety of stimuli including electrical shock, pressure (pinching), heat, cold, and pH changes. Once it is initiated, the impulse is propagated along the cell without dependence on a continuing stimulus. The velocity with which an impulse travels along the nerve fiber is not dependent on the strength of the stimulus; that is, it does not travel faster if initiated by a stronger stimulus.

Dromic vs. Anti-dromic Conduction

If the long process of a nerve fiber is exposed and experimentally stimulated (e.g., by electrical shock) at a point midway along its length, the fiber will simultaneously conduct an impulse away from the point of stimulation in *both* directions. One of these directions is the direction that leads to the terminal branches of the axon, whereas the other di-

rection leads to the terminal branches of the dendrite(s). The impulse that travels in the direction of the axon's terminals is the "physiologically correct" direction (i.e., the direction in which the process normally conducts an impulse). This "physiologically correct" direction is called the **dromic** direction. The impulse that travels in the direction of the dendrite's terminals is "physiologically incorrect" (i.e., it is not the direction in which the cell normally conducts an impulse) and is called the **antidromic** direction. Antidromic conduction can be elicited experimentally, but it does not normally occur in the body. However, the ability to experimentally elicit dromic and antidromic conduction is a source of useful information regarding the nature of the nerve impulse and the properties of neurons.

Nature of the Nerve Impulse

A nerve impulse is an electrochemical phenomenon and is not the same as the conduction of electrical current in a copper wire. Electricity travels through copper wires at speeds approaching the speed of light, namely 186,000 miles per second. The movement of an impulse along a nerve fiber rarely occurs at speeds greater than about 100 miles per hour. Thus, electricity travels many hundreds of thousands times faster than a nerve impulse.

Observations Using A Galvanometer. Some of the electrical characteristics of impulse conduction may be learned from a simple experiment in which a nerve fiber is exposed and is connected to a **galvanometer**. A galvanometer is a device that measures the flow of electrical current. It consists of a meter and two measuring electrodes that can be placed on the surface of an object suspected of carrying electrical charge. For example, if the measuring electrodes of a galvanometer are attached to the positive and negative poles of a battery, the needle of the galvanometer deflects to one side, indicating that electrons are flowing through the galvanometer from the negative pole of the battery to the battery's positive pole (fig. 5.4).

If the measuring electrodes of a galvanometer are placed at two distant points on the surface of an exposed (unmyelinated) nerve fiber and the fiber is not conducting an impulse, the galvanometer needle remains undeflected (fig. 5.5). This implies that the two points on the surface of the nerve fiber have the same electrical potential. However, changes in the position of the galvanometer needle will be observed if a stimulus (e.g., an electrical shock) of sufficient intensity to cause the fiber to conduct an impulse is applied to the fiber (fig. 5.6). When such a stimulus is applied, an impulse travels along the fiber (fig. 5.6A). The galvanometer needle transiently deflects to one side as the impulse passes the first electrode. The deflection that is observed (fig. 5.6B) indicates that the surface of the fiber below the first electrode (i.e., electrode-1) is electrically negative with respect to the surface below the second electrode (i.e., electrode-2). It is as though the nerve cell surface below electrode-1 were the negative pole of a battery and the surface below electrode-2 were the positive pole of a battery. Because of the polarity differences between these two points, electrons flow through the galvanometer from electrode-1 to electrode-2.

5-2

The deflection of the galvanometer needle is only temporary, because as soon as the impulse passes electrode-1, the needle

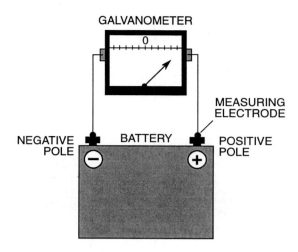

Figure 5.4
A galvanometer measures the flow of electrons between two differently charged points. When the galvanometer's two measuring electrodes are attached to the negative and positive poles of a battery, the galvanometer's needle deflects to one side. The needle's deflection indicates that electrons are flowing through the galvanometer from the negative pole of the battery to the battery's positive pole. ("Current" is said to flow from positive pole to negative pole.)

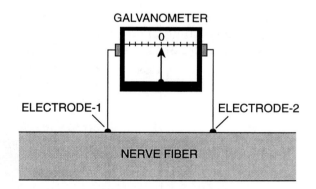

Figure 5.5
When a galvanometer's measuring electrodes are placed at two positions on the surface of a nonconducting nerve fiber, there is no deflection of the galvanometer needle. This indicates that in the nonconducting fiber, all points along the fiber's surface have the same electrical charge.

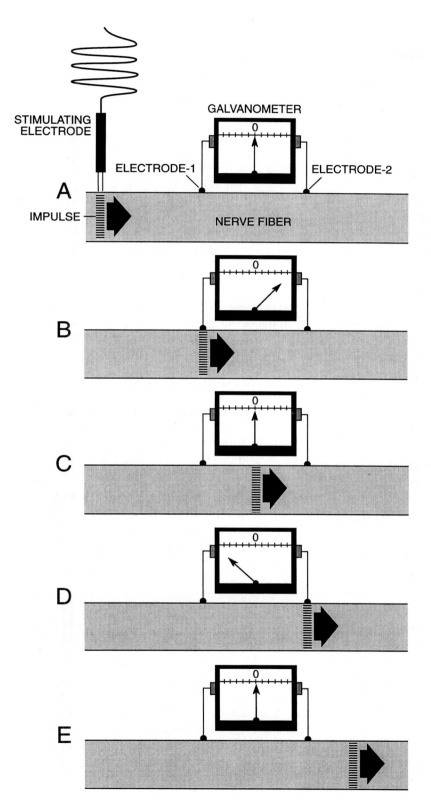

Figure 5.6
Movements of a galvanometer needle that accompany the conduction of an impulse. In A, the fiber is stimulated electrically in front of the first of the two measuring electrodes that have been placed on the surface of the nerve fiber. In B, the impulse sweeps past electrode-1, deflecting the galvanometer needle. In C, the impulse is between the two electrodes and the galvanometer needle returns to the zero position. In D, the impulse sweeps past electrode-2, again deflecting the galvanometer needle (but toward the opposite side of the scale). Finally, in E, the impulse proceeds beyond electrode-2, and the galvanometer returns to the zero position. (See the text for an explanation of the galvanometer needle's movements.)

returns to its initial position (fig. 5.6C). This indicates that once again the surfaces of the fiber below electrodes 1 and 2 have the same electrical potential.

As the impulse passes below electrode-2, the galvanometer needle is again deflected. However, this time the needle moves in the *opposite* direction (fig. 5.6D), indicating that now the surface of the fiber below electrode-2 is electrically negative relative to the surface below electrode-1 (i.e., electrons are now flowing through the galvanometer from electrode-2 to electrode-1). Finally, as the impulse proceeds beyond electrode-2, the galvanometer needle returns to its initial position. This indicates that the surfaces of the fiber below both electrodes once again have the same electrical potential.

What do these observations mean? This experiment with the galvanometer implies that (1) *the surface of a nerve fiber is electrically the same everywhere, except in the region of the nerve impulse*, and (2) *in the region representing the transient position of the impulse, the fiber's surface temporarily becomes negative relative to other regions of the surface*.

An additional informative observation may be made with a galvanometer if one of its electrodes is inserted *through* the nerve fiber's surface into the underlying cytosol, while the other electrode remains on the outside surface (fig. 5.7). When this is done, the galvanometer needle is again deflected, the deflection indicating that the inside surface of the fiber (i.e., the surface that faces the cytosol) is electrically negative relative to the outside surface.

Distributions of Ions Across the Nerve Fiber's Surface. All of the observations described above may be explained by examin-

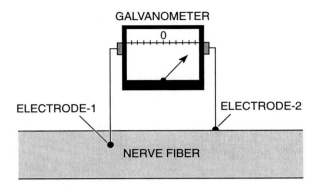

Figure 5.7
When one measuring electrode of a galvanometer is inserted through the cell surface into the cytosol below and a second electrode is placed on the fiber's surface, electrons flow through the galvanometer from the internal electrode to the external electrode. This shows that in a neuron at rest, the outside surface of the cell membrane is electrically positive relative to the inside surface.

ing and comparing the distributions of ions across the nerve cell's surface when the fiber is "resting" (i.e., when the neuron is *not* conducting an impulse) and when it is conducting an impulse. The fluid bathing the surface of a nerve fiber and the fluid within the fiber (i.e., the cytosol) are rich in a number of ions. To make our discussion somewhat simpler, we will focus on two ions of particular importance, namely Na^+ (sodium ions) and K^+ (potassium ions). The fluid bathing the membrane's outer surface contains a greater concentration of Na^+ than does the cytosol, whereas the cytosol contains a greater concentration of K^+ than does the extracellular fluid. In addition to the Na^+ and K^+ concentration differences across the nerve cell membrane, the combined concentration of Na^+ and K^+ in the fluid bathing the outside surface of the membrane is greater than in the fluid bathing the inside surface. Another way of saying this is

(1) $[Na^+]_{outside} > [Na^+]_{inside}$

(2) $[K^+]_{inside} > [K^+]_{outside}$

and

(3) $\{ [Na^+]_{outside} + [K^+]_{outside} \} >$

$\{ [Na^+]_{inside} + [K^+]_{inside} \}$

As a result of these ion distributions, the outside surface of the nerve cell membrane is positive relative to the inside surface (fig. 5.8). This is why electrical current flows through the galvanometer when one measuring electrode is inserted through the nerve cell membrane to the other side, while the other electrode remains on the outside surface of the fiber (fig. 5.7). Because the outside surface of the fiber is electrically positive relative to the inside surface, the nerve fiber is said to be **polarized**.

The nerve cell membrane is permeable to Na^+ and K^+ and these ions are continuously diffusing from the side of the membrane where they are at higher concentration to the side where they are at lower concentration. The ions pass along their respective concentration gradients from one side of the membrane to the other through channels in the membrane; these channels are protected by "gates," referred to as the **sodium gate** and the **potassium gate**. In order to maintain the concentration differences of these two cations, some of the cell's metabolic energy is used to pump inwardly-diffusing Na^+ out of the cell and outwardly-diffusing K^+ into the cell. The mechanism that accomplishes this is known as the **sodium/potassium pump** and involves an enzyme located in the nerve cell membrane. Sodium ions, together with ATP, are bound to an enzyme site exposed at the inside membrane surface, and K^+ is bound to the outer surface. Binding to the enzyme is followed by a conformational change in the enzyme that acts to move the ions through the membrane. The ATP is converted to ADP and phosphate in the process (see chapter 4), and the sodium and potassium ions are released.

Movements of Na^+ and K^+ During Conduction. The conduction of an impulse by a nerve fiber is associated with specific changes

Figure 5.8
The difference in electrical charge that exists between the outside and inside surfaces of a nerve fiber's cell membrane is due principally to the differential distributions of sodium and potassium ions in the fluid that bathes the surface of the cell and the cytosol.

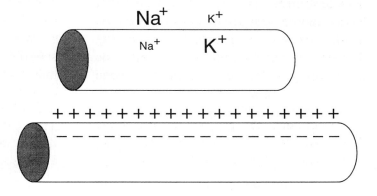

in the distributions of Na⁺ and K⁺ across the nerve cell membrane. These changes are summarized in figure 5.9.

Stimulation of a nerve fiber at a point on its surface serves to **depolarize** the membrane in that region, and this widens the sodium gates (figure 5.9, stage A). As a result, Na⁺ can diffuse more rapidly through the membrane along its concentration gradient (figure 5.9, stage B). For purposes of easy reference, let's call this region of the nerve cell "region-1." The inward movement of just a small amount of Na⁺ in region-1 is sufficient to reverse the membrane's polarity in that region, thereby causing the inside surface to become electrically positive relative to the outside surface (figure 5.9, stage B). Thus, in a localized portion of the nerve fiber (i.e., in region-1), the membrane's polarity has been reversed by this action. The action is referred to as **reverse polarization** and one of its consequences is the widening of the potassium gates.

The localized reversal of the membrane's polarity creates neighboring regions on the outside (and inside) surface(s) where the electrical polarity differs. This state is depicted in figure 5.9, stage C1, in which region-1 is negative externally and positive internally, whereas the neighboring region (let's call it "region-2") is still positive externally and negative internally. This condition is unstable and leads to the flow of electrons from each negative region to the neighboring positive region. As electrons flow into the neighboring positive region on the outside surface (i.e., region-2), that region is depolarized, thereby widening the sodium gates and permitting an influx of sodium ions (figure 5.9, stage C2). Meanwhile, in the region of the fiber where the polarity had already been reversed (i.e., back in region-

1), potassium ions diffuse outward in sufficient quantity to again reverse the membrane's polarity (i.e., to restore the original polarity of the membrane, in which the outside is positive relative to the inside; figure 5.9, stage C2). Restoration of the membrane's original polarity (called **repolarization**) narrows the sodium and potassium gates, at which time the sodium/potassium pump is able to restore the original distribution of Na⁺ and K⁺ across that region of the membrane (i.e., the Na⁺ and K⁺ concentrations across the membrane in region-1 are returned to their initial states).

Only a small percentage of the sodium and potassium ions initially present outside and inside the cell need to pass through the membrane in order to cause the reverse polarization and the ensuing repolarization. Indeed, a nerve fiber can conduct a number of successive impulses without appreciably diminishing the ion concentration gradients across the membrane. Between successive conductions, the sodium and potassium gates narrow and the sodium/potassium pump restores the original ion concentrations. It is estimated that the normal distribution of Na⁺ and K⁺ across the membrane can be restored in less than 0.005 seconds.

Returning to region-2 of stage C2 in figure 5.9, the influx of Na⁺ has reversed the membrane's polarity in that region, again creating conditions in which the flow of electrons into the next region (say, "region-3") perpetuates the entire cycle (figure 5.9, stage D). Thus, the conduction of an impulse may be thought of as a repeating series of events that progressively occurs along the nerve fiber's surface (figure 5.9, stage E). The events may be summarized as follows:

1. depolarization of the outer surface of the membrane,

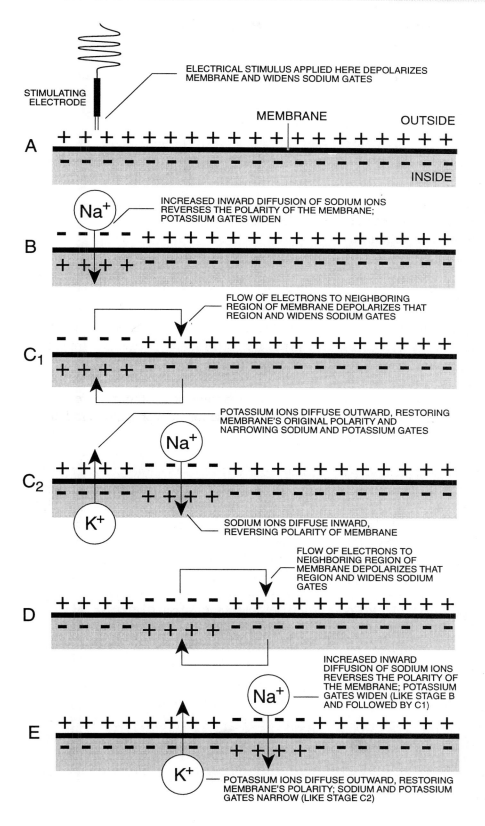

Figure 5.9
Movements of Na⁺ and K⁺ through the nerve fiber membrane during the conduction of an impulse. See text for explanation.

STIMULATING ELECTRODE

ELECTRICAL STIMULUS APPLIED HERE DEPOLARIZES MEMBRANE AND WIDENS SODIUM GATES

MEMBRANE   OUTSIDE

A

INSIDE

Na⁺   INCREASED INWARD DIFFUSION OF SODIUM IONS REVERSES THE POLARITY OF THE MEMBRANE; POTASSIUM GATES WIDEN

B

FLOW OF ELECTRONS TO NEIGHBORING REGION OF MEMBRANE DEPOLARIZES THAT REGION AND WIDENS SODIUM GATES

C₁

POTASSIUM IONS DIFFUSE OUTWARD, RESTORING MEMBRANE'S ORIGINAL POLARITY AND NARROWING SODIUM AND POTASSIUM GATES

Na⁺

C₂

K⁺

SODIUM IONS DIFFUSE INWARD, REVERSING POLARITY OF MEMBRANE

FLOW OF ELECTRONS TO NEIGHBORING REGION OF MEMBRANE DEPOLARIZES THAT REGION AND WIDENS SODIUM GATES

D

INCREASED INWARD DIFFUSION OF SODIUM IONS REVERSES THE POLARITY OF THE MEMBRANE; POTASSIUM GATES WIDEN (LIKE STAGE B AND FOLLOWED BY C1)

Na⁺

E

K⁺   POTASSIUM IONS DIFFUSE OUTWARD, RESTORING MEMBRANE'S POLARITY; SODIUM AND POTASSIUM GATES NARROW (LIKE STAGE C2)

2. widening of the sodium gates in the depolarized region of the membrane (e.g., region-1),

3. inward movement of Na^+, thereby reversing the membrane's polarity in region-1 and widening the potassium gates,

4. flow of electrons to next region of the membrane (e.g., region-2), thereby depolarizing that region of the membrane and widening the sodium gates,

5. outward movement of K^+ in previous region of the membrane (i.e., region-1), thereby restoring the original polarity of the membrane in that region and narrowing the sodium and potassium gates,

6. restoration of original Na^+ and K^+ distributions across the membrane by the sodium/potassium pump,

7. return to step 3, except that it is the next region of the membrane that is changing.

SYNAPTIC TRANSMISSION

As noted earlier, the passage of an impulse from one nerve fiber to another (i.e., "transmission") is a different phenomenon than the passage of an impulse from one end of a fiber to the other end of the same fiber (i.e., "conduction"). Unlike impulse conduction, which is an electrochemical process, impulse transmission is wholly chemical. The transmission of an impulse from one nerve fiber to another occurs at specific junctions called **synapses** (fig. 5.10). These synapses usually occur between axonal endings of one fiber (the so-called *presynaptic* fiber) and either the dendritic endings of the next fiber (the so-called *postsynaptic* fiber) or the postsynaptic fiber's cell body (figs. 5.10 and 5.11). Synapses between axonal and dendritic endings (fig. 5.11A) are called **axodendritic** synapses. Synapses between

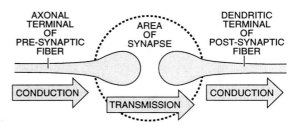

Figure 5.10
A synapse is a junction between two (or more) nerve fibers. At a synapse, nerve impulses are transmitted from the presynaptic fiber to the postsynaptic fiber.

axonal endings of one neuron and the cell body of the next neuron are called **axosomatic** synapses (fig. 5.11B). Also possible (but rarer) are synapses between axonal endings of one neuron and the axon of the next neuron; these are called **axoaxonic** synapses (fig. 5.11C). The surfaces of the two nerve fibers at a synapse are separated from one-another by a narrow, fluid-filled gap called a **synaptic cleft**.

The transmission of an impulse from one nerve fiber to another generally takes the following form. The arrival of the impulse at the axonal terminals of the presynaptic fiber causes a localized widening of Ca^{++} gates in the terminal's membrane. This is quickly followed by an influx of Ca^{++} from the surrounding extracellular fluid. The rise in intracellular Ca^{++} serves to activate enzymes that promote the fusion of **neurotransmitter**-rich vesicles (called **synaptic vesicles**) with the plasma membrane, thereby releasing the neurotransmitter into the synaptic cleft (fig. 5.12). The most common neurotransmitter of the *peripheral* nervous system (see later) is **acetylcholine**; less common neurotransmitters are **epinephrine** and **norepinephrine**. Neurotransmitters of the *central* nervous system (again, see later) are more varied, several dozen having been

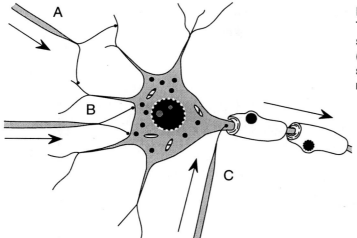

Figure 5.11
Types of synapses. (A) axodendritic synapse, (B) axosomatic synapses, and (C) axoaxonic synapse. The arrows show the direction of conduction in each nerve fiber.

identified. The most common are **glutamate**, acetylcholine, **dopamine**, and **serotonin**.

The neurotransmitter molecules released into the synaptic cleft quickly diffuse across the space separating the surfaces of the two fibers and bind to neurotransmitter receptors in the plasma membrane of the postsynaptic fiber. In the case of acetylcholine, the neurotransmitter binds to **acetylcholine receptors**. The ultimate effect of neurotransmitter binding is to widen the sodium gates in that region of the membrane and in so doing to trigger the events of conduction in the postsynaptic fiber (as described in detail earlier in the chapter). The process is repeated at the next synapse.

As illustrated in figure 5.12, acetylcholine bound to acetylcholine receptors in the plasma membrane of the postsynaptic nerve fiber is broken down into *choline* and *acetate* by the enzyme *acetylcholinesterase*. This enzyme is located on the surface of the postsynaptic fiber. Shortly afterwards, the acetate and choline are taken up by the presynaptic fiber, which recycles these substances so that new molecules of acetylcholine are formed.

In the simplest case, the transmission of an impulse involves only two fibers, namely the presynaptic fiber and postsynaptic fiber. However, many fibers may form a synapse, so that signals may pass from the axonal endings of one or more of the fibers to the dendritic endings of other fibers forming the synapse. Moreover, different neurohumors may be released by different cells forming a synapse, and the neurohumor released by one cell may act to block the transmission of an impulse that would otherwise be achieved through the action of a neurohumor released by another cell.

MAJOR SUBDIVISIONS OF THE NERVOUS SYSTEM

The nervous system may be subdivided into 3 major portions: (1) the **central nervous system** (or **CNS**), (2) the **peripheral nervous system** (or **PNS**), and (3) the **autonomic nervous system** (or **ANS**).

The central nervous system is comprised of the **brain** and the **spinal cord**. The peripheral nervous system is comprised of the nerves that arise from the brain and the spinal cord, namely the **cranial nerves** (arising from the brain) and the **spinal nerves**

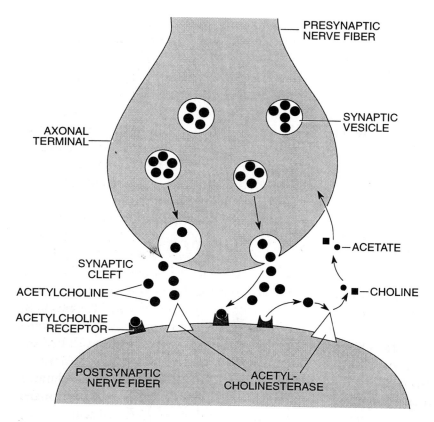

Figure 5.12

Synaptic transmission. Acetylcholine released from synaptic vesicles into the synaptic cleft that separates the surfaces of two neurons at a synapse binds to receptor molecules in the surface of the postsynaptic cell. This causes the initiation of a nerve impulse in the postsynaptic cell. Acetylcholine molecules are broken down into acetate and choline by *acetylcholinesterase* enzymes in the surface of the postsynaptic cell; the acetate and choline are absorbed by the presynaptic cell and used to form new acetylcholine.

(arising from the spinal cord). The autonomic nervous system includes a number of nerves that regulate the activities of the body's internal organs (e.g., heart, digestive organs, and excretory organs). The autonomic nervous system has two divisions of its own; these are called the **sympathetic** division and the **parasympathetic** division. We will proceed with a discussion of the nervous system by considering the system's divisions in the order that has just been outlined.

THE CENTRAL NERVOUS SYSTEM

Grey vs. White Matter

Before we proceed any further into our discussion of the nervous system, it is important to explain what is meant by "grey matter" and "white matter," because both of these terms will now be cropping up quite regularly. The myelin of Schwann cells and oligodendrocytes that form sheaths around many nerve cell processes is white in appearance. Consequently, nerve tissue that is rich in myelinated processes looks white and is called **white matter**. Even when a nerve cell is myelinated, the myelin sheath does not extend over the cell body, which therefore retains its grey appearance. As a result, nerve tissue that is rich in nerve cell bodies and unmyelinated processes appears grey and is called **grey matter**.

The Brain

The brain of the average adult contains about 100,000,000,000 (i.e., 100 billion!) neurons, virtually all of which are produced during embryonic development. (Very few new neurons are produced after birth, so that any neurons that are lost through disease or injury are not replaced; in contrast, the neuroglial cells of the brain, which support the activities of the neurons, do retain the capacity for growth and division.) There are three anatomically and functionally different parts of the brain (fig. 5.13). The main part of the brain is the **cerebrum**, which is further divided into left and right halves (i.e., the left and right cerebral hemispheres).

Underneath the rear of the cerebrum is the second major portion of the brain, the **cerebellum**. Connecting the cerebrum to the cerebellum and also linking both of these to the top of the spinal cord is the so-called **brain stem**, consisting of the **midbrain**, **pons**, and **medulla**. Supporting the three parts of the brain from below and also encapsulating the brain at its surface (thereby physically protecting the brain) is a thick sheet of bone called the **skull**. The cranial nerves (see below) that arise in the brain emerge through small apertures in the skull.

The Cerebrum. The cerebrum is characterized by the many grooves that run across its surface. The deeper grooves are called **fissures**, while the shallower grooves are called **sulci** (singular = **sulcus**). The elevations between neighboring fissures and sulci are called **gyri** (singular = **gyrus**) or **convolutions**. A very deep fissure, called the **longitudinal fissure**, which runs from the front of the cerebrum to the rear separates the left and right **cerebral hemispheres**.

Each cerebral hemisphere is divided into four **lobes** (fig. 5.13); these are (1) the **frontal lobe**, (2) the **parietal lobe**, (3) the **occipital lobe**, and (4) the **temporal lobe**. If a cut is made into the surface of a cerebral lobe, one notes that the outer layer (called the **cerebral cortex**) is grey, indicating that it is comprised predominantly of neuron cell bodies. Below the cortex, the cerebral tissue is white, indicating the predominance of myelinated processes. Although the grey cortex of the cerebrum is only about 5 mm thick, its surface area is very large due to the great amount of folding created by the fissures and sulci.

5-4

A great deal is known about the actions of different regions of the cerebrum. Although much of this information has been derived from studies with experimental animals, quite a lot has also been learned by studying humans who have suffered injuries to different parts of the brain and who therefore lose certain capabilities. An important anatomical and physiological landmark of each cerebral hemisphere is the **central sulcus**, which separates the frontal and parietal lobes (figure 5.13, top). The frontal lobe gyrus that lies in front of the central sulcus is an important area initiating nerve impulses that bring about muscle movements. For this reason, the *precentral gyrus* of the frontal lobe is known as

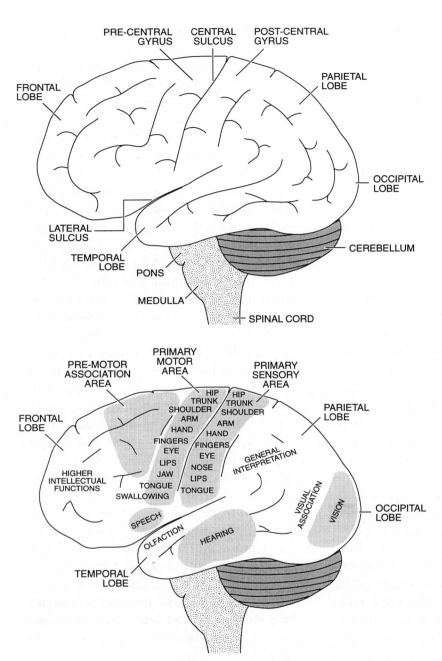

Figure 5.13
A superficial view of the brain as seen from the left side. The top illustration highlights the major parts of the brain and distinguishes the different lobes of the left cerebral hemisphere. Not seen (because it is hidden by the temporal lobe) is the midbrain. The midbrain, pons, and medulla comprise the brain stem. The bottom illustration shows the point-for-point relationship between the primary motor and sensory areas of the pre- and post-central gyri and the areas of the body with which they physiologically interact. With some exceptions, the right side of the brain is symmetric with the left side shown here.

the **primary motor area**. By applying electrical stimuli to different parts of the cortex of the pre-central gyrus and observing the body movement that results, it has been possible to establish the topological distribution of control centers along the pre-central gyrus

(see figure 5.13, bottom and also figure 5.15). Accordingly, movements of the toes, feet, legs, and hips are initiated from the end of the pre-central gyrus closest to the longitudinal fissure. Moving laterally along the pre-central gyrus from the longitudinal fissure toward the temporal lobe, we find the centers that initiate movements in the body's trunk, shoulders, arms, hands, and fingers. Finally, the most lateral areas of the pre-central gyrus initiate movements of the eyes, face, jaws, lips, and tongue. Although there is a clear spatial (anatomical) relationship between the order of control centers in the pre-central gyrus and the regions of the body that are controlled by these centers, the *amount* of cerebral cortex devoted to different regions of the body bears no such anatomical relationship. For example, a disproportionately larger amount of cerebral cortex is devoted to the control of the tongue, eyes, and fingers than to the body trunk and legs. This disproportionality is consistent with the much finer control we are able to exercise over the movements of the tongue, eyes, and fingers than over the movements of our body trunk and legs.

Immediately behind the central sulcus is the **post-central gyrus** of the parietal lobe; this gyrus contains the cerebrum's **primary sensory area** (figs. 5.13 and 5.15). It is into this area that nerve impulses flow from different regions of the body, informing us of stimuli that arise in these regions. For example, pressure or heat applied to the skin of the fingers will elicit *sensory* nerve impulses that arise in the fingers and soon reach the area of the post-central gyrus corresponding to sensation in the fingers. The topological distribution of the sensory areas of the post-central gyrus is shown in figures 5.13, bottom and 5.15. The sensory areas for vision, hearing and olfaction (i.e., the sense of smell) are not in the primary sensory area but are represented individually in the occipital and temporal lobes (see figure 5.13, bottom). The amount of cortex devoted to sensations arising from a given area of the body varies

with that area's sensory importance. Hence, much more sensory cortex is devoted to sensations arising from the fingers, face and lips than from the arms, body trunk, and legs.

| 5-5 |

It should be noted at this point that stimuli applied to the *left* side of the body (e.g., squeezing the fingers of the left hand) produce sensory impulses that reach the primary sensory area of the *right* cerebral hemisphere. By the same token, stimuli applied to the *right* side of the body produce nerve impulses that are carried to the primary sensory area of the *left* cerebral hemisphere. This is because nerve fibers conducting these impulses cross from one side of the central nervous system to the other during their ascent toward the cerebrum. (The same is true for impulses arising in the primary motor areas. That is to say, stimulating the "fingers region" of the pre-central gyrus of the *left* cerebral hemisphere leads to movements of the fingers of the *right* hand.) Most of the fibers that cross from the right to the left (or from the left to the right) sides of the central nervous system cross in the brain stem.

Motor association areas occupy a portion of the frontal lobes just anterior to the right and left primary motor areas. These association areas appear to be involved in the learning and control of skilled body movements. Other parts of the frontal lobe are believed to be involved in higher intellectual functions such as learning, memory and emotion.

The translation of thoughts into speech occurs in the **speech area** (also called **Broca's area**) located on the frontal side of the **lateral sulcus**, which separates the frontal and temporal lobes (fig. 5.13). Impulses are sent from the speech area to the primary motor area, which then controls the throat muscles, vocal cords, and respiratory muscles which make speech possible. Interestingly, the speech areas of the cerebrum are not bilaterally symmetric. Rather, in a very high percentage of right-handed people, the speech area is in the left frontal lobe; whereas in about half of all left-handed people, the speech area is in the right frontal lobe.

Just posterior to the two primary sensory areas of the cerebrum's parietal lobes are the **sensory association areas**. These areas play a role in the interpretation of sensations, such as determining the texture of an object that is touched or determining the shape of an object or identifying it by its feel.

A principal concern of the left and right occipital lobes is vision (figure 5.13, bottom). The **visual cortex** at the rear of each occipital lobe is concerned with the evaluation of incoming visual signals concerning colors, shapes, and motion. The **visual association area** that lies just anterior to each primary visual area is involved with the recognition and interpretation of what is being seen.

The temporal lobes house the centers for hearing and olfaction. The **hearing center** is responsible for our ability to distinguish and identify sounds and to understand speech. The **olfactory center** makes it possible to distinguish and identify different odors.

The interior of the cerebrum is revealed in the midsagittal section shown in figure 5.14 and the midfrontal section shown in figure 5.15. As noted earlier, much of the grey matter of the cerebrum is represented by a thin layer at the surface (i.e., the **cortex**). Below this lies the more expansive area of white matter. There are, however, a number of "islands" of grey matter distributed within the white matter; these include the **thalamus,**

hypothalamus, and **basal ganglia** (discussed below).

A large bridge of white matter at the base of the frontal and parietal lobes connects the left and right cerebral hemispheres. This structure is called the **corpus callosum** and consists of nerve fibers that conduct impulses from one cerebral hemisphere to the other. Below the corpus callosum in each lobe there is a fluid-filled chamber called a **lateral ventricle** (seen in figure 5.15, but not seen in figure 5.14). The two lateral ventricles merge medially and inferiorly to form the so-called **third ventricle** (figs. 5.14 and 5.15); a **fourth ventricle** lies just anterior to the cerebellum. The fluid that fills the ventricles of the brain, called **cerebrospinal fluid** (or CSF), also fills the **central canal** of the spinal cord (see below). Altogether, the ventricles of the brain and the central canal of the spinal cord contain about 140 c.c. of CSF. CSF is derived from blood circulating through capillaries of the brain called **choroid plexuses**, about 800 c.c. being formed daily and eventually returning to the bloodstream. Cerebrospinal fluid plays a protective role and also serves in the removal of chemical wastes produced during | 5-6 | the metabolism of the tissues of the central nervous system.

The upper portion of the side walls of the third ventricle contains masses of grey matter that form an important part of the brain called the thalamus; the lowermost portion of the wall, together with the third ventricle's floor, form the grey matter of the hypothalamus. The thalamus serves as a major relay center containing many axosomatic synapses in which impulses that are passing upward to higher brain centers or downward into the brain stem and spinal cord | 5-7 | are transmitted from one nerve fiber

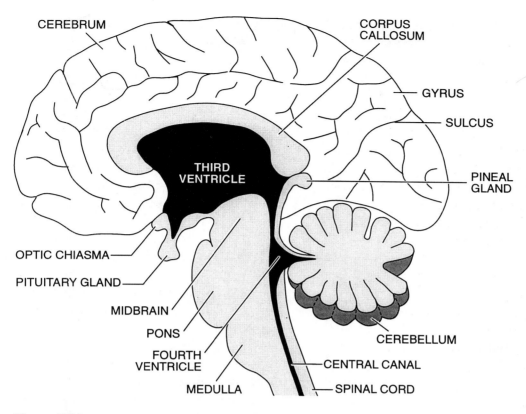

Figure 5.14
Midsagittal section through the brain (cut parallel to the longitudinal fissure). Seen in this section is the medial surface of the right cerebral hemisphere. The two hemispheres are connected anatomically and functionally by the corpus callosum. Below the corpus callosum are three chambers (or ventricles) filled with cerebrospinal fluid. Seen here is the third ventricle (the first and second ventricles, also called lateral ventricles, are behind and in front of the plane seen in this section but are depicted in fig. 5.15). The fourth ventricle lies anterior to the cerebellum.

to another. The hypothalamus contains centers that are very important in the regulation of such physiological processes as (a) body temperature, (b) the body's water and salt balance, (c) hunger, and (d) sexual activity. The hypothalamus is also the source of chemicals that regulate the production and secretion of hormones by the *anterior lobe* of the pituitary gland. Additionally, the hypothalamus manufactures two important hormones of its own (**antidiuretic hormone**

and **oxytocin**). These hormones are conveyed by axonal processes to the *posterior lobe* of the pituitary, where they are temporarily stored prior to secretion. Antidiuretic hormone is concerned with the regulation of water loss from the body through the kidneys. Oxytocin influences the contractions of the walls of the uterus during childbirth and also facilitates the release of milk from the mammary glands). (The functions of the body's **endocrine glands** and the diversity

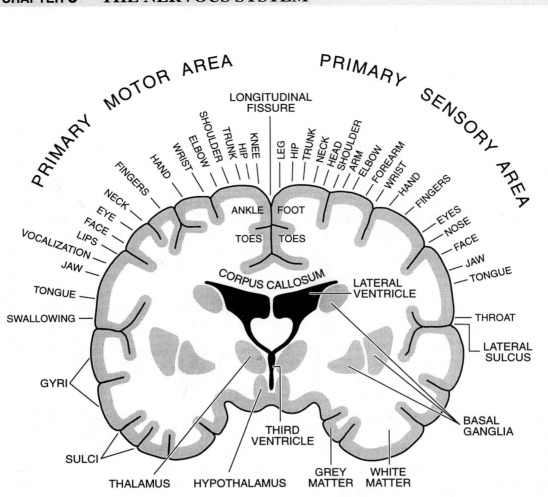

Figure 5.15
Midfrontal section through the cerebrum. On the left side of the figure, the plane of the section passes just anterior to the central sulcus, so that the primary motor area and its centers are revealed. On the right side of the figure, the plane of the section passes just posterior to the central sulcus, so that the primary sensory area and its centers are revealed.

of hormones that they secrete are considered in some detail in chapter 14.)

Like the thalamus and hypothalamus, the basal ganglia are areas of grey matter that are surrounded by the cerebrum's expansive white matter. The basal ganglia serve as relay centers in which impulses that are on their way *up* to the cerebrum (*from* lower brain centers or the spinal cord) or are on their way *down* from the cerebrum (*to* lower brain centers or the spinal cord) are transmitted from pre- to postsynaptic fibers.

Extending downward and forward from the floor of the third ventricle into a small pocket in the underlying skull bone is the **pituitary gland** (or **hypophysis**; figure 5.14). The pituitary is divided into two anatomically distinct and functionally different lobes. As has already been mentioned, the posterior lobe (known as the **neurohypophysis**) temporarily stores and then secretes two hormones (manufactured in the overlying hypothalamus). In response to chemical messages received from the hypothalamus, the anterior lobe (known as the **adenohypophysis**) manufactures and secretes a variety of hormones, including **human growth hormone, thyroid-stimulating hormone, prolactin, adrenocorticotropic hormone, follicle-stimulating hormone,** and **luteinizing hormone**.

Human growth hormone contributes to the control of the body's growth, especially the growth of the skeleton. Thyroid-stimulating hormone regulates the production and secretion of the hormones of the thyroid gland. Prolactin promotes the production of milk in the mammary glands of the mother in the weeks and months that follow childbirth. Adrenocorticotropic hormone controls the production and secretion of hormones by the outer layer (i.e., the "cortical layer") of the adrenal glands. Follicle-stimulating hormone and luteinizing hormone affect the body's reproductive organs (i.e., the ovaries in females and the testes in males). In females, follicle-stimulating hormone regulates the development of the ovarian follicles and the egg cells that the follicles contain. In males, the same hormone influences the production of sperm cells. Luteinizing hormone plays an important part in promoting ovulation (i.e., the release of an egg cell from the ovaries during each menstrual cycle). In males, luteinizing hormone promotes sex hormone production by the testes. An in-depth discussion of the hormones of the pituitary gland (as well as other endocrine glands) is presented in chapter 14, which deals with the physiology of the endocrine system.

Extending posteriorly from the rear wall of the third ventricle is the **pineal gland** (or **epiphysis**; fig. 5.14). The functions of the pineal gland are somewhat unclear; what is known is that the gland secretes the hormone **melatonin** and that melatonin secretion is indirectly influenced by the amount of light that enters the eyes. Apparently, during the daytime, light entering the eyes results in the passage of nerve impulses to the pineal gland and this inhibits the gland's melatonin synthesizing activity. Thus, it is not surprising to find that in most people the blood's melatonin level is very low between 7:00 A.M. and 11:00 P.M. but rises dramatically between 11:00 P.M. and 7:00 A.M. In humans, the effects of melatonin are uncertain, although there is evidence that melatonin inhibits development of and hormone production by the ovaries.

The Cerebellum. At the rear of the brain, lying underneath the two occipital lobes of the cerebrum is the **cerebellum** (figs. 5.13 and 5.14). Like the cerebrum, the cerebellum is divided into right and left hemispheres. Also like the cerebrum, the thin cortex of the cerebellum consists of grey matter, while most of the underlying tissue is white. The cortex folds inward to create sulci and gyri like those of the cerebrum, but in the cerebellum these folds and rises are less prominent and are organized in a parallel fashion. Large bundles of nerve fibers called *cerebellar peduncles* link the cerebellum to the cerebrum and to the brain stem.

The principal function of the cerebellum is the subconscious regulation of muscle activity in such common body actions as maintaining posture, standing upright, walking, and running. The cerebellum is in continuous communication with the muscles and joints of the body, receiving signals from these remote regions, as the muscles contract, relax, or are stretched. Acting through the primary motor area of the cerebrum, the cerebellum responds to this sensory information by ensuring smooth and coordinated muscle activity.

The Brain Stem. The **brain stem** is the third major part of the brain. It connects the cerebrum above with the spinal cord below and is also joined posteriorly to the cerebellum. The brain stem consists of three major portions called the **midbrain, pons,** and **medulla** (figures 5.13 and 5.14). The midbrain lies immediately below the cerebrum and contains descending nerve fibers that connect the cortex of the cerebrum with the pons and ascending fibers that connect the spinal cord with the thalamus. The midbrain is also the origin of nerve fibers that comprise cranial nerves III and IV (see later). Within the midbrain are centers that control the movements of the eyes and head. The two *superior colliculi* of the midbrain control reflexes in which the eyes are moved in order to fix upon a particular object as the head is turning. The paired *inferior colliculi* control reflexive movements of the head in the direction of a source of sound. Together, the superior and inferior colliculi comprise the *corpora quadrigemina.*

The pons lies immediately below the midbrain and just above the medulla (fig. 5.14). Acting in concert with the **inspiratory center** of the medulla, the pons regulates breathing. Depending on physiological conditions, the pons' **apneustic center** has a stimulatory effect on the inspiratory center (leading to inflation of the lungs), whereas the **pneumotaxic center** has an inhibitory effect on the inspiratory center (leading to lung deflation). The actions of the respiratory centers of the pons and medulla are considered in detail in chapter 10. The pons is also the origin of nerve fibers that become part of cranial nerves V, VI, VII, and VIII (nerve VIII also has roots in the medulla).

In addition to housing several control centers, the medulla contains numerous ascending and descending nerve fibers that provide communication between the brain and spinal cord. Interestingly, most of the descending nerve fibers cross over from one side of the medulla to the other, with the result that (for the most part) actions taken by the right side of the body are controlled by the left side of the brain, and actions taken by the left side of the body are controlled by the right side of the brain. Ascending fibers carrying *proprioceptive* signals (i.e., information concerning the amount of stretch or contraction taking place in a muscle) also cross from one side to the other in the medulla. Consequently, proprioceptive sensations arising from the muscles of the right side of the body are perceived in the left side of the brain (and vice-versa). (Nerve fibers carrying signals that give rise to sensations of heat, cold and pain also cross from one side of the body to the other *but not in the medulla*; rather, these fibers cross from one side to the other *within the spinal cord.*)

There is a grey region in the medulla called the **reticular formation**, which contains centers that regulate heart rate (i.e., the **cardioaccelerator** and **cardioinhibitory centers**), breathing (i.e., the **inspiratory** and **expiratory centers**), and blood pressure (i.e., the **vasomotor center**).

Cranial nerves VIII, IX, X, XI, and XII arise in the medulla, although VIII also has roots in the pons. As will be discussed later, the medulla controls much of the activity of the **autonomic nervous system.**

The Spinal Cord

Descending from the brain stem through the neck, chest, and upper abdominal regions of the body is the **spinal cord** (fig. 5.16). In the average adult male, the spinal cord is about 17 inches long and is protected by a chain of 33 bony enclosures called **vertebrae** (fig. 5.17). Collectively, the vertebrae form the **vertebral column**. Successive vertebrae are separated from one another by pads of cartilage called **disks**; this arrangement of vertebrae and disks provides protection while also allowing the spinal cord to bend or twist according to body movement and body position. The bottom-most nine vertebrae are fused with one another and form the *sacrum* and *coccyx*. Arising from the left and right sides of the spinal cord (usually between successive vertebrae) are 31 pairs of **spinal nerves**.

Figure 5.18 shows a cross section through the spinal cord. As you examine this figure, it is important to note that the inner portion of the spinal cord consists of **grey matter** (i.e., neuron cell bodies), whereas the outer region consists of **white matter** (i.e., myelinated processes). This is the reverse of the condition that exists in the brain, where the outer portion (i.e., the cortex) is grey and the inner portion is predominantly white. The grey matter of the left and right sides of the spinal cord is divided into three **horns**: the *posterior* horn, the *anterior* horn, and the *lateral* horn. At the center of the spinal cord is the narrow **central** (or **spinal**) **canal**. The central canal is filled with cerebrospinal fluid, which serves in the nourishment of the surrounding tissue. (Remember that as you follow the central canal upward through the spinal cord into the brain, the canal leads into the brain's ventricles; see figure 5.14).

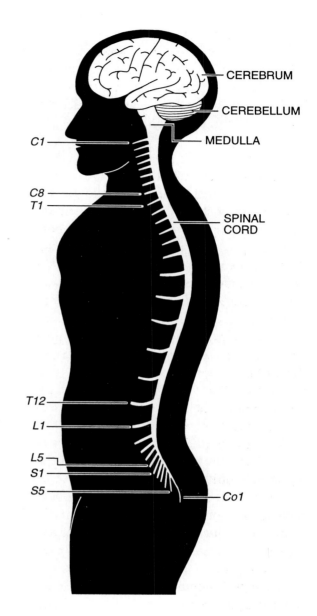

Figure 5.16
The spinal cord. Arising from the spinal cord (usually between successive vertebrae) are 31 pairs of spinal nerves, identified by a letter and number denoting the body region in which they arise and their order of occurrence. There are 8 pairs of cervical (C1-C8), 12 pairs of thoracic (T1-T12), 5 pairs of lumbar (L1-L5), 5 pairs of sacral (S1-S5), and one pair of coccygeal (Co1) spinal nerves. Not shown is the chain of vertebrae protecting the spinal cord (see fig. 5.17).

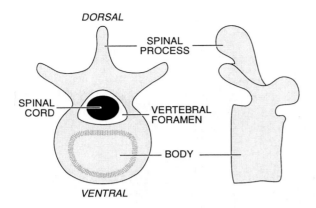

DORSAL

SPINAL PROCESS

SPINAL CORD

VERTEBRAL FORAMEN

BODY

VENTRAL

Figure 5.17
A vertebra, seen from above (left) and from the side (right). Although the structure of the vertebrae vary somewhat depending on their location in the vertebral column, they generally take the form shown here. The spinal cord passes vertically downward through the vertebral foramens of successive vertebrae. Skeletal muscles are attached to the spines of the vertebrae. Neighboring vertebrae are cushioned by cartilaginous disks seated on the body portion of each vertebra. (Dorsal = "rear" and ventral = "front".)

The white matter of the spinal cord contains **spinal tracts** (fig. 5.19); these are bundles of neuronal processes that conduct impulses up or down the spinal cord. Tracts containing fibers that conduct impulses up

the spinal cord are called *ascending tracts*, whereas tracts containing fibers that conduct impulses down the spinal cord are called *descending tracts*.

THE PERIPHERAL NERVOUS SYSTEM

Cranial and Spinal Nerves

Emerging from the brain and spinal cord are the **cranial nerves** and **spinal nerves**. The internal organization of a representative nerve is shown in the cross section depicted in figure 5.20. The nerve is enclosed by a band of connective tissue called **epineurium**. Internally, the individual nerve fibers are arranged in large bundles called **fascicles**. (The fascicles of a cranial or spinal nerve are analogous to the *tracts* of the spinal cord.) Each fascicle is covered by a layer of connective tissue called **perineurium**. Within each fascicle, groups of nerve fibers are embedded in yet another form of connective tissue called **endoneurium**. The endoneurium is separate and distinct from the myelin sheaths that may envelope specific fibers.

There are 31 pairs of spinal nerves, one

Figure 5.18
The appearance of the spinal cord in cross section.

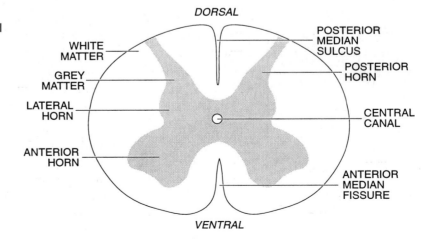

DORSAL

WHITE MATTER

GREY MATTER

LATERAL HORN

ANTERIOR HORN

POSTERIOR MEDIAN SULCUS

POSTERIOR HORN

CENTRAL CANAL

ANTERIOR MEDIAN FISSURE

VENTRAL

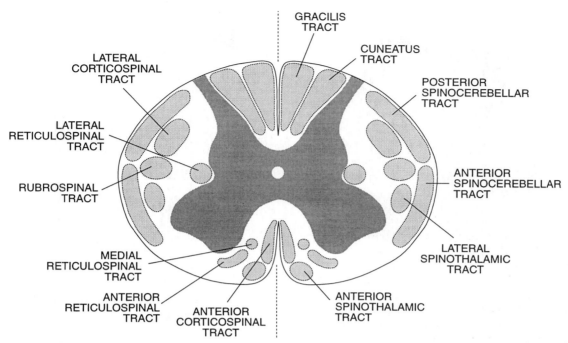

Figure 5.19
A cross section of the spinal cord showing the locations of the spinal tracts. The tracts whose labels appear on the left side of the diagram are descending tracts (their fibers conduct impulses down the spinal cord); the tracts whose labels appear on the right side of the diagram are ascending tracts (their fibers conduct impulses up the spinal cord). (Note that ascending and descending tracts are symmetrically distributed on both sides of the spinal cord.)

member of each pair exiting from the right half of the spinal cord and the other member of the pair exiting from the left half of the spinal cord. A single spinal nerve may contain several hundred fibers. The fibers that comprise a spinal nerve are a mixture of **sensory fibers** (i.e., fibers that conduct impulses toward the spinal cord) and **motor fibers** (i.e., fibers that conduct impulses away from the spinal cord). An individual fascicle usually contains a mixture of sensory and motor fibers.

There are 12 pairs of cranial nerves. However, unlike the spinal nerves some cranial nerves consist exclusively of sensory fibers (i.e., fibers that conduct impulses toward the

brain) and are called "sensory nerves"; others consist exclusively of motor fibers (i.e., fibers that conduct impulses away from the brain) and are called "motor nerves." The remaining cranial nerves are like all of the spinal nerves; that is, they are "mixed nerves" (i.e., they contain both sensory and motor fibers). Table 5.1 lists the various cranial nerves; as you examine the table, note that it is customary to identify the nerves using Roman numerals. The numbering of the cranial nerves is based on their anatomical position, cranial nerve number I arising uppermost in the brain, and cranial nerve number XII arising lowermost in the brain (fig. 5.21).

5-8

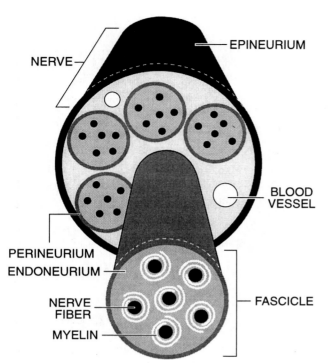

Figure 5.20
A spinal or cranial nerve seen in cross section.

EPINEURIUM

NERVE

BLOOD VESSEL

PERINEURIUM
ENDONEURIUM

NERVE FIBER

MYELIN

FASCICLE

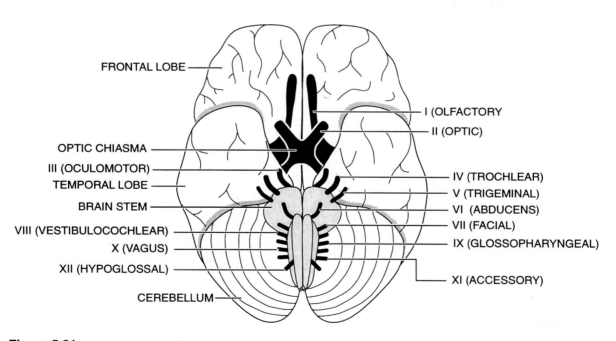

FRONTAL LOBE

I (OLFACTORY

II (OPTIC)

OPTIC CHIASMA

III (OCULOMOTOR)

TEMPORAL LOBE

BRAIN STEM

VIII (VESTIBULOCOCHLEAR)

X (VAGUS)

XII (HYPOGLOSSAL)

CEREBELLUM

IV (TROCHLEAR)

V (TRIGEMINAL)

VI (ABDUCENS)

VII (FACIAL)

IX (GLOSSOPHARYNGEAL)

XI (ACCESSORY)

Figure 5.21
The 12 pairs of cranial nerves. In this view, the brain is seen from below so that the origins of the cranial nerves are apparent. See table 5.1 for the nerve functions.

TABLE 5.1   THE CRANIAL NERVES

NUMBER	NAME	ORIGIN	TYPE	FUNCTION
I	Olfactory	Cerebrum	Sensory	Olfaction (sense of smell)
II	Optic	Cerebrum	Sensory	Vision
III	Oculomotor	Midbrain	Mixed	Movements of the eyes; accommodation of the lens; movements of eyelids; varies pupil size; proprioception
IV	Trochlear	Midbrain	Mixed	Movements of the eyes; proprioception
V	Trigeminal	Pons	Mixed	Sensations from face, nose, mouth, and teeth; chewing
VI	Abducens	Pons	Mixed	Eye movements; proprioception
VII	Facial	Pons	Mixed	Sense of taste; salivation; facial muscles of expression
VIII	Vestibulocochlear (Acoustic)	Pons and medulla	Sensory	Sense of hearing; equilibrium and balance
IX	Glossopharyngeal	Medulla	Mixed	Swallowing; salivation; sense of taste; blood gas monitoring
X	Vagus	Medulla	Mixed	Sensory information from and motor control of many thoracic and abdominal organs (e.g., heart, lungs, stomach, and intestine)
XI	Accessory	Medulla and spinal cord	Motor	Head and shoulder movement; proprioception
XII	Hypoglossal	Medulla	Motor	Movements of tongue

Cranial nerves I and II arise from the cerebrum and are comprised exclusively of sensory fibers. As we will see in chapter 6, the **olfactory** and **optic** nerves are concerned with the senses of smell and vision. Cranial nerves III and IV (**oculomotor** and **trochlear**) arise from the midbrain and are primarily concerned with the control of movements of the eyes. Cranial nerves V, VI, and VII (**trigeminal**, **abducens**, and **facial**) arise from the pons and are concerned with sensations arising from the face and mouth and the control of eye movement and chewing. The remaining cranial nerves (VIII through XII) arise from the medulla. The **vestib-ulocochlear** (VIII) is a sensory nerve concerned with the senses of hearing and balance. The **glossopharyngeal** (IX) provides for the sense of taste and controls swallowing. The **vagus** (X) controls and monitors the actions of many of the thoracic and abdominal organs. The **accessory** (XI) is comprised of motor fibers only and controls head and shoulder movements. Finally, the **hypoglossal** (XII), also motor, controls the movements of the tongue.

INTERACTIONS BETWEEN THE NERVOUS SYSTEM'S CENTRAL AND PERIPHERAL DIVISIONS

Spinal Reflexes

Using the foregoing description of the organization of the nervous system as a foundation, we may now proceed to a discussion of interactions between the nervous system's central and peripheral divisions. Among the fundamental interactions are the **spinal re-** **flexes**—interactions that involve the spinal cord and the spinal nerves. As noted earlier, the spinal nerves arise from the spinal cord between successive vertebrae. As seen in figure 5.22, the nerves emerge from the cord as two **roots**: a **dorsal root** (arising at the rear of the cord; i.e., dorsal = "rear" or "back") and a **ventral root** (arising at the "front" of the cord; i.e., ventral = "front"). The dorsal and ventral roots of each nerve merge to form the nerve proper a short distance from the spinal cord.

Simple (Monosynaptic) Spinal Reflexes

During a physical examination, it is not uncommon for a physician to perform a test in which you are asked to cross your legs, following which you are gently struck below the knee with a rubber mallet. The normal response is a sudden (and involuntary) extension (or "jerk") of the crossed leg. This reaction is an example of a **simple spinal reflex** (called the **knee-jerk reflex**). "Postural reflexes" are another illustration of simple spinal reflexes. In a postural reflex, contractions of the musculature of the back and/or neck maintain the body and head in an

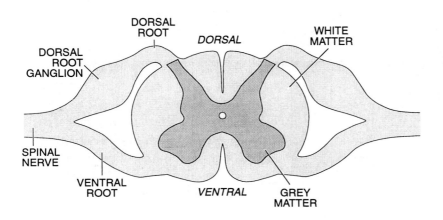

Figure 5.22
Cross section through the spinal cord, showing the dorsal and ventral roots of the spinal nerves and the dorsal root ganglia.

DORSAL ROOT

DORSAL

WHITE MATTER

DORSAL ROOT GANGLION

SPINAL NERVE

VENTRAL ROOT

VENTRAL

GREY MATTER

upright position. For example, you may have experienced a situation in which you were seated upright watching television or listening to a lecture and began to drift off to sleep. As your head started to tilt sideways or forward, the postural reflex suddenly brought your head back to its upright position, perhaps startling you in the process. These and other simple spinal reflexes are founded on the following interactions.

The typical simple spinal reflex (fig. 5.23) begins when **receptor cells** sense physical or chemical changes within the body or near to the body surface. For example, in the knee-jerk reflex, the blow of the rubber mallet stretches the **patellar tendon**, which anchors the *quadriceps femoris* muscle to the "knee bone" (i.e., patella) and tibia. The stretching of the tendon stimulates receptor cells in the tendon. The receptors are associated with the dendritic nerve endings of

sensory (or **afferent**) **nerve fibers**, so that activation of the receptors is quickly followed by stimulation of these dendritic endings. The result is the propagation of nerve impulses along the sensory fibers toward the spinal cord.

The dendrites of the sensory fibers become part of a spinal nerve. Each fiber's cell body is housed in the small, bulbous enlargement of the spinal nerve's dorsal root (the *dorsal root ganglion*). The axonal process of each sensory nerve fiber extends from the dorsal root ganglion into the grey matter of the spinal cord. Within the ventral horn of the spinal cord's grey matter, the axonal processes of the sensory fibers synapse with either the dendritic endings or the cell body of **motor** (or **efferent**) **nerve fibers**. The nerve impulses conducted along the sensory fibers are transmitted to the motor fibers in the grey

5-9

Figure 5.23
A simple (monosynaptic) spinal reflex. The arrows show the direction in which the impulses are conducted and transmitted. See text for explanation.

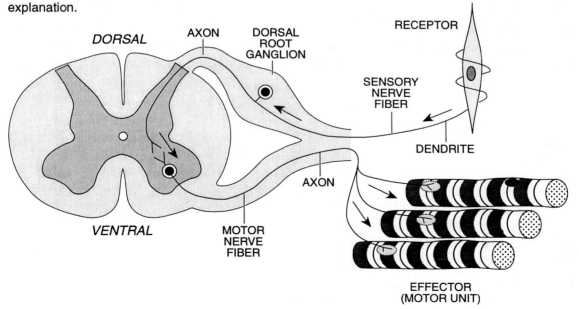

matter. The axonal processes of the motor fibers exit the grey and white matter of the spinal cord through the ventral root and become a part of the spinal nerve. The motor fibers extend into the leg musculature and branch into a number of fine endings that form junctions with an effector; in this case, the effector is a number of striated muscle fibers (i.e., the fibers of a *motor unit*) in the quadriceps femoris muscle. Impulses reaching these striated cells cause contraction, which elevates the lower leg in a quick, jerky motion. The arrangement of receptor, sensory fiber, motor fiber, and effector is depicted in figure 5.23.

| 5-10 |

There are several important points to be kept in mind regarding a simple spinal reflex and its illustration in figure 5.23. First of all, note that there are only *two types* of nerve fibers involved in a simple spinal reflex: *sensory* fibers and *motor* fibers. However, this is not intended to imply that the completion of a simple spinal reflex employs only two fibers. Indeed, depending upon the intensity with which the mallet strikes the leg, several dozen sensory fibers may be caused to conduct impulses into the spinal cord. By the same token, several dozen motor fibers may conduct impulses out of the spinal cord and on toward several dozen motor units in the leg muscle that jerks the leg upwards. Notice also that the pathway is **monosynaptic**; that is, there is a single *type* of synapse–the type that occurs between sensory fibers and motor fibers. Finally, it is to be emphasized that the reflex is mediated through the spinal cord and does not require participation of the brain in order to be completed (i.e., a knee jerk reflex can be elicited from an unconscious person). This does not mean that one is unaware of the impact of the mallet, it simply means that the reflexive

action occurs whether or not the brain is "aware" of what has taken place.

Polysynaptic Spinal Reflexes

We will now turn to some spinal reflexes that are somewhat more complex. A good illustration is the sudden reflexive withdrawal of one's hand from the surface of a hot object that is inadvertently touched. The pathways involved in this reflex are shown in figure 5.24.

Again, the reflex begins with the stimulation of receptors. In this case, the receptors are cells (or nerve endings) in the skin that are sensitive to heat. The stimulated receptors respond by triggering nerve impulses in the sensory fibers with which the receptors are coupled. As in a simple spinal reflex, the impulses are conducted into the grey matter of the spinal cord through the spinal nerve's dorsal root.

Within the grey matter of the spinal cord, the impulses are transmitted to nerve fibers called **association fibers** (or **interneurons**). Some of these fibers then conduct impulses to the dendritic endings (or cell bodies) of motor fibers; the latter complete the reflex by conducting impulses to the effector cells (e.g., muscle cells in the arm that cause you to lift your hand off the hot object). As in the simple spinal reflex described earlier, the motor impulses are conducted from the spinal cord via the ventral root. Notice that a major difference between this reflex and the one described earlier is the involvement of a third class of nerve fibers, namely *association* fibers.

The interposition of association fibers between sensory fibers and motor fibers has a number of consequences. One of these is the ability of the brain to easily override the

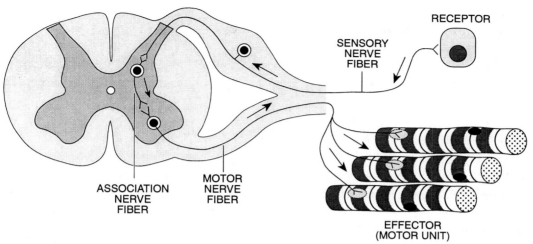

Figure 5.24
A polysynaptic spinal reflex, such as withdrawing your hand from a hot surface. The arrows show the direction of the nerve impulses (see text for explanation).

reflex. For example, suppose that you place your hand on an object that is hot but is not so hot that it could injure (i.e., burn) the skin. As the skin receptors respond to the sudden rise in temperature, the spinal reflex is triggered and you begin to withdraw your hand. However, sensing that the object is not really too hot to touch, you can return your hand to the warm surface and keep it there. In such a scenario, what you are doing is "overriding" the spinal reflex. This is achieved physiologically by sending nerve impulses down through tracts in the white matter of the spinal cord from the brain to the junctions between association and motor fibers of the reflex arc. These impulses serve to *inhibit* the transmission of impulses from association to motor fibers and therefore allow you to leave your hand on the hot object. In other words, you can exercise conscious control over this kind of spinal reflex.

Flexion and Extension. The two reflexes described above illustrate two alternative

classes of muscle actions: **flexion** and **extension**. By flexion is meant contraction of a muscle that serves to move a body part such as an arm or leg closer to the torso (e.g., closer to the head, chest, or abdomen). For example, when you pull your hand away from the surface of a hot (or sharp) object, you pull your hand toward the body (i.e., you don't simply move it to one side or the other). By extension is meant a muscle contraction that serves to move a body part further away from the torso. This is illustrated by the knee jerk, in which muscle contraction extends the lower leg, so that the foot is further away from the torso.

Reciprocal Innervation

Flexion of a muscle is accompanied by stretching of the *antagonistic* muscle. For example, when the biceps muscle of the upper arm contracts, the arm is flexed (i.e., the lower arm is raised toward the shoulder).

111

As a result, the triceps muscle of the upper arm is stretched. By the same token, extension of the lower arm brought about by contraction of the triceps muscle serves to stretch the biceps muscle. Because they have opposite effects, the biceps and triceps muscles are said to be antagonistic. Clearly, it would be counter-productive if a pair of antagonistic muscles were to contract at the same time. When one of a set of antagonistic muscles is stimulated by motor pathways and caused to contract, stimulation of the antagonistic muscle is inhibited (and thereby prevented from contracting). This important physiological mechanism is called **reciprocal innervation**.

The Crossed-Extensor Reflex

An interesting spinal reflex that involves both flexion and extension is the **crossed-extensor reflex**. This reflex may be illustrated as follows. Suppose that a person is walking barefoot and steps on a thumbtack with his left foot. Puncturing of the skin of the sole of the foot and injury to underlying receptors and nerve endings triggers a reflex in which there is *flexion of the left leg* as flexor muscles contract and extensor muscles are allowed to stretch (through reciprocal innervation). This raises the left foot off the ground, and by taking the weight of the sole of the foot prevents further penetration of the thumbtack. At the same time, there is *extension of the right leg* as extensor muscles contract and allow the body weight to be supported by one leg. The nerve pathways involved in this reflex are shown in figure 5.25.

The reflexive actions of the left leg are said to be **ipsilateral**, which means that the incoming sensory signals (from the sole of the left foot) and outgoing motor signals (to the flexor and extensor muscles of the left leg) involve only one side of the body and one side of the nervous system (i.e., the left side). In contrast, the reflexive actions of the right leg are said to be **contralateral** (i.e., they involve the opposite side of the body and opposite side of the nervous system).

In all of the reflexes that we have considered so far, the incoming sensory impulses and the outgoing motor impulses involve the same level of the spinal cord. For example, the knee jerk reflex involves only the pelvic region of the spinal cord, and withdrawing one's hand from a hot object involves only the shoulder level of the spinal cord. Some reflexive actions, however, involve sensory input and motor output at different levels of the spinal cord. For example, when you are abruptly pushed on the shoulder, displacing your body to one side, you reflexively extend the leg on the opposite side of the body in order to avoid falling. (Note that the above action is *contralateral*.) Whenever either contralateral or ipsilateral reflexes involve two (or more) levels of the spinal cord, the conduction of impulses between the different levels is mediated by association fibers whose long processes are part of a spinal tract.

THE AUTONOMIC NERVOUS SYSTEM

The autonomic nervous system includes a number of motor pathways that control the actions of many of the body's internal organs. These motor pathways are divided anatomically and physiologically into two groups: (1) pathways of the **sympathetic**

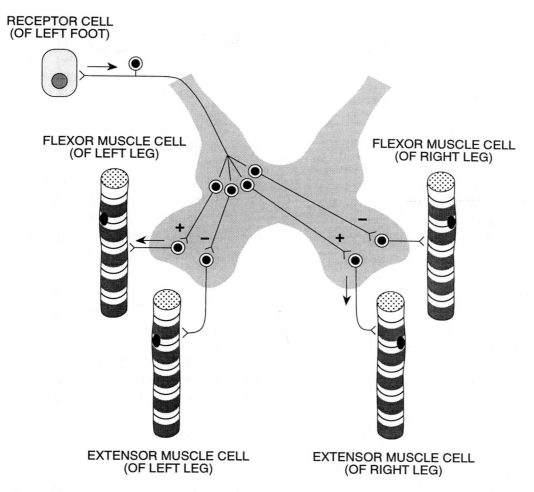

RECEPTOR CELL
(OF LEFT FOOT)

FLEXOR MUSCLE CELL
(OF LEFT LEG)

FLEXOR MUSCLE CELL
(OF RIGHT LEG)

EXTENSOR MUSCLE CELL
(OF LEFT LEG)

EXTENSOR MUSCLE CELL
(OF RIGHT LEG)

Figure 5.25
Reciprocal innervation and the crossed-extensor reflex. The spinal reflex
depicted here involves flexion of the left leg and extension of the right leg.
The association fibers that are marked (+) have a stimulatory effect on the
motor fibers with which they synapse; association fibers marked (-) have an
inhibitory effect. See the text for explanation of the reflex.

division, and (2) pathways of the **parasympathetic division**. It is, perhaps, easiest to
understand the overall functions of these two
divisions by considering examples of what
each division does. Let us begin by listing
some of the physiological changes that occur
in the body when the sympathetic division of
the autonomic nervous system becomes
active.

Actions of the Sympathetic Division

The comprehensive action of the sympathetic
division of the autonomic nervous system
becomes obvious if we consider a series of
individual effects on the body's physiology.

1. The amount of light entering each eye is

regulated by a smooth muscle called the **iris**. The iris muscle is shaped like a donut, the opening at the center forming the eye's **pupil** (i.e., the central black spot); the iris musculature also contains pigments that determine one's eye color (i.e., brown eyes, blue eyes, etc.). The muscle tissue of the iris is arranged in two ways: *circumferentially* and *radially*. Contraction of the circumferential muscle tissue makes the pupil smaller and reduces the amount of light entering the eye. Contraction of the radial muscle tissue dilates the pupil, allowing more light to enter the eye. Motor fibers of the sympathetic division innervate the radial muscle tissue of the iris, thereby causing dilation of the pupil.

2. Motor fibers of the sympathetic division also innervate the **salivary glands** (glands in the walls of the mouth that secrete saliva). The effect of impulses reaching the glands over these sympathetic pathways is to reduce the blood flow to the glands. This reduces the water content of the saliva, making the saliva thicker (more concentrated). The accompanying sensation is a dryness of the mouth.

3. Motor fibers of the sympathetic division also innervate the **sweat glands** (glands in the skin that secrete a watery fluid onto the skin's surface, thereby cooling the skin). The effect of impulses reaching the sweat glands over these sympathetic pathways is *stimulatory*, thereby causing perspiration.

4. The walls of the digestive organs (stomach, small intestine, etc.) are rich in smooth musculature, the contractions of which help to digest the food and propel it through the digestive tract. The musculature of the digestive tract receives nerve impulses over autonomic pathways. In the case of the sympathetic division, the effect on this musculature is *inhibitory*, bringing a halt to such digestive activity.

5. The walls of the digestive organs are also rich in blood vessels and the walls of these vessels contain smooth muscle tissue. The role of this muscle tissue is to regulate the diameter of the blood vessel, decreasing the diameter by contracting and increasing the diameter by relaxing. Increasing the vessel's diameter increases the flow of blood through the vessel and decreasing the diameter reduces the flow of blood through the vessel. Smooth muscle tissue in the walls of these blood vessels is innervated by the autonomic nervous system, and the effect of the sympathetic division is to cause *contraction*. This reduces the diameter of the vessels, thereby reducing the flow of blood to the digestive tract. (Note that this action is consonant with number 4.)

6. The autonomic nervous system also regulates the flow of blood to the skeletal musculature by either dilating or constricting the blood vessels that carry blood to these organs. In the case of the sympathetic division, the action is to inhibit smooth muscle cells in these blood vessels so that the vessels dilate, thereby conducting more blood into the skeletal musculature. Consequently, blood that is diverted from the digestive organs by the action of the sympathetic division (see number 5, above) is carried instead to the skeletal muscles.

7. The passageways that carry air to the lungs (i.e., the bronchi and bronchioles) have smooth muscle tissue in their walls. Motor impulses reaching this muscle tissue via

fibers of the sympathetic division inhibit these cells, thereby causing dilation of the air passageways.

8. Finally, the control centers of the heart (i.e., the sinoatrial node and atrioventricular node) also receive impulses from the autonomic nervous system. The effect of the sympathetic division is stimulatory, causing an increase in heart rate.

By considering these eight effects together, it is clear that the sympathetic division of the autonomic nervous system can bring about major changes in the body's physiological state: the pupils dilate, the mouth becomes dry, the skin perspires, digestion of food ceases, additional blood flows to the skeletal muscles, large amounts of air are readily conveyed to the lungs, and the heart begins to beat faster. All of these events serve to switch the body's level of physiological and metabolic activity from a normal or "idling" state to one in which there can be a sudden and massive expenditure of energy. The change is often referred to as the **flight-or-fight reaction**, in the sense that the body can either quickly flee from a threatening situation or "take a stand" and expend the energy in combat. Of course, some or all of these physiological changes may take place in the absence of a genuine threat (such as watching a frightening film), but it is the action of the motor fibers of the sympathetic division that bring about the physiological changes.

Actions of the Parasympathetic Division

As you might suspect at this point, the action of the parasympathetic division of the autonomic nervous system is to return the body

from a high level of metabolic and physiological activity to a resting or idling state. The parasympathetic division's motor fibers conduct impulses to the same effectors as the sympathetic division's fibers; however, *the effect on the target tissue is just the opposite*. Whereas the sympathetic division dilates the pupils, the parasympathetic division constricts the pupils; whereas, the sympathetic division increases heart rate, the parasympathetic division reduces heart rate; and so on. Indeed, nearly every effect of the sympathetic division may be reversed by the parasympathetic division. Acting cooperatively, the sympathetic and parasympathetic divisions of the autonomic nervous system adjust the levels of activity of the various organs in order to meet the body's prevailing needs.

Organization of the Sympathetic Division

Let us now turn to a consideration of the manner in which the motor fibers of each division of the autonomic nervous system are arranged and organized. Unlike the motor pathways that we have considered previously (in which a single motor neuron extends the entire distance from the central nervous system to the effector), motor pathways of the autonomic nervous system always consist of *two successive fibers*. In the case of the sympathetic division, the cell body of the first fiber lies in the grey matter of the spinal cord. Its axonal process exits the spinal cord through the ventral root of a spinal nerve. The process soon leaves the nerve and enters the nearest **lateral ganglion** in a chain of 22 **sympathetic** (or **paravertebral**) **ganglia** that lie one on each side of the vertebral column in the thoracic and lumbar regions of the body

115

(this is why the sympathetic division is also known as the **thoracolumbar division**; figure 5.26). The first fiber may synapse in that ganglion with the second fiber of that pathway (with the second fiber temporarily re-entering the spinal nerve) or the first fiber may ascend or descend to another ganglion in the chain before synapsing. The first fiber may pass through the ganglion of the sympathetic chain (without synapsing) and synapse in one of several **collateral** (or **prevertebral**) **ganglia** that lie several inches away from the sympathetic chain (fig. 5.27). The first of the two successive motor fibers

is called the **preganglionic fiber**, whereas the second fiber of the pathway is called the **postganglionic fiber**. The axonal process of the postganglionic fiber terminates at its junction with the effector.

Organization of the Parasympathetic Division

Like the sympathetic division, the motor pathways of the parasympathetic division consist of two successive fibers (a preganglionic fiber and a postganglionic fiber). The preganglionic fibers emerge from two different regions of the central nervous system: the brain stem and the bottom tip of the spinal cord. (This is why the parasympathetic division is also known as the **craniosacral division**). Those preganglionic fibers that emerge from the brain stem enter certain cranial nerves (most enter the cranial nerve X, the vagus nerve).

Preganglionic fibers that originate in the grey matter at the bottom of the spinal cord exit the cord through the ventral root of a spinal nerve. In the parasympathetic division, the ganglia in which pre- and postganglionic fibers synapse lie close to the organs that are regulated (fig. 5.28). Thus, preganglionic fibers are long, whereas postganglionic fibers are very short. The ganglia in which synapsis occurs are called **terminal ganglia**.

As noted above, the autonomic nervous system regulates the activities of many of the body's internal organs. The relationship between the sympathetic and parasympathetic divisions and their pre- and postganglionic fibers is illustrated for several organs in figure 5.29. For simplicity, only 5 lateral and 2 collateral ganglia are shown.

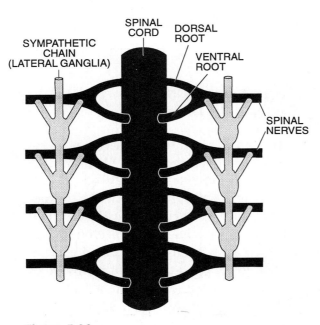

Figure 5.26
Organization of the sympathetic chains. The sympathetic chains are two chains of interconnected (lateral) ganglia that descend through the body on each side of the spinal cord, just ventral to the spinal nerves. The ganglia are connected to the spinal nerves just distal to the points where the dorsal and ventral roots merge. Only 3 of the 22 pairs of lateral ganglia are shown here.

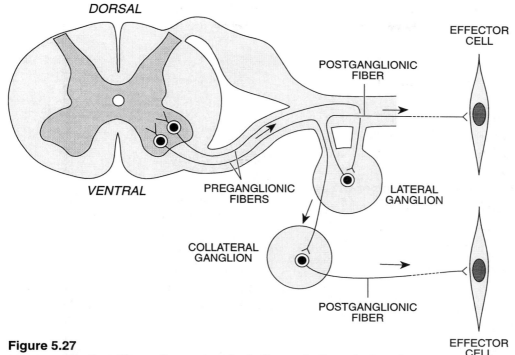

Figure 5.27
In a sympathetic pathway, the preganglionic fiber exits the spinal cord through the spinal nerve's ventral root. It then passes from the spinal nerve into a lateral ganglion where it may synapse with a postganglionic fiber; alternatively, the preganglionic fiber may pass through the lateral ganglion and ascend or descend the sympathetic chain without synapsing. In such an instance, the preganglionic fiber synapses with the postganglionic fiber in a collateral ganglion.

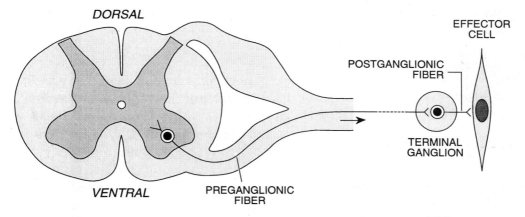

Figure 5.28
In a parasympathetic pathway, the preganglionic fiber exits the spinal cord through a ventral root and later synapses with a postganglionic fiber in a terminal ganglion, located in or near the surface of the effector.

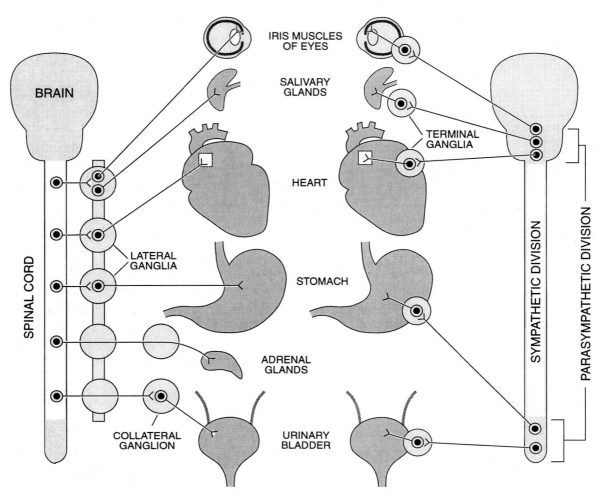

Figure 5.29
A comparison of the motor pathways (preganglionic fibers, postganglionic fibers, and ganglia) of the sympathetic (left) and parasympathetic (right) divisions of the autonomic nervous system.

Neurotransmitters of the Autonomic Fibers

In both divisions of the autonomic system, the axonal endings of preganglionic fibers release the neurotransmitter acetylcholine (discussed earlier in this chapter). Acetylcholine is also released at the junctions between the postganglionic fibers of the parasympathetic division and the effector (e.g., at the junction with muscle or gland tissue). However, in the sympathetic division, most postganglionic endings release the neurotransmitter **norepinephrine**. Since the sympathetic and parasympathetic divisions of the autonomic nervous system have opposite effects on the tissues and organs with which they communicate, it is not surprising that the neurotransmitters released at the effector junctions are different.

The Autonomic Nervous System and Heart Rate

The opposite effects of the sympathetic and parasympathetic divisions are clearly illustrated in the case of the regulation of heart rate (fig. 5.30). As you will learn in a later chapter, the rate at which the heart beats can be altered in order to meet varying demands for blood that are presented by the body's organs. Heart rate is increased by the autonomic nervous system's sympathetic division and decreased by the parasympathetic division. Changes in heart rate are initiated by nerve centers in the brain stem that communicate with the **sinoatrial node** of the heart. Slowing the heart down is achieved by increasing the frequency with which impulses are sent via pre- and postganglionic fibers of the parasympathetic division, the fibers be-

ing constituents of the tenth cranial nerve (i.e., the vagus nerve).

An increase in heart rate is effected by increasing the frequency of impulses reaching the sinoatrial node over sympathetic pre- and postganglionic fibers. As seen in figure 5.30, this is achieved by first sending impulses down through the spinal cord to the level of the heart via association fibers that are components of a descending tract. At the level of the heart, the impulses are transmitted to the cell bodies of preganglionic fibers in the spinal cord's grey matter. The pathway is then completed as the pre- and then postganglionic fibers relay the nerve impulses to the sinoatrial node. The sinoatrial node either increases or decreases the heart rate according to the relative balance of acetylcholine and norepinephrine released at the node by the postganglionic fibers.

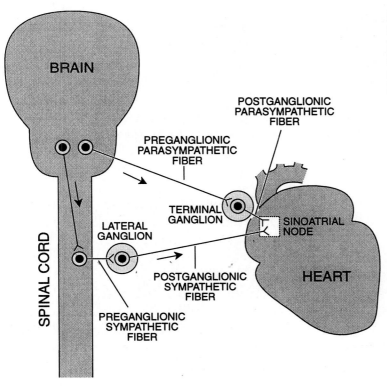

Figure 5.30
Sympathetic and parasympathetic pathways linking the brain with the heart's sinoatrial node. Arrows show the direction in which the impulses travel.

The Adrenal Medulla

Among the tissues innervated by the sympathetic division, is the *medulla* (i.e., the central portion) of the **adrenal glands**. Only preganglionic fibers innervate the adrenal medulla (i.e., there are no postganglionic fibers; see figure 5.29). In response to these sympathetic impulses, the adrenal glands release two closely-related hormones into the bloodstream; these are **epinephrine** and **norepinephrine** (i.e., the same substance that is released by the postganglionic fibers of the sympathetic division). These two hormones serve to augment the responses of organs that are also receiving sympathetic innervation, and they also influence tissues not receiving any sympathetic innervation at all. Moreover, the hormones circulate in the bloodstream for some period of time. As a result, adrenal secretion has a much longer lasting effect than does the discharge of nerve impulses by the postganglionic fibers of the sympathetic division.

SELF TEST*

True/False Questions

1. An axon is defined as a nerve cell process that conducts impulses away from the central nervous system (i.e., away from the brain and spinal cord).

2. Under normal circumstances, a single neuron conducts an impulse in one direction; that direction is referred to as the "dromic" direction.

3. Nerve impulses are conducted more rapidly by some nerve fibers than by other nerve fibers.

4. The surface of a myelinated nerve fiber is exposed at numerous points along its length; these points are called nodes of Ranvier.

5. The "knee-jerk" is a simple, ipsilateral spinal reflex.

6. The medulla of the adrenal glands secretes hormones that mimic the effects of the sympathetic division of the autonomic nervous system.

Multiple Choice Questions

1. The fatty substance that acts as an insulator around the long processes of neurons is called (A) actin, (B) myelin, (C) myosin, (D) cholesterol, (E) neurilemma.

2. Saltatory conduction takes place (A) in nerve cells that are enveloped by a layer of Schwann cells, (B) only in the brain, (C) only in motor nerve fibers, (D) only in sensory nerve fibers, (E) only in association nerve fibers.

3. If one measuring electrode of a galvanometer is placed on the surface of a non-conducting nerve fiber and the other measuring electrode placed inside the fiber, then (A) the galvanometer would show no flow of current between the two electrodes, (B) the galvanometer would show the flow of current between the two

---

* *The answers to these test questions are found in Appendix III at the back of the book.*

electrodes, (C) it is not possible to place a measuring electrode inside a nerve fiber, (D) flow of current through the galvanometer would only occur if the fiber were conducting an impulse.

4. Clusters of cell bodies of nerve fibers would not be found in (A) spinal tracts, (B) grey matter, (C) dorsal root ganglia; (D) lateral ganglia, (E) the cerebral cortex.

5. When one member of a pair of antagonistic muscles is stimulated by motor pathways and the other is inhibited, this is known as (A) a crossed-extensor reflex, (B) a simple spinal reflex, (C) reciprocal innervation, (D) crossed innervation.

6. The flight-or-fight response is intimately related to the actions of (A) the cerebellum, (B) the sympathetic division of the autonomic nervous system, (C) the parasympathetic division of the autonomic nervous system; (D) the crossed-extensor reflex.

7. In a nerve fiber that is not conducting a nerve impulse, (A) there are no sodium ions inside of the cell, (B) there are no potassium ions inside of the cell, (C) the fluid that bathes the outside surface of the cell is richer in positive ions than the fluid that bathes the inside surface of the cell, (D) the fluid that bathes the inside surface of the cell is richer in positive ions than the fluid that bathes the outside surface of the cell.

CHAPTER 6

THE RECEPTOR SYSTEM

Our perception of ourselves and the universe around us relies on the ability of the body's receptor system to transform chemical and physical stimuli into nerve impulses that are conducted to the central nervous system for proper interpretation. The cells and tissues that comprise the receptor system may be divided into classes according to the nature of the stimuli that they detect. Accordingly, there are (1) **mechanoreceptors** (which respond to *touch*, *pressure*, and *vibration*), (2) **thermoreceptors** (which respond to changes in *temperature*), (3) **photoreceptors** (which respond to *light*), (4) **chemoreceptors** (which respond to a variety of *chemical* *stimuli*), and (5) **nocioceptors** (which respond to *physical* or *chemical damage*).

Receptors may also be classified according to their location in the body. Using this criterion, three major classes may be identified: (1) **interoceptors**, (2) **proprioceptors**, and (3) **exteroceptors**. The interoceptors (also called **visceral receptors**) are responsible for sensing stimuli that arise deep within the body. For example, sensations of pressure arising from the organs of the digestive tract or from the urinary bladder are the result of stimulation of interoceptors in these organs. On a more subtle scale, interoceptors in the walls of certain blood vessels serve to sense blood pressure or the oxygen and carbon dioxide contents of the blood. Sensations of hunger and thirst are also considered visceral senses.

Proprioceptors are located in the skeletal muscles and also in joints and are responsible for our awareness of muscle position and movement. Finally, the exteroceptors are responsible for our awareness of stimuli at the body surface or at some distance from the body. Included here are the receptors of the skin (i.e., the **cutaneous receptors**) and the so-called **special receptors** that are responsible for the senses of **gustation** or taste (i.e., the tongue), **olfaction** or smell (i.e., the nose), **vision** (i.e., the eyes), and **hearing** (i.e., the ears).

Receptors can be organized in several different ways. Often the receptor consists of individual, highly specialized cells that respond to a particular kind of stimulus. For example, the **cone** cells and **rod** cells of the eyes are receptor cells that selectively respond to a light stimuli (see later in chapter). Sometimes the receptors are **encapsulated nerve endings** (i.e., the dendritic endings of one or more sensory nerve fibers enclosed within a shell of other [often connective tissue] cells). The Pacinian corpuscles of the skin (see later) are examples of this type of receptor organization. Finally, receptors can be **free nerve endings** (i.e., non-encapsulated dendritic endings of sensory nerve fibers). Pain receptors (see later) have this type of structural organization.

KINESTHESIA

If you were asked to close your eyes and slowly raise up your hand to touch your nose with your index finger, you probably could perform this task with relative ease. Your ability to do this rests with your constant awareness of the position of the different parts of your body (in this case your arm, hand, and nose) and their motion with respect to one another. This sense is known as **kinesthesia** and it relies on the properties and actions of the body's **proprioceptors**.

There are two major kinds of proprioceptors; these are the **Golgi tendon organ** and the **muscle spindle apparatus**. Both are mechanoreceptors. Golgi tendon organs are located in tendons (tendons join one muscle to another and also join muscle to bone; figure 6.1). Each time that a muscle contracts, the tendon tissue at the ends of the muscle is stretched, and this stretching acts as a mechanical stimulus to the Golgi tendon organs. The stimulated receptors respond by triggering nerve impulses in the sensory neurons with which each Golgi tendon organ is associated. The long dendritic processes of these fibers are myelinated, so that the sensory impulses reach the central nervous system very quickly.

Each muscle spindle apparatus consists of several tapered receptor cells, called **intrafusal fibers**, enveloped in a thin layer of connective tissue. These assemblies are

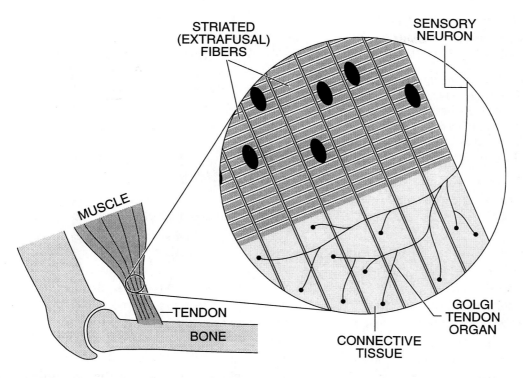

STRIATED (EXTRAFUSAL) FIBERS

SENSORY NEURON

MUSCLE

TENDON

BONE

CONNECTIVE TISSUE

GOLGI TENDON ORGAN

Figure 6.1
The Golgi tendon organ. Contraction of a muscle stretches the muscle's tendons and mechanically stimulates the Golgi tendon organs.

buried among the "normal" striated cells (or **extrafusal fibers**) near the ends of muscles, where the muscle fibers give way to the elastic tissue of the attached tendons. Within each spindle apparatus, the receptor cells are oriented with their long axis parallel to the axis of muscle contraction (fig. 6.2). The dendritic endings of sensory nerve fibers are twisted around each of the spindle cells to form a helical belt. The stretching of a muscle (for example, as you carry a briefcase or book at your side) stretches and thereby stimulates the intrafusal cells. The receptors then stimulate the spiral belt of dendritic endings of the associated sensory nerve fibers. The result is the flow of nerve impulses over the sensory fibers to the central nervous system, where the number and frequency of the incoming impulses is used as a measure of the amount of muscle stretching that is taking place. The stretching of the muscle is followed by reflexive contraction in order to return the muscle to its normal length.

Proprioceptors also play an important role in determining the extent of muscle contraction that is necessary in order to lift or move an object. For example, far less contraction of the biceps muscle is needed in order to lift a magazine off the table than is needed to lift a heavy book. As the muscle attempts to raise the object off the table, the resistance offered by the object's weight brings about an initial stretching of the muscle and this stretching is sensed by the proprioceptors. The greater the weight (and resistance) of the object, the greater is the degree of initial

Figure 6.2
The muscle spindle apparatus. These receptors are located in the tendons that link muscles to one another and to bone. The receptor consists of spindle fibers (intrafusal fibers) around which are twisted the dendritic endings of sensory neurons. When a muscle is stretched, the spindle fibers are mechanically stimulated; they then stimulate the associated sensory neurons.

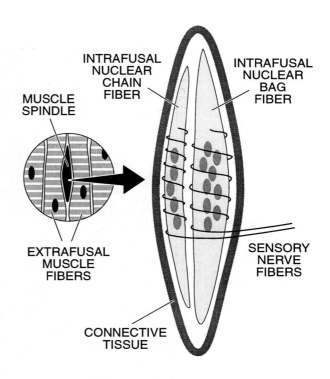

stretching, and the greater are the number and frequency of sensory impulses arising in the sensory fibers associated with the proprioceptors. This sensory information is used as a gauge to determine the intensity of the motor output necessary to bring about sufficient contraction in the muscle to lift the entire weight. Thus, the degree of contraction is no less or greater than that actually needed.

EXTEROCEPTION

Most thoroughly studied and best understood among the receptors of the body are those involved in exteroception, that is sensing stimuli that occur at the surface of the body or that originate at some distance away from the body surface. The senses that are provided by the exteroceptors are divided into two groups: the **cutaneous senses** and the **special senses**.

CUTANEOUS SENSES

There are three classes of cutaneous receptors. Those that give rise to sensations of touch and pressure are called **tactile receptors**; those that give rise to sensations of heat or cold are called **thermoreceptors**; and those that give rise to sensations of pain are called **pain receptors**.

Tactile Receptors

The skin is a complex tissue (fig. 6.3) and contains a variety of different **tactile receptors**. Those receptors that respond to mild disturbances of the skin (such as that caused by the weight of an insect or a small piece of paper) include **free nerve endings**, **Meissner's corpuscles** (cell-encapsulated nerve endings), and **hair-end-organs** (figures 6.3 and 6.4). Hair-end-organs are the

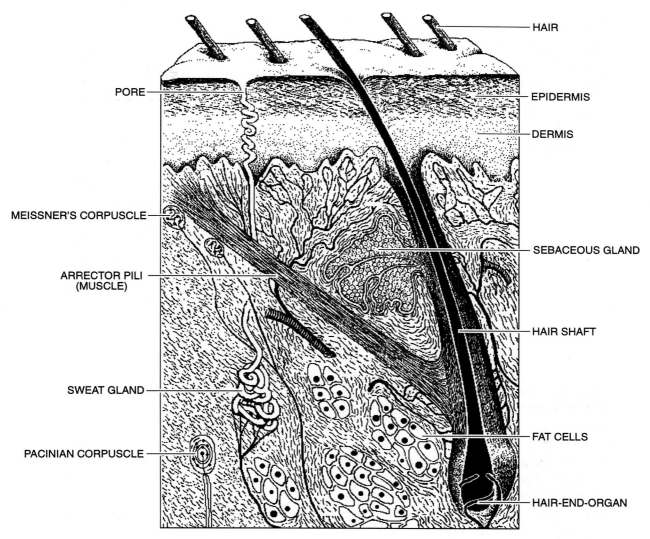

Figure 6.3
A section through the skin showing some of the cutaneous receptors that respond to touch and pressure. Also shown are hairs that arise from the sub-epidermal layers and the sebaceous (i.e., oil) and sweat glands.

dendritic endings of sensory nerve fibers that are wound around the shaft of a hair, near the hair's root. Movement of the hair mechanically stimulates these nerve endings and gives rise to a sensation of touch. (Use the tip of a pencil to move a hair on the back of your hand and you will sense this mild disturbance.)

Pacinian corpuscles are also tactile receptors. However, Pacinian corpuscles respond to more vigorous deformation of the skin, such as that which occurs when a heavier weight (e.g., a book) is placed on the skin. Pacinian corpuscles are therefore said to respond to *pressure* rather than to touch. Like Meissner's corpuscles, Pacinian corpuscles

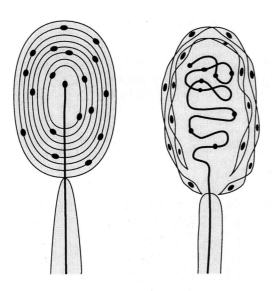

Figure 6.4
Pacinian (left) and Meissner's (right) corpuscles. These receptors (which respectively
respond to pressure and to touch) consist of
dendritic terminals of sensory neurons that are
encapsulated in one or more layers of connective tissue cells. As shown here, the long
dendritic processes are usually myelinated.

are encapsulated nerve endings (fig. 6.4);
however, the encapsulation takes the form of
several successive layers of connective tissue
cells, so that the organization is much like the
concentric arrangement of layers in an on-
ion. As you might expect, tactile receptors
are not distributed uniformly over the body
surface. Instead, some regions of the skin
(e.g., the fingertips and lips) contain a much
higher concentration of receptors than do
other regions. Consequently, these receptor-
rich areas are much more sensitive to touch
and pressure.

Thermoreceptors

There are two major types of **thermorecep-**
tors: those that respond to *decreasing* skin
temperature (usually called **cold receptors**)
and those that respond to *increasing* skin
temperature (usually called **heat receptors**).
Structurally, the two types of thermoreceptors are represented for the most part by the
free endings of the dendritic branches of
sensory nerve fibers.

Pain Receptors

Most pain receptors are free dendritic endings of sensory nerve fibers originating in the
skin. These endings are stimulated by injury
to the surrounding tissue. The release of the
substances **bradykinin** and **histamine** by
damaged cells has been linked to the stimulation of pain receptors. In this sense, the
pain receptors are acting as chemoreceptors.
Intense pressure applied to the skin and un-
usually high (or low) temperatures also give
rise to sensations of pain and are believed to
result from the direct stimulation of the free
nerve endings.

Adaptation

You probably have had the experience of
entering a swimming pool and finding the
water uncomfortably cold at first; but after a
few minutes, the water did not seem as cold.
This phenomenon is known as **adaptation**.
What happened was that the initial submer-
sion of the skin in the pool water stimulated
cold receptors and gave rise to sensations of
cold. Although the pool temperature did not
change, the skin's cold receptors ceased to
be stimulated. As a result, sensory nerve
impulses apprising you of the coldness of the
water were halted. A similar effect is noticed

when you place a watch on your wrist or a ring on your finger. At first, you are aware of the presence of the watch or ring, as tactile receptors in the skin are stimulated and trigger sensory nerve impulses. After a few minutes, they cease being stimulated and your awareness of the watch and ring subsides. You may have heard stories about near-sighted people who raise their eyeglasses up to their forehead while they read a newspaper, and then later search everywhere for their glasses. (After several minutes they no longer feel the weight of their glasses on their forehead and are unaware of the presence of the object.)

The phenomenon of adaptation is important. Because adaptation does not begin until the stimulus has been applied continuously (and unchanged) for a short period of time, you are made aware of the stimulus' presence and have the option to react to it in some way. For example, sensing that a fly has settled onto your skin, you may elect to brush it away. Sensing that the pool water is quite cool, you may decide to get out of the pool. However, if the stimulus persists unchanged for some time, your awareness of it abates because the receptors adapt and fail to generate sensory signals. This frees the central nervous system to attend to the multitude of other sensory signals. Were this not so, you would be preoccupied continuously by the weight and feel of your clothing, eyeglasses, rings, watch, shoes, and so on.

Different receptors adapt at different rates. As indicated above, tactile and thermoreceptors adapt very quickly. On the other hand, pain receptors (and also proprioceptors) adapt very slowly. It is important to remember that the adaptation that takes place involves the receptors–*not the central nervous system.*

THE SPECIAL SENSES

There are four special senses: **gustation**, the sense of taste; **olfaction**, the sense of smell; **vision**, the sense of sight; and **audition**, the sense of hearing.

GUSTATION

Gustation (the sense of *taste*), and olfaction (the sense of *smell*) are closely related to one another. In fact, it is often difficult to distinguish between taste and smell. For example, when eating a meal, the "flavor" that is sensed is the result of stimulation of *both* the **gustatory receptors** and the **olfactory receptors**. Moreover, you have probably noticed that when you have a head cold and your nasal passageways are blocked, food seems to lose much of its taste.

The gustatory receptors are examples of chemoreceptors and are stimulated by chemical substances that dissolve in the **saliva**. The gustatory receptors are clustered together to form a number of small **taste buds** distributed primarily over the surface of the tongue where they line the walls of the tongue's **papillae** (fig. 6.5). A number of taste buds are also present in the palate and the walls of the pharynx. Each taste bud contains a mixture of gustatory receptor cells and supportive cells and opens onto the surface of the papillae through a small pore (fig. 6.6). Gustatory hairs (*cilia*) project from the apical ends of the receptor cells through the pores of the taste buds and into the oral cavity. The gustatory receptor cells are innervated by the dendritic endings of sensory nerve fibers. These fibers conduct impulses to the brain stem via the *facial* and *glossopharyngeal* nerves (cranial nerves VII and

Figure 6.5
The surface of the tongue is littered with papillae. The walls of the papillae (shown in vertical section in the enlarged view) are lined with taste buds.

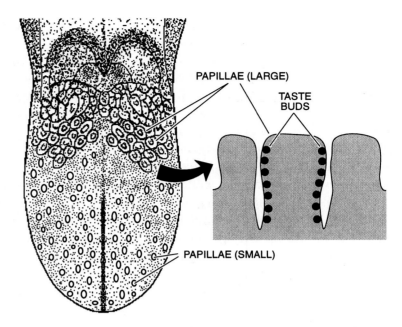

PAPILLAE (LARGE)

TASTE BUDS

PAPILLAE (SMALL)

IX; see table 5.1). Impulses reaching the brain stem from gustatory receptors activate the autonomic nervous system, which reflexively increases the secretion of saliva. Other fibers conduct impulses from the brain stem to the cerebral cortex where the taste is perceived.

It is generally agreed that there are four different "taste qualities": *sweet*, *salty*, *sour*, and *bitter*. Chemicals that taste sour are acids, the degree of sourness varying in relation to the hydrogen ion concentration or pH

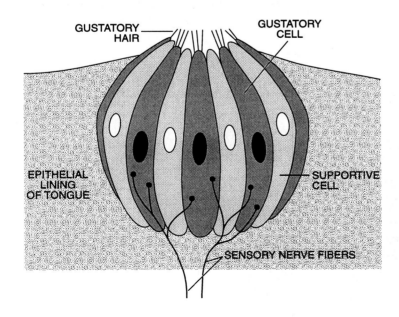

GUSTATORY HAIR

GUSTATORY CELL

EPITHELIAL LINING OF TONGUE

SUPPORTIVE CELL

SENSORY NERVE FIBERS

Figure 6.6
Organization of gustatory receptor cells and supportive cells to form a taste bud. The taste bud is surrounded by epithelial tissue and opens at the surface of the tongue through a small pore.

that is created in the saliva by the sour sub-stance. As you might expect, chemicals that taste salty are usually salts, such as sodium chloride and potassium chloride. In addition to sugars, a variety of other chemical sub-stances give rise to sweet sensations. Bitter tastes are also produced by a variety of chemical substances. Many poisonous sub-stances taste bitter, a fact that has obvious survival value.

Not all regions of the tongue are equally sensitive to the four taste qualities. Rather, different regions respond to different sub-stances; the regions of the tongue that are most sensitive to each of the four tastes are shown in figure 6.7. Not all individuals show equal sensitivity to specific substances, and this appears to be genetically determined. Individual variations are best illustrated in the case of PTC (phenylthiocarbamide). About 80% of the population finds that PTC tastes sour, whereas the other 20% finds that PTC has no taste at all. Although individual gustatory receptors are most sensi-tive to substances that produce a specific taste, receptors can be stimulated by more than one type of taste.

6-1

OLFACTION

Like the gustatory receptors, the **olfactory receptors** are chemoreceptors. The sub-stances detected by the olfactory receptors dissolve first in the air that is inspired through the nose and then dissolve in the thin layer of **mucus** that covers the olfactory receptors.

The receptor cells form a small patch of olfactory membrane in a narrow cleft at the top of the nasal cavity (fig. 6.8). During quiet breathing, the inhaled air does not flow over these receptors, although they can be reached by diffusion. More forceful inspiration or sniffing does bring the air into contact with the receptors and is followed by a strong olfactory sensation. The olfactory sense is more complex than the gustatory sense, and most individuals can distinguish thousands of different odors.

As seen in figure 6.8, the receptors are elongate cells with small projections (again cilia) extending from the apical ends of the cells into the nasal cavity. At their basal ends, the receptors synapse with sensory fibers of the *olfactory* nerve (cranial nerve number I). Impulses are conducted by the olfactory nerve to the olfactory region of the

Figure 6.7
The surface of the tongue is not equally sensitive to all four taste qualities. Shown here are the areas of the tongue's surface that are most sensitive to sweet, salty, sour, and bitter substances.

SWEET   SALTY   SOUR   BITTER

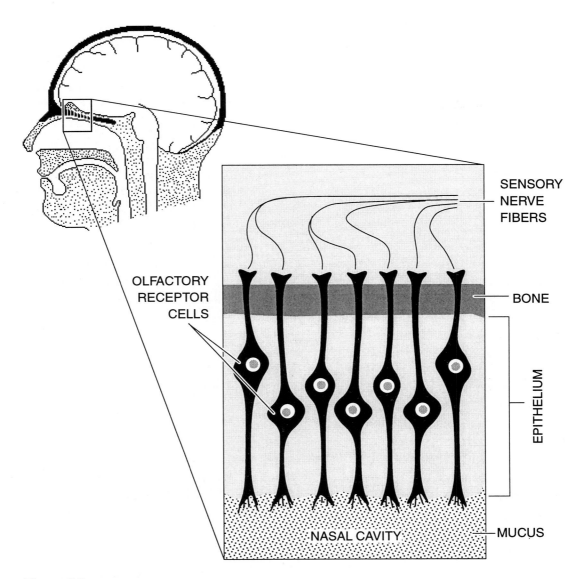

Figure 6.8
The olfactory membrane. In a narrow cleft at the top of the nasal cavity (seen in the small insert) is the olfactory membrane–the receptor apparatus that detects odors. Forceful inspiration and sniffing act to draw air onto the exposed hairlike endings (cilia) of the olfactory receptor cells. Chemicals that are dissolved in the inspired air stimulate these cells, thereby giving rise to sensory impulses and, ultimately, the sense of smell.

cerebral cortex, where the odor is perceived. Like cutaneous receptors, olfactory receptors exhibit adaptation. As a result, an odor is noticed for only a short period of time.

VISION

Perhaps the most important of the special senses in humans is the sense of vision. To

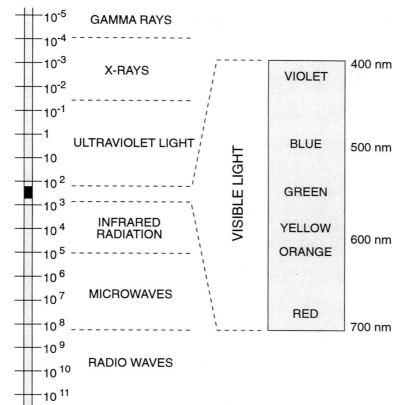

Figure 6.9
Part of the electromagnetic energy spectrum, showing ray wavelengths in nanometers (1 nm = 1 billionth of a meter). The visible light spectrum comprises only a small portion of the total energy spectrum, extending from about 400 nm (violet light) to 700 nm (red light).

understand and appreciate this remarkable sense, it is first necessary to know a little bit about the properties of light and the manner in which light rays are affected by the substances through which the light passes.

Light

Light is one of a number of different forms of **electromagnetic radiation** (fig. 6.9) and consists of a stream of units of energy called **photons**. In air, light travels at a speed of 186,000 miles per second. The stream of photons is called a **ray** and has the character of a **wave** (fig. 6.10) in which the distance from any point on the wave to an equivalent point on an adjacent wave is the **wavelength**.

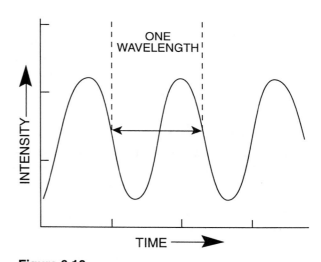

Figure 6.10
The wavelike character of electromagnetic radiation. A wavelength is equal to the distance from any point on the wave to an equivalent point on an adjacent wave.

Among the differences between the radiations of the electromagnetic spectrum are differences in wavelength. At one extreme are the long waves (e.g., radio and television waves), while at the opposite end of the spectrum are the short waves (e.g., X rays and gamma rays).

That part of the electromagnetic spectrum that the human eye can detect is called *visible light* and extends from a wavelength of about 400 nanometers (nm) (corresponding to the color *violet*) to about 700 nm (corresponding to the color *red*; figure 6.9). Within this range, different wavelengths of light are perceived as different colors.

In addition to varying wavelength, visible light of a given color can vary in intensity (fig. 6.10). Intensity corresponds to the brightness of the light. Increasing a colored

light's brightness does not alter its color; it simply alters the light's intensity.

Refraction of Light

Light readily passes through transparent media such as air, water, and glass. However, when light passes from one transparent medium into another, the light's speed is altered and it is also bent or **refracted**. In the case of light passing from a low density medium into one that has greater density (e.g., from air into glass or from air into water), the speed of the light diminishes and the light rays are bent "toward the **normal**" (a normal to a surface at a given point is a line drawn perpendicular to the surface at that point; figure 6.10). In contrast, when light passes from a

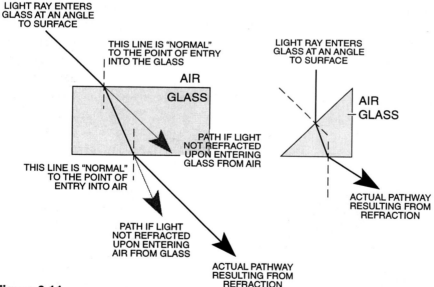

Figure 6.11
Refraction of light as it passes between glass and air. Shown on the left is the path taken by the light as it enters one side of a rectangular piece of glass and exits on the opposite side. On passing from the air into the glass, the light rays are bent toward the normal. On exiting the glass and reentering the air, the rays are bent away from the normal. Notice that the entry and exit rays are parallel. Shown on the right is the path taken through a triangular piece of glass (such as a prism).

dense medium into one that is less dense (e.g., from glass into air or from water into air) the speed of the light increases and the light rays are bent "away from the normal" (fig. 6.11). An appreciation of these rules regarding the refraction of light is fundamental to your understanding of how the eye is able create a clear (properly focused) field of view.

Anatomy of the Eye

The eyes are responsible for the sense of vision and consist of an elaborate apparatus that ultimately converts the energy of light rays into nerve impulses. To understand how the eyes function, let's begin by considering the organ's anatomy, noting the individual functions of the eye's various components.

Figure 6.12 is an exterior view of the eye as seen from the front, and figure 6.13 shows the internal parts as seen in a side view. The eye is a hollow, ball-shaped structure, the surface of which is formed by three coats of tissue. The outermost coat is the **sclera**; the middle coat is the **choroid**; and the innermost coat is the **retina**. The sclera, which covers the entire outer surface of the eye, is

formed from connective tissue and serves as a protective capsule. At the front of the eye, the sclera is modified so that it is transparent. This region of the sclera is called the **cornea**. A fine transparent membrane called the **conjunctiva** covers the front of the cornea and attaches to the undersurfaces of the upper and lower eyelids. When the eyes are open (which is most of the time), the conjunctiva and cornea are exposed to the surrounding air. To prevent these delicate tissues from becoming dry, their surfaces are kept moist by the flow of tears from the **lacrimal glands** (fig. 6.12). The tears flow across the front of the eyes and are drained by the **nasolacrimal ducts**, which empty the fluid into the nasal cavity.

When looking at the front of the eye, the donut-shaped **iris** muscle can be seen behind the cornea. This muscle, which is rich in pigment and gives rise to the characteristic eye colors (e.g., brown eyes, blue eyes, green eyes, hazel eyes, etc.), is a part of the middle coat of the eye–the **choroid** coat. The hole at the center of the iris is called the **pupil**; the size of this opening can be decreased by contracting those iris smooth muscle cells that are arranged in a *circular* manner. Contraction of the iris

6-2

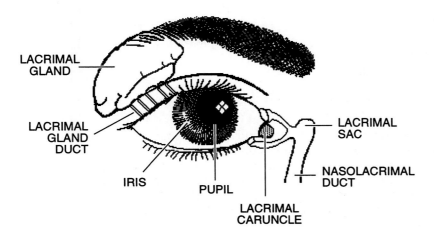

Figure 6.12
A front view of the right eye. The secretions of the lacrimal gland sweep across the surface of the eye and are drained into the nasolacrimal duct. The pupil (black spot) is an opening at the center of the pigment-rich iris muscle.

LACRIMAL GLAND

LACRIMAL GLAND DUCT

IRIS

PUPIL

LACRIMAL CARUNCLE

LACRIMAL SAC

NASOLACRIMAL DUCT

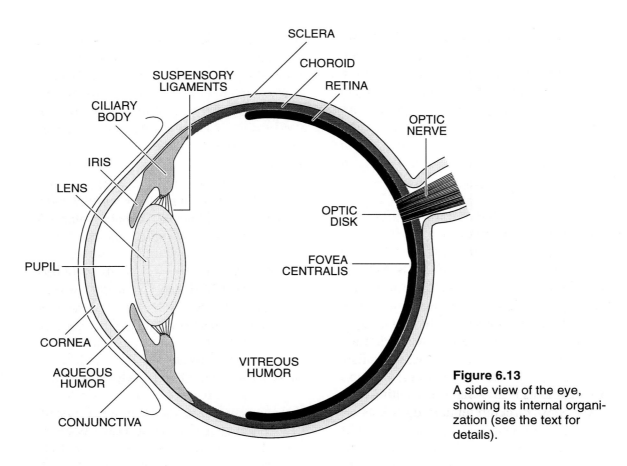

Figure 6.13
A side view of the eye, showing its internal organization (see the text for details).

cells that are arranged *radially* increases the size of the pupil. The size of the pupils determines the amount of light that enters the eyes.

6-3

The space between the rear of the cornea and the front of the **lens** is filled with a watery fluid called the **aqueous humor**. The aqueous humor is secreted by the ciliary body and makes its way through the pupil into the eye's anterior cavity.

Pupillary Reflexes

The status of the circular and radial musculature of the iris is controlled by the autonomic nervous system. Contraction of the radial muscles is caused by motor nerve impulses from the sympathetic division and acts to dilate (widen) the pupil (thereby letting more light into the eye). Contraction of the circular muscles is caused by the parasympathetic division and acts to constrict the pupil (thereby reducing the amount of light entering the eye). Three autonomic pupillary reflexes may be described: the **near reflex**, the **light reflex**, and the **dilator reflex**.

The near reflex is the constriction of the pupil that occurs when looking at an object that is very close to you. This restricts the light that enters the eye to the center portion of the eye's **lens** (see below), where image focus and sharpness is best. The light reflex is the pupillary constriction and dilation that occur in response to the brightness of the light entering the eye. In very bright light, the pupil constricts, whereas in dim light the pupil dilates. In the dilator reflex, the pupil dilates in reaction to fright and is part of the "flight-or-fight" response described in chapter 5.

The Lens and Accommodation

Behind the iris, the choroid coat forms a ring of smooth muscle tissue called the **ciliary body** (or **ciliary muscle**; figure 6.13). Tiny, nonelastic ligaments, called **suspensory ligaments** or **zonules of Zinn**, extend axially from the ciliary body and suspend the eye's **lens** in position behind the pupil (figs. 6.13 and 6.14). The lens is a flexible structure whose natural shape is round, much like a ball. However, the tension in the suspensory ligaments is high and the resulting pull on the lens' circumferential edges acts to flatten the lens, altering its shape to that of a gently biconvex disk. Contraction of the ciliary muscles pulls the choroid coat forward a tiny distance, and this serves to introduce slack into the suspensory ligaments. Under these conditions, the natural elasticity

of the lens causes it to become rounder, a phenomenon called *elastic recovery*. This relationship between the shape of the lens and the status of the ciliary muscles is fundamental to the mechanics of **accommodation** (fig. 6.14); therefore, the essential points bear repeating:

(1) by pulling the choroid coat forward, contraction of the ciliary muscles loosens the suspensory ligaments, and allows the lens to become rounder;

(2) relaxation of the ciliary muscles allows the choroid coat to slide backwards, increasing the pull on the suspensory ligaments and thereby flattening the lens.

Lens Shape and Focusing Power

A simple experiment that you can quickly perform illustrates the relationship between lens shape, ciliary muscle action, and the focusing of light rays entering the eye. Look out of the window or across the room at an object that is 20 or more feet away. Now close your eyes for several seconds and

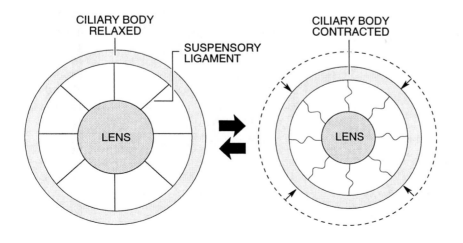

Figure 6.14
Contraction of the ciliary body acts to pull the choroid coat forward and thereby reduce the tension of the suspensory ligaments. As a result, the natural elasticity of the lens causes it to change its shape from flat (thin) to round (fat).

reopen them. You should note that immediately upon opening your eyes, the object is in focus. Now, hold this page about eight inches from your eyes and focus on the print. Close your eyes, and after several seconds open them. Notice that this time the object (i.e., page of print) was *not* immediately in focus and that it takes a small fraction of a second to bring the print into focus. You might also be able to sense the muscle strain that accompanies focusing on such a close object.

What do these observations demonstrate? Focusing on a distant object requires a flattened lens, whereas focusing on a near object requires a round (biconvex) lens. Indeed, the nearer the object, the more biconvex the lens must be (fig. 6.15). Focusing on a distant object occurs without ciliary muscle contraction because the tension in the suspensory ligaments flattens the lens. In contrast, focusing on a near object requires contraction of the ciliary muscles; their contraction

slackens the suspensory ligaments, thereby allowing the lens to assume a biconvex shape. The process by which the shape of the lens is changed in order to keep an image properly focused on the photoreceptor layer at the rear of the eye is called **accommodation**.

The lens of the eye is rich in a protein called **alpha-crystallin**, produced by a layer of cells at the lens' surface by secretion. As a person ages, the capacity for accommodation diminishes, and by the time one reaches an age around 45 it is lost altogether. The resulting defect in vision is called **presbyopia** (a form of **farsightedness**, see later). It remains uncertain whether presbyopia results from an age-related loss of the lens' elasticity or the progressive accumulation in the lens of insoluble alpha-crystallin complexes. In either case, although contraction of the ciliary muscles introduces slack into the suspensory ligaments, the lens fails to become as round as it once did. The result is a

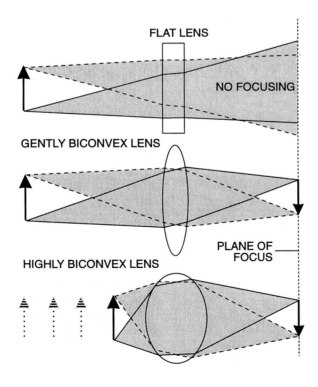

Figure 6.15
Effects of curvature on the focusing power of a lens. In this figure, light that is reflected from the surface of an object (i.e., the arrows to the left) is directed into each of three lenses. Whereas a flat lens has no focusing power at all (*top*), a biconvex lens focuses the reflected light on a plane located some distance behind the lens (*middle*). The more biconvex the lens (i.e., the greater its curvature; *bottom*), the closer the object may be to the lens while still remaining in focus. Notice that the image of the object created on the focal plane is inverted (i.e., the arrows to the right).

progressively deteriorating ability to focus on objects that are close to the eyes. Thus, most people over the age of 45 require eyeglasses that place an additional biconvex lens in front of each eye.

The Retina

The lens of the eye focuses incoming light on the innermost coat of the eye—the **retina**. The retina contains the complex of receptor cells (called **rods** and **cones**, figure 6.16) and nerve cells that convert light rays into nerve impulses. The organization of the several layers of cells that comprise the retina is shown in figure 6.17. At the surface of the retina that is flush against the choroid coat is the **pigment layer**. This is a layer of cells

rich in the dark pigment **melanin**. The role of the pigment layer is similar to that played by the black coat of paint that covers the inside surfaces of a camera–namely, to prevent the reflection of light rays. Thus, any light not absorbed by the photoreceptor cells themselves is absorbed by the pigment layer.

Seated on the pigment layer is the layer of photoreceptor cells. Two classes of photoreceptors are present; they are called **rods** and **cones**. The names of the photoreceptors describe the general shapes of the cells, the rods being rodlike and the cones being conical. It is estimated that the retina contains about 3,000,000 cones and about 100,000,000 rods.

The rods and cones form junctions with the layer of **bipolar nerve cells**. In the dark, there is a continuous outward leakage of

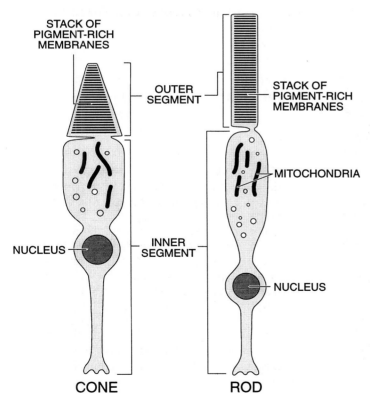

STACK OF PIGMENT-RICH MEMBRANES

OUTER SEGMENT

STACK OF PIGMENT-RICH MEMBRANES

MITOCHONDRIA

NUCLEUS

INNER SEGMENT

NUCLEUS

CONE

ROD

Figure 6.16
The photoreceptor cells of the eyes: cones and rods. The cells are stimulated when pigment in the membranes of the outer segments of the cells absorbs light that has entered the eyes. The cones are responsible for our ability to distinguish different colors of the visible spectrum.

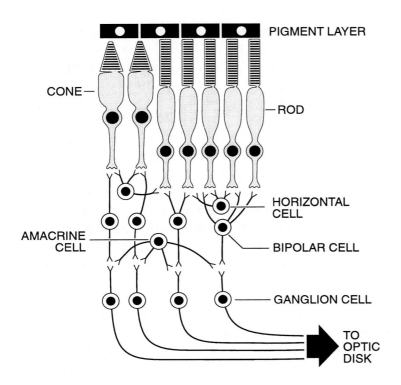

PIGMENT LAYER

CONE

ROD

HORIZONTAL CELL

AMACRINE CELL

BIPOLAR CELL

GANGLION CELL

TO OPTIC DISK

Figure 6.17
The retina. The pigment layer is appressed against the choroid coat. Seated on the pigment layer is the photoreceptor layer containing the cones and rods. The photoreceptors stimulate bipolar nerve cells, which then transmit impulses to the layer of ganglion cells. The long axonal processes of the ganglion cells converge on the optic disk and become part of the optic nerve. It should be noted that the light rays must pass through the ganglion cells and bipolar cells before reaching the photoreceptor layer.

sodium ions from the receptor cells that acts to keep their plasma membranes in a depolarized state. The loss of Na^+ is accompanied by the release of a chemical that acts to inhibit the bipolar cells. Thus, in the dark, no nerve impulses are generated in the bipolar nerve cell layer. However, when the photoreceptors absorb light, sodium gates in the photoreceptor membranes close and the membranes quickly become hyperpolarized. This causes a reduction in the release of inhibitory chemical by the photoreceptor. This reduction releases the bipolar nerve cells from inhibition and they now conduct impulses to their synapses with the **ganglion nerve cell** layer of the retina (fig. 6.17). The impulses are transmitted from the bipolar cells to the ganglion cells, which then conduct the impulses onward to the brain (see later).

The rods and cones have different properties and function differently in vision. The rods are the more sensitive photoreceptors and function in dim light, although they do not provide much image detail or information about color. In contrast, the cones function in bright light and provide detailed images that possess color.

The light-absorbing substance of the rods is a purple material called **rhodopsin**. It is comprised of the pigment **11-cis-retinal** and the protein **opsin**. When rhodopsin absorbs light rays, it breaks down into its component parts, retinal and opsin. This chemical reaction leads to the changes in membrane polarity and the reduced secretion of the inhibitory chemical just described. Some of the retinal recombines with opsin, thereby regenerating rhodopsin. The remainder of the retinal is reduced to **vitamin A**, some of which enters the bloodstream and is eventually degraded. Retinal lost in this way must be replaced, and under normal circumstances is resynthesized from vitamin A in the diet.

Bright light causes the rapid breakdown of rhodopsin and temporarily renders the rods functionless. This explains why a person entering a darkened theater from daylight finds it extremely difficult to see. Gradually, the rhodopsin is replenished and vision returns. This phenomenon is called "dark adaptation."

When reentering daylight from a darkened theater, the reverse effect is noted. Initially, the daylight seems painfully bright, but as the rhodopsin produced during dark adaptation is degraded, normal daylight vision returns. As you might expect, this phenomenon is called "light adaptation."

Although the rods provide sensitivity to low levels of light, they do not provide any information about color. Color vision is a function of the cones. Although most individuals can distinguish hundreds of different colors and shades, there are only three different kinds of cones: *red-sensitive* cones, *green-sensitive* cones, and *blue-sensitive* cones. Each type of cone possesses a different kind of light-absorbing pigment.

The red-sensitive family of cones contains the pigment **erythrolabe**; the green-sensitive family of cones contains the pigment **chlorolabe**; and the blue-sensitive family of cones contains the pigment **cyanolabe**. Erythrolabe, chlorolabe, and cyanolabe are similar to rhodopsin in that they are comprised of 11-cis-retinal and the protein opsin. Whereas retinal is chemically identical in all four pigments, the protein portion of the molecule differs in each of the four pigments and is responsible for their different light-absorbing properties.

Our ability to distinguish red, green, and blue objects depends on which family of cones is stimulated by the light reflected by the object. For example, light that is reflected into the eyes from a red object is differentially absorbed by the pigment molecules of the red-sensitive cones; light that is reflected into the eyes from a green object is differentially absorbed by the pigment molecules of the green-sensitive cones, and so on. As shown in figure 6.18, there is some overlap in the color sensitivities of the three cone families and rhodopsin of the rods. Our perception of colors other than red, green, and blue stems from the stimulation of combinations of red-, green-, and blue-sensitive cones. For example, when yellow light falls on the retina, both red- and green-sensitive cones are stimulated. The blue-sensitive cones are not stimulated at all by the yellow light. If light falling on the retina stimulates all three types of cones equally, the color is interpreted as being white.

Color-Blindness. Color-blindness is an abnormality in which a person either fails to see color altogether or has difficulty distinguishing certain colors. "Complete color-blindness" is very rare, whereas "red-green color-blindness" is quite common (about eight percent of males, but less than one percent of females are red-green color-blind). The abnormality (which is inherited via the sex chromosomes) occurs when one of the three families of cones is either absent or poorly functioning. Usually, the defect involves either the red-sensitive or the green-sensitive cones. In red-green color-blindness due to malfunctioning (or absent) red-sensitive cones, red light falling on the retina is absorbed by and weakly stimulates the green-sensitive cones. For such a person, a red object is difficult to distinguish from a green object.

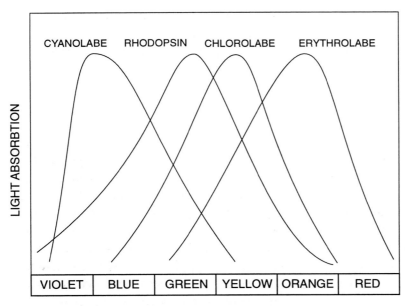

Figure 6.18
Absorption spectra of the three visual pigments found in the cones and rhodopsin of the rods. Red-sensitive cones contain the pigment erythrolabe; green-sensitive cones contain chlorolabe; and blue-sensitive cones contain cyanolabe. Note the overlap among the spectra.

Distribution of Rods and Cones in the Retina. The photoreceptor layer of the retina consists of millions of cones and rods; however, these cells are not uniformly distributed in the retina. Instead, the cones are concentrated in a small area at the center of the retina called the **fovea centralis**. As seen in figure 6.13, the fovea centralis forms a small depression in the retina and lies directly in line with the center of the cornea and the center of the lens. As one moves away from the fovea, one finds fewer and fewer cones and relatively greater numbers of rods. At the margins of the retina, the photoreceptors are all rods; there are no cones at all.

As noted above, the rods are more sensitive to light than the cones, but the cones provide color vision. Therefore, it is not surprising that when you look straight ahead (and do not turn your eyes to either side), it is difficult to distinguish the colors of the objects at the edges of your field of view. By the same token, in very dim light, one can

more easily detect objects at the edges of the field of view than at the center.

Nerve Cells of the Retina. The cones and rods form junctions with a layer of nerve cells called **bipolar cells** (see figure 6.17). When the photoreceptors are stimulated by light, they cause nerve impulses to be propagated in these bipolar cells. Whereas each cone forms a junction with a different bipolar cell, two or more rods may form a junction with a single bipolar cell. Because there is a 1:1 relationship between cones and bipolar cells, the cones provide greater visual *acuity*. The bipolar cells are quite short, and within the retina, they transmit their nerve impulses to a second layer of nerve cells called **ganglion cells**. The long axons of the ganglion cells run over the surface of the retina and converge upon a particular point known as the **optic disk** (fig. 6.13). At the optic disk the axons are bundled together to form the **optic nerve**, which exits at the rear of the eye. The optic disk is also the entry point for blood vessels that nourish the eye.

Because the optic disk contains no cones or rods, any light that falls on this region of the retina goes undetected. Consequently, the optic disk is also referred to as the eye's **blind spot**. You can readily demonstrate the blind spot to yourself using figure 6.19. Hold the page six to ten inches in front of your right eye with the black spot centrally aligned (and the *plus sign* over to the right. With your left eye closed, stare *through* the black spot while you slowly move the figure in a circular path (keeping the page the same distance in front of the eye). You should find that there is a particular position in which you can no longer detect the plus sign, even though you can still see the surrounding white paper. In that position, the image of the plus sign is falling on the blind spot. Normally, a person is unaware of the existence of the blind spot because both eyes are open and it is not possible for an image to fall on the blind spots of both eyes *at the same time*. Even with one eye closed, you cannot normally discern the blind spot because the brain "fills in" the missing visual information.

The surface of the retina is bathed by a viscous fluid called the **vitreous humor**, which fills the space between the rear of the lens and the retina's surface (see figure 6.13).

Visual Areas of the Brain

Figure 6.20 shows the nerve pathways that lead from the retinas of the two eyes to the visual areas of the brain. As seen in the figure, the axons of the ganglion cells that originate in the nasal half of each of the two retinas cross over in the **optic chiasma** to the opposite side of the brain. In contrast, the axons of the ganglion cells that originate in the marginal (i.e., outer) half of each retina enter the optic chiasma but do not cross over to the opposite hemisphere. (In figure 6.20, the ganglion cells of the retina of the left eye are collectively depicted by a thick, solid black line, whereas the ganglion cells of the right eye are depicted by a broken black line.) Impulses are transmitted in the **lateral geniculate bodies** to fibers that will comprise the left and right **optic tracts**. The optic tract fibers then conduct impulses to the **visual cortex** of the left and right **occipital lobes** of the cerebrum for evaluation and

Figure 6.19
Demonstrating the blind spot (see the text for an explanation of how to use this figure to demonstrate the blind spot of your right eye).

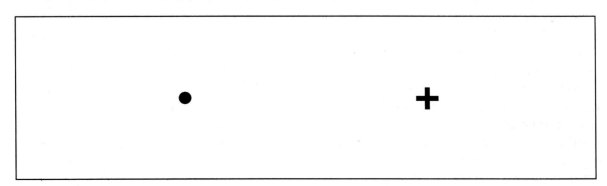

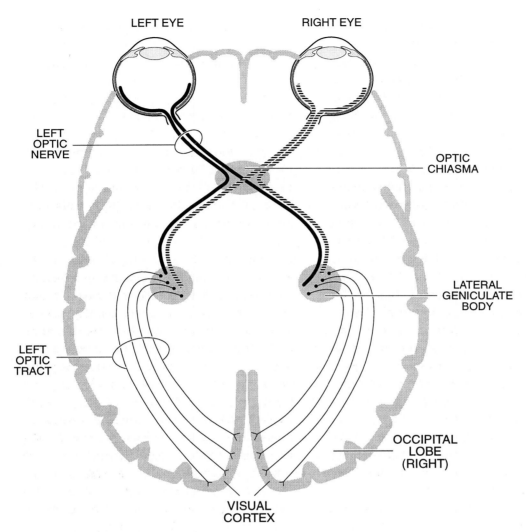

LEFT EYE

RIGHT EYE

LEFT
OPTIC
NERVE

OPTIC
CHIASMA

LATERAL
GENICULATE
BODY

LEFT
OPTIC
TRACT

OCCIPITAL
LOBE
(RIGHT)

VISUAL
CORTEX

Figure 6.20
Nerve pathways that lead from the retinas of the two eyes to the brain. The
fibers that originate in the nasal half of each retina cross over in the optic
chiasma to the opposite side of the brain. In contrast, fibers that originate in the
outer half of each retina remain on the same side of the brain. Therefore, each
side of the brain receives visual signals from both eyes.

interpretation. Note that each optic tract conducts signals that arose in the retinas of *both*
eyes. For example, the fibers of the
left optic tract conduct signals that
originate in the outer half of the left
eye and the nasal half of the right
eye.

6-4

The crossing over that occurs in the optic
chiasma has several implications. Clearly, if
the left optic nerve is destroyed by injury or
disease, vision in the left eye is totally lost.
However, destruction of the left optic tract
leaves a person with partial vision in both
eyes. That is, the affected person would

retain vision in the nasal half of the left eye and the outer half of the right eye. (Similarly, if the right optic tract were destroyed, vision would be retained in the nasal half of the right eye and the outer half of the left eye.)

Common Eye Abnormalities

In a person who has normal vision (**emmetropia**), the cornea and lens of each eye are smooth and clear, and the changes in lens shape that occur during accommodation ensure that the entire image that is being viewed is properly focused onto the retina. There are, however, a number of common eye abnormalities in which the image (in either one or both eyes) is not properly focused on the retina; among these defects are the conditions known as **farsightedness** (or **hyperopia**), **nearsightedness** (or **myopia**), and **astigmatism**.

Farsightedness. The correct focusing of light onto the retina is illustrated in the *top* section of figure 6.21. However, in many individuals (especially people who are over 40 years old) who view an object that is only a short distance from the eyes, the focusing power of the lens is too low to focus the image on the retina. Instead, the image created on the retina is "not yet" in focus; that is, the image is focused on a hypothetical plane that lies a short distance *behind* the retina (*middle* section of figure 6.21). This condition usually results from one of the following two problems; either (1) the eye is too short (i.e., the distance between the rear of the lens and the surface of the retina is too short), or (2) the lens has lost some of its elasticity and fails to become sufficiently round when the ciliary muscles contract (a condition called **presbyopia**, see earlier). The

latter is usually the case in people who are over 40.

The condition just described is called **farsightedness** (also known as **hyperopia**) because such individuals easily focus distant images. A distant scene does not require that the lens become so round in order for the image to be properly focused on the retina. Far-sightedness can be corrected using eyeglasses whose lenses are *convex* (fig. 6.21). Such lenses provide for the preliminary focusing of the light rays before the rays enter the eye.

Nearsightedness. A person who suffers with **nearsightedness** (or **myopia**) does not see distant images clearly. Usually, this is because the eye is too long (i.e., the distance between the rear of the lens and the surface of the retina is too great), with the result that the image is focused on a plane that lies *in front of* the retina (*bottom* section of figure 6.21). Nearsightedness can be corrected with eyeglasses whose lenses are *concave*. Such lenses provide for the divergence of light rays before the rays enter the eye.

Astigmatism. A common eye abnormality is **astigmatism**. In astigmatism, certain features of the image are in focus while other features are not. For example, vertical features may be focused on the retina, whereas horizontal features may be focused in front of or behind the retina. The reverse may also occur (i.e., horizontal features are in focus but vertical features out of focus). The most common sources of this abnormality are defects in the curvature of the cornea and irregularities in the smoothness or clarity of the lens. Like hyperopia and myopia, many forms of astigmatism can be corrected using eyeglasses having appropriately shaped lenses.

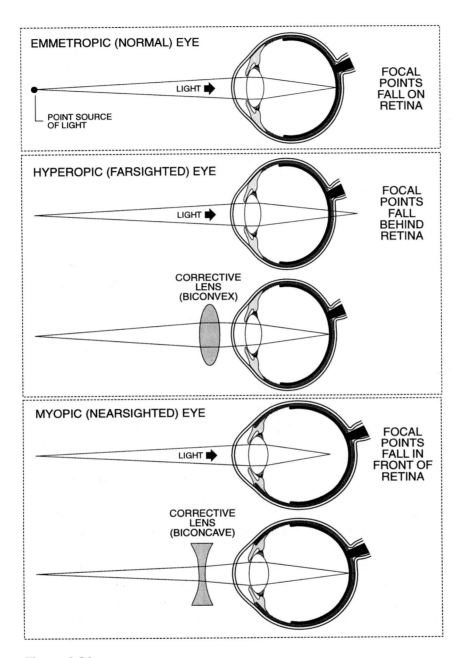

Figure 6.21

In a normal (emmetropic) eye, light rays are focused on the retina. In a farsighted (hyperopic) eye, the rays are focused on a plane that lies behind the retina (and the image is blurred). This condition may be corrected by placing a biconvex lens in front of the eye; this lens performs some preliminary focusing of the light rays before they reach the eye. In the nearsighted (myopic) eye, light rays are focused before they reach the retina (again, the image is blurred). This condition may be corrected by placing a biconcave lens in front of the eye; this lens performs some preliminary diverging of the light rays before they reach the eye.

HEARING AND BALANCE

The ears are the source of several special senses. While the most obvious of these is the sense of hearing, the ear also provides the brain with information about head position and head movement; the latter is called the **vestibular sense**. There are three major parts to the ear: the **outer ear**, the **middle ear,** and the **inner ear**. Of these, it is the inner ear that contains the receptor apparatus for hearing (called the **cochlea**) and for head position and movement (called the **vestibular apparatus**).

Sound

To properly understand and appreciate the sense of hearing, it is necessary to understand what sound is. Sound is produced by the vibration of an object (e.g., a guitar string, a metal tuning fork, or your own vocal cords). As the object vibrates back and forth, it pushes against and then pulls back the adjacent layers of air molecules. The motion of the air is then transferred to neighboring molecules which repeat the cycle, thereby causing a *wave* that moves in all directions away from the original source of vibration. (A good analogy to this phenomenon are the ripples that are produced when a small stone is dropped into a very still pond.) In the parlance of physics, a **pressure wave** is created radiating in all directions from the sound source. In the case of sound travelling through air, the waves typically travel at speeds around 1,100 feet per second (about 760 miles per hour). Thus, a sound made 2,200 feet away takes about 2 seconds to reach your ears. Note that the speed of a sound wave is hundreds of thousands of times slower than the speed of light

(which is 186,000 miles per second). Also, whereas light travels through a vacuum, sound requires a medium of some sort for its conduction. When travelling through materials other than air, the speed of the sound is altered (for example, the speed of sound in water is more than four times as great as in air).

In addition to velocity, sound waves have several other special properties. Among these is **frequency** (or **pitch**), which is a measure of how rapidly the air is vibrating back and forth and is measured in *cycles per second* (each cycle is a complete back and forth movement) or **hertz (Hz)**. The higher the frequency, the higher the pitch. Humans are capable of detecting sound waves having frequencies as low as about 20 Hz and as high as about 20,000 Hz. Within this range, the ears can differentiate about 2,000 different pitches.

6-5

Two sounds having the same frequency may be perceived quite differently. For example, the musical note called "middle A" which has a frequency of 440 Hz sounds quite differently when produced by a piano than when produced by a flute. These "qualitative" differences among sounds of the same frequency are called "tonal" differences or differences in **timbre**.

Another property of a sound wave is its **amplitude** (or **intensity** or **loudness**), which is measured in **decibels**. For example, a ticking watch has an amplitude of about 20 decibels, ordinary conversation a value of about 60 decibels, and the sound of a car horn only 15 feet away about 100 decibels (fig. 6.22). The scale is logarithmic, with each 10 decibel increase in amplitude corresponding to a ten-fold increase in loudness. Thus, a sound 100 decibels loud is 10,000 times louder (i.e., 10^4) than a sound that is 60 decibels loud. Sounds louder than about 150

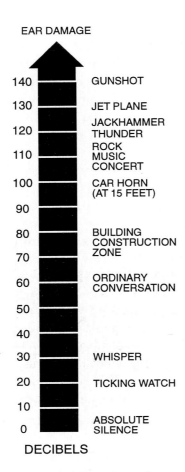

EAR DAMAGE

DECIBELS	
140	GUNSHOT
130	JET PLANE
120	JACKHAMMER / THUNDER
110	ROCK MUSIC CONCERT
100	CAR HORN (AT 15 FEET)
90	
80	BUILDING CONSTRUCTION ZONE
70	
60	ORDINARY CONVERSATION
50	
40	
30	WHISPER
20	TICKING WATCH
10	
0	ABSOLUTE SILENCE

Figure 6.22
The loudness of a sound is measured in decibels. The decibel scale is logarithmic with each 10 decibel increase corresponding to a ten-fold rise in loudness.

decibels can be harmful to the ears. The subjective sensation of the loudness of a sound depends on the frequency of the sound as well as its amplitude, the ear being more sensitive to some frequencies than to others.

The Outer Ear

The outer (or external) ear (fig. 6.23) consists of the **pinna** (or **auricle**) and the **au-**ditory canal (also known as the **external auditory meatus**). In humans, the pinna is not particularly large or highly developed, whereas in many other mammals it serves as a horn to trap sound waves and can even be directed toward the source of the sound. In humans, the pinna is a flap of cartilage and skin that is usually flush (or nearly flush) against the side of the head; therefore, its ability to trap sound is considerably diminished. Sound passing into the pinna travels along the auditory canal (which passes slightly upward and backward for a distance of about one inch) and reaches the **tympanic membrane** (or **eardrum**). The tympanic membrane seals off the end of the auditory canal and serves to separate the outer ear from the middle ear. Sound waves striking the tympanic membrane cause the membrane to vibrate with the same frequency as the impinging sound. By its vibration, the tympanic membrane transmits the sound to the middle ear.

The auditory canal and the tympanic membrane are lined by skin. The skin of the auditory canal is rich in glands that secrete wax (called **cerumen**) onto the canal's surface. The wax lubricates the canal and rarely accumulates in amounts sufficient to interfere with normal hearing. Thus, one should not attempt to remove the wax by inserting objects into the canal and running the risk of damaging the tympanic membrane. Hairs in the auditory canal act to trap airborne particles.

6-6

The Middle Ear

The middle ear is a small chamber in the skull (fig. 6.24) containing three tiny bones called **ossicles**. The chamber is not completely

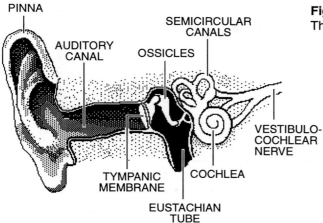

Figure 6.23
The outer ear, middle ear, and inner ear.

sealed, since the floor of the chamber contains a narrow opening that leads via the **eustachian tube** into the rear of the mouth (i.e., the **pharynx**). Air can enter or leave the middle ear through this channel, and this permits equilibration of the gas pressure of the middle ear with the atmospheric pressure around us. Equilibration of pressure pre-

vents undue distention (or even rupturing) of the tympanic membrane when there is a change in external pressure (as when travelling by plane at great heights). Normally, the eustachian tube is closed; however, swallowing and chewing bring about a temporary opening of the channel, thereby allowing pressure equilibration. The eustachian tube

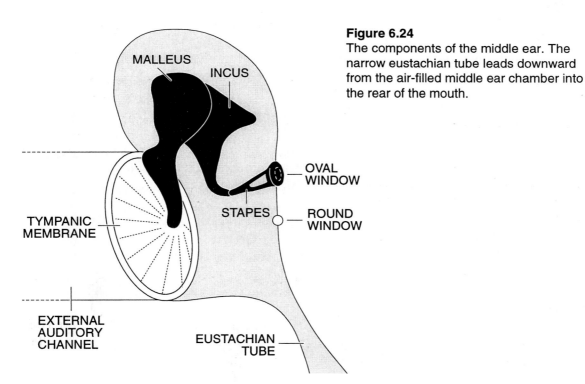

Figure 6.24
The components of the middle ear. The narrow eustachian tube leads downward from the air-filled middle ear chamber into the rear of the mouth.

can be an avenue for passage of bacteria from the mouth into the middle ear, causing middle ear infections.

The far wall of the middle ear also contains two openings; these openings, however, are covered by thin membranes. The upper opening is the **oval window**, and the lower is the **round window**. The ossicles are arranged to form a series of levers that transmit the vibrations of the tympanic membrane across the middle ear to the oval window. As we will see, the inner ear contains fluid that is set in motion by sound. Much more force is required to set liquid in motion than to set air in motion. The lever-like arrangement of the ossicles provides the means for the 20-fold increase in force needed to cause vibration of fluid in the inner ear (i.e., the ossicles act in a manner similar to a set of gears). Excessive movements of the ossicles are prevented by small muscles and ligaments that connect the ossicles to the surrounding skull bone. The ossicle attached to the tympanic membrane is the **malleus**; the malleus articulates with the second ossicle, called the **incus**; finally, the incus moves the **stapes** back and forth, pushing it in and out of the oval window. The oval window leads into a long, fluid-filled, twisted channel that eventually leads back to the round window. Thus, as the oval window is pushed inward by movement of the stapes, the round window bulges outward into the middle ear.

The Inner Ear–Cochlea

The inner ear consists of the **cochlea, saccule, utricle**, and **semicircular canals** (fig. 6.25); these structures occupy a complex labyrinth in the temporal bone of the skull. The cochlea is concerned with the sense of hearing, whereas the other components of the inner ear allow us to sense head position and head movement; the latter senses are known as the **vestibular senses**.

6-7

The cochlea is a spiral-shaped chamber similar in appearance to a snail shell. Within the chamber are three fluid-filled channels called the **scala vestibuli**, the **scala media**, and the **scala tympani**. In figure 6.26, the three channels are shown in an untwisted arrangement, so that the relationships among them and their relationship to the oval and round windows are more easily understood. Also indicated in the figure are the ranges of frequency sensitivity in different regions of the cochlea. Note that the apical end of the cochlea responds to low frequency sounds, whereas the basal end of the cochlea responds to high frequency sounds.

The scala vestibuli begins at the oval window near the base of the cochlea and winds its way up to the cochlea's apex, where it leads into the scala tympani; the scala tympani then winds its way back down to the cochlea's base, ending at the round window. The scala vestibuli and scala tympani make about two and one-half turns as they wind their ways between the base and apex of the cochlea. The fluid that fills these channels is called **perilymph**. Vibrations of the stapes are transmitted across the oval window so that the perilymph is set into motion.

The third fluid-filled channel of the cochlea, the scala media, is sandwiched between the scala vestibuli and scala tympani and is filled with a fluid called **endolymph** (fig. 6.26). The floor of the scala vestibuli (which is also the roof of the scala media) is a thin membrane called the **vestibular membrane** (also known as **Reissner's membrane**). A thicker

6-8

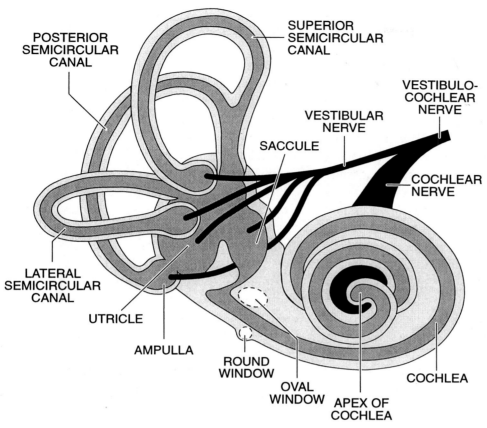

Figure 6.25
The inner ear. The inner ear consists of the cochlea, saccule, utricle, and semicircular canals. The cochlea is concerned with the sense of hearing, whereas the saccule, utricle, and semicircular canals are concerned with the sense of balance.

structure called the **basilar membrane** serves as the floor of the scala media (which is also the roof of the scala tympani). The scala media houses the delicate machinery of sound detection called the **organ of Corti**. Figure 6.27, which depicts a cross section through the entire cochlea, shows the three channels as they wind their way upward from the base to the apex.

The organ of Corti (shown in detail in fig. 6.28) contains the auditory receptor cells or **phonoreceptors**. Because hairlike projections (cilia) emerge from the upper sur-

faces of these cells, the cells are also referred to as **hair cells**. The end of the hairlike projections are embedded in the overhanging **tectorial membrane** which "floats" freely in the endolymph. The hair cells, which are estimated to be about 15,000 in number, are seated on the basilar membrane and form a succession of rows that begin at the base of the cochlea and extend up to the apex. In the basilar membrane, the hair cells form junctions with the dendritic endings of sensory nerve fibers. Collectively, these fibers form the **cochlear nerve**, which, following its exit

151

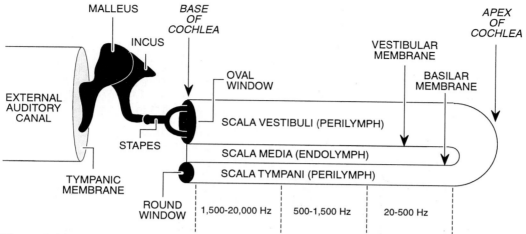

Figure 6.26

The three channels of the cochlea. In this diagram, the cochlea has been unwound so that the relationships among the scala vestibuli, scala media, and scala tympani may be more readily understood.

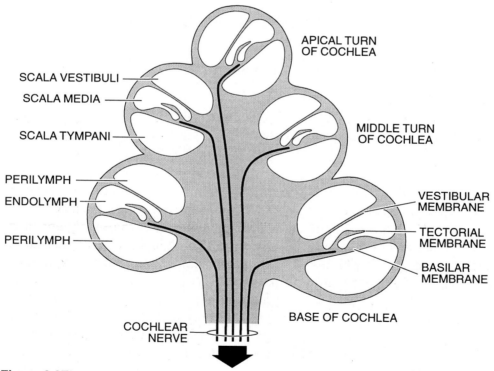

Figure 6.27

A cross section through the cochlea, showing the scala vestibuli, scala media, and scala tympani as they wind their way up from the cochlea's base to apex. The cochlear nerve runs upward through the central axis of the cochlea, giving rise to nerve fibers that form a spiral and radial (spokelike) pattern.

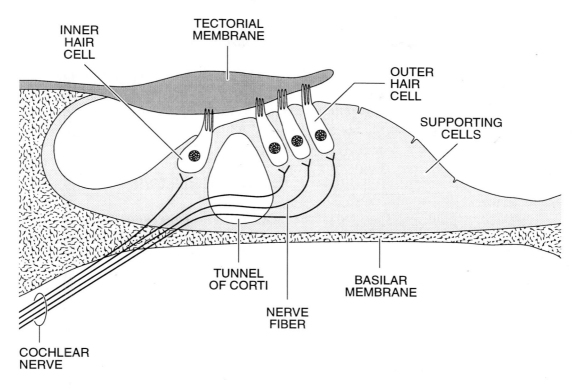

Figure 6.28
A magnified view of the organ of Corti (compare with figure 6.27).

from the cochlea, becomes a part of the **vestibulocochlear** (or **acoustic**) nerve (i.e., cranial nerve VIII).

Movements of the oval window at the base of the cochlea cause the perilymph of the scala vestibuli to vibrate and this sets the vestibular membrane into vibration. These vibrations are transmitted across this thin membrane, so that the endolymph vibrates in sympathy. The vibrations of the endolymph then cause the basilar membrane to vibrate. This membrane contains thousands of **stiff fibers** that radiate through the membrane from the cochlea's axis much like the spokes that radiate from the hub of a bicycle wheel to the wheel's rim. These stiff fibers are believed to have vibrational properties that are similar to the reeds of a musical instru-

ment. The lengths of the stiff fibers increase progressively as the basilar membrane winds its way up to the cochlea's apex. Although the basilar membrane is capable of vibration over its entire length, it vibrates *maximally* at a specific point that is determined by the vibration frequency of the endolymph (and, therefore, the original source of the sound). The point of maximum vibration of the basilar membrane corresponds to the region of the membrane containing stiff fibers of the appropriate **resonant frequency**. Low frequency sounds vibrate the basilar membrane over its entire length, with the vibration peak occurring near the apical end of the scala media, where the stiff fibers are longer and have lower resonant frequencies. Movements of the basilar membrane caused by high frequency sounds peak closer to the basal end of the scala media where the stiff fibers are shorter and have higher resonant frequencies.

At the point of maximum movement of the basilar membrane, motion of the hair cells causes their stimulation, with the result that nerve impulses are generated in the sensory nerve fibers associated with these receptors. As seen in figure 6.26, most of the basilar membrane's length is sensitive to frequencies between 20 and 1,500 Hz (a range within which most of the sounds that we hear are produced). Indeed, below 1,000 Hz, the human ear can distinguish frequencies differing by as little as 3 Hz.

Impulses arising from a particular region of the organ of Corti are conducted to a specific region of the auditory portion of the cerebral cortex and are interpreted as sound of a specific pitch. Most sounds are a complex mixture of pitches that stimulate several regions of the organ of Corti. The resulting pattern of impulses received by the brain is sorted out, so that the interpretation matches the complexity of the original sound. The intensity or loudness of the original sound determines the amplitude of the vibrations of the tympanic membrane, the ossicles, the oval window, the cochlear fluids, and the basilar membrane. The greater the amplitude, the more intense is the sensation. Some of the sensory neurons arising from the cochlea (and other regions of the inner ear) cross over in the brain stem. Therefore, both sides of the cerebral cortex receive input from each of the ears.

The Inner Ear–Vestibular Apparatus

The **vestibular apparatus** consists of the **saccule, utricle,** and **semicircular canals** (fig. 6.25) and is important in the maintenance of balance. The apparatus provides information about the *position* of the head and also its *motion*. Like the cochlea, the interior of the vestibular apparatus is filled with fluid.

The Saccule and Utricle

The **saccule** and **utricle** are the source of sensory information about (1) the head's *static position* in space and (2) *linear acceleration* of the head in the horizontal (saccule) and vertical (utricle) planes. Both structures contain oval patches of hair cells (receptor cells) similar to those of the organ of Corti (fig. 6.29); the oval patches are called **maculae**. The projections (i.e., "hairs") are of two types, consisting of a single large cilium called a **kinocilium** and several smaller cilia called **stereocilia** (fig. 6.30). The cilia ascend from the apical surface of the receptor cells into a jellylike block of material whose upper region is littered with crystals of calcium carbonate called **otoliths** (or **otoconia**). The gelatinous layer, together with the otoliths, is called the **otolithic membrane** (fig. 6.29). At their basal ends, the hair cells are associated with the dendritic endings of sensory nerve fibers that form a branch of the vestibular nerve. (The vestibular nerve and the cochlear nerve form the vestibulocochlear [or acoustic] nerve.)

In a person who is standing upright, the macula in each utricle is oriented horizontally (i.e., as depicted in figure 6.29). In each saccule, however, the macula is oriented in a vertical plane (i.e., at right angles to that depicted in figure 6.29). Consequently, the hair cells of the utricle are directed verti-

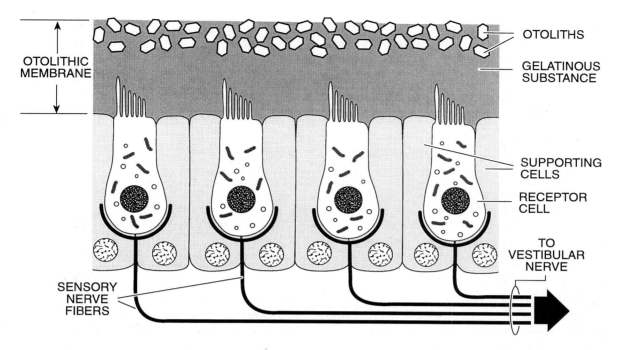

OTOLITHS

GELATINOUS SUBSTANCE

OTOLITHIC MEMBRANE

SUPPORTING CELLS

RECEPTOR CELL

TO VESTIBULAR NERVE

SENSORY NERVE FIBERS

Figure 6.29
Organization of the saccule and utricle. In a person who is standing upright, the utricle is oriented as shown in the diagram, whereas the saccule is oriented at right angles to that shown here. When standing motionless, the weight of the otoliths pushes down on the cilia of the receptor cells uniformly stimulating the entire horizontal carpet of cells (i.e., the entire macula). Tilting the head acts to flop the otolithic membrane to one side, thereby applying a differential stimulus to the hair cells.

cally, while the hair cells of the saccule are directed horizontally.

When standing upright and motionless, gravity pulling downward on the otoliths acts to compress the otolithic membrane of the utricles. This applies a uniform and equal stimulus to the cilia of the hair cells. Under these conditions, the hair cells stimulate the dendritic endings of the nerve cells with which they are associated, causing a continuous and uniform flow of nerve impulses to the brain. However, when the head is tilted, the weight of the otoliths flops the otolithic membrane toward the lowered side of the head. This pulls the cilia in a particular

direction and acts to apply a *differential* stimulus to the hair cells (fig. 6.30).

Depending upon the direction of tilt, the membranes of the hair cells will either hyperpolarize (when the stereocilia are pulled *away from* the kinocilium) or depolarize (when the stereocilia are pulled *toward* the kinocilium). *Hyperpolarization* of the hair cell membranes leads to a *decrease* in the flow of nerve impulses to the brain over the sensory nerve fibers, whereas *depolarization* leads to an *increase* in the frequency. The brain interprets the changing frequency with which it receives these sensory impulses as a tilt of the head in a particular direction. In a

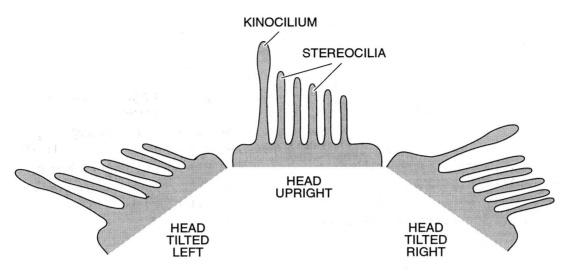

KINOCILIUM

STEREOCILIA

HEAD
UPRIGHT

HEAD
TILTED
LEFT

HEAD
TILTED
RIGHT

Figure 6.30
Actions of the hair cells of the utricles (and saccules). Tilting the head differentially stimulates the utricles. When the head is tilted one way (toward the right
in the diagram above), the stereocilia are pulled away from the kinocilium.
When the head is tilted the other way (toward the left in the diagram above),
the stereocilia are pulled toward the kinocilium. These movements differentially
stimulate the hair cells and their associated sensory nerve fibers.

similar manner, the hair cells of the saccule are differentially stimulated, depending upon the direction in which the cilia of the hair cells are pulled.

The saccule and utricle are also stimulated by linear acceleration of the head. The utricle responds to vertical linear acceleration, whereas the saccule responds to horizontal linear acceleration. For example, if a person standing upright on a chair suddenly jumps down to the floor (i.e., this would be vertical linear acceleration), the inertia of the otoliths in the otolithic membranes of the utricles causes the otolithic layers to lag behind as the head decelerates downward. This acts to pull upward on the cilia of the hair cells. Jumping upward from the floor onto a chair would have the opposite effect, causing an initial compression of the otolithic mem-

brane. Similar effects in the utricles allow you to discern whether the elevator you are in is accelerating downward or upward.

The saccules respond to acceleration in the horizontal plane (as when you accelerate forward in an automobile when the traffic light turns green or come to a halt at a traffic light that is red). During sudden forward acceleration, the inertia of the otoliths in the saccules causes them to lag behind, thereby stretching the otolithic membranes and pulling on the cilia. When you brake to a sudden halt the inertia of the otoliths attempts to keep the otolithic membranes moving forward and this acts to compress the otolithic membranes. As in the case of the utricle, the stretching or compression of the cilia of the hair cells provides the differential stimulus that alters the flow of nerve impulses to the brain; these differences are perceived as acceleration or deceleration in the horizontal plane.

The Semicircular Canals

Arising from the end of the vestibular apparatus are the three **semicircular canals** (figs. 6.25 and 6.31). These structures provide information about rotational movements of the head (or *dynamic acceleration*). Each canal is like a ring oriented in one of the three planes of space (i.e., "upward/downward" [the **superior** or **anterior canal**], "forward/backward" [the **posterior canal**], and "side-to-side" [the **lateral** or **horizontal canal**]). Close to its point of origin, each canal possesses an expanded portion (called an **ampulla**) that houses the canal's receptor apparatus (known as a **crista**; figure 6.31). The receptor consists of a carpet of hair cells whose hairlike projections (cilia) ascend into a gelatinous tuft called the **cupula**.

The cristae of the canals are stimulated by rotational acceleration within that canal's

Figure 6.31
The crista inside an ampulla of one of the three semicircular canals. As the head is rotated in one (or a combination) of the three planes in space, the inertia of the endolymph inside the canal causes a relative rotation in the opposite direction. Movement of endolymph against the surface of the cupula pushes the cupula to one side. Movement of the cupula acts as a stimulus for the hair cells (receptors).

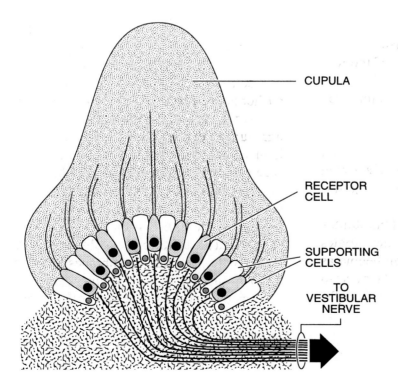

CUPULA

RECEPTOR CELL

SUPPORTING CELLS

TO VESTIBULAR NERVE

plane. For example, when you rotate your head clockwise (to the right) in the horizontal plane, the inertia of the endolymph in each of your lateral canals causes the endolymph to lag behind the canal's movement. Relative to the moving canals, the endolymph rotates in the opposite direction (i.e., counter clockwise, or to the left). Rotation of the endolymph pushes on the cupulas in the canals' ampullae disturbing the hair cells and their cilia. Clockwise rotation of the head stimulates hair cells in the right horizontal semicircular canal and inhibits

hair cells in the left horizontal semicircular canal. This leads to corresponding changes in the frequencies with which sensory impulses are conducted to the brain by sensory nerve fibers associated with the hair cells in each ampulla. When you stop rotating your head, the reverse effect takes place (left horizontal canal is stimulated by inertial movements of the endolymph, while the right horizontal canal is inhibited). Forward/backward and upward/downward rotation of the head produce corresponding effects in the other two pairs of semicircular canals.

SELF TEST*

True/False Questions

1. Golgi tendon organs are stimulated by being stretched when a skeletal muscle contracts.

2. Pacinian corpuscles are receptors that respond to mild compression of the skin.

3. Adaptation is slowest in thermoreceptors and fastest in pain receptors.

4. The front tip of the tongue is especially sensitive to bitter tastes.

5. The front of the eye's choroid coat is transparent in order to allow light to enter the eye.

6. The tension in the ligaments that suspend the lenses of the eyes is greater when reading a book than when looking at a distant mountain top.

7. By acting very much like a set of minia-

ture gears, the ossicles are able to convert low frequency vibrations into high frequency vibrations.

8. The semicircular canals respond to rotational movements of the head.

Multiple Choice Questions

1. Your awareness of the positions in space of the various parts of your body is known as (A) consciousness, (B) vertigo, (C) kinesthesia, (D) exteroception.

2. The term "gustation" has to do with (A) the spinal cord, B) the ears, (C) the tongue, (D) the nose, (E) none of these choices is correct.

---

* *The answers to these test questions are found in Appendix III at the back of the book.*

3. The fovea centralis contains (A) large numbers of cones but very few rods, (B) large numbers of rods but very few cones, (C) approximately equal numbers of rods and cones.

4. Which one of the following structures is responsible for creating the retina's "blind spot?" (A) the fovea centralis, (B) the optic nerve, (C) rods, (D) cones, (E) the optic chiasma.

5. In the condition known as "farsightedness," (A) light rays reach the retina before they are brought into focus, (B) light rays are focused before they reach the retina, (C) distant objects are not seen with clarity.

6. Suppose that you are outside on a bright, sunny day and stare at the center of long, tall wall that has been painted blue. Under such circumstances, your eyes' (A) blue-sensitive cones would be stimulated, (B) blue-sensitive rods would be stimulated, (C) blue-sensitive cones and blue-sensitive rods would be stimulated, (D) rods and blue-sensitive cones would be stimulated.

7. The eustachian tube (A) leads from the pinna to the tympanic membrane, (B) connects the apex of the cochlea with the oval window, (C) connects the pharynx (rear of the mouth) with the middle ear, (D) connects the semicircular canals with the round window, (E) connects the pharynx (rear of the mouth) with the inner ear.

8. Which one of the following ossicles pushes in and out of the round window? (A) the pinna, (B) the incus, (C) the stapes, (D) the malleus, (E) the round window is not in contact with the ossicles.

9. Which of the following is responsible for sensing high frequency sounds? (A) the basal end of the cochlea, (B) the apex of the cochlea, (C) the saccule, (D) the scala tympani.

10. Which one of the following receptor organs is stimulated when you come to an abrupt stop at a traffic light? (A) the semicircular canals, (B) the utricle , (C) the saccule, (D) the cochlea.

11. At the base of each semicircular canal there is an enlargement that contains the canal's receptor organ; these enlargements are called (A) otolithic membranes, (B) rotational bodies, (C) saccules, (D) ampullae, (E) lobi concentricus.

THE CIRCULATORY SYSTEM

The circulatory system is the system of organs responsible for the circulation of the blood (and the lymph). Several major physiological roles are played by this system; these include the following:

1. the transport of oxygen, nutrients, and raw materials to the body's tissues,

2. the removal of carbon dioxide and metabolic wastes from the body's tissues,

3. the transport of hormones and other secretory substances from one tissue (or organ) to another, and

4. the distribution of the body's heat of metabolism.

MAJOR COMPONENTS OF THE CIRCULATORY SYSTEM

The circulation of the blood requires a mechanism that keeps the blood in motion and an array of vessels through which the blood may be carried (fig. 7.1). The motion

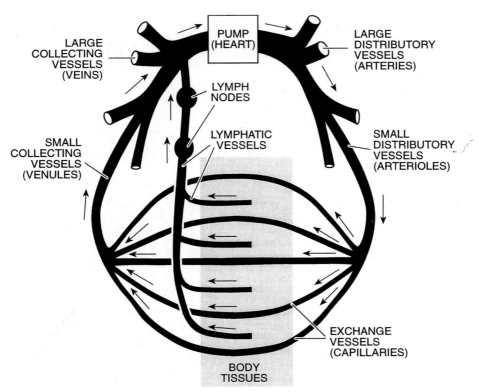

Figure 7.1
Major components of the circulatory system. The circulatory system requires a pump (i.e., heart), distributory vessels (i.e., arteries and arterioles), vessels for exchange (i.e., capillaries), and collecting vessels (i.e., venules and veins). In humans, an auxiliary system of collecting vessels exists (lymphatics), which empty their contents into major veins.

of the blood is created by the pumping action of the **heart**. Blood is carried away from the heart by major vessels called **arteries**; indeed, an artery is defined as any major vessel that conducts blood *away from the heart* (note that the definition is unconcerned with the blood's composition, such as whether it is rich in oxygen or is oxygen-poor). It is the arteries that carry the blood into the body's major organs.

Arteries branch into a number of smaller distributory vessels called **arterioles**, and the arterioles then branch into a **network** of still finer vessels called **capillaries**. Whereas the

walls of arteries and arterioles contain smooth muscle whose state of contraction or relaxation determines the blood pressure and regulates the flow of blood through the vessel, the capillary walls have no musculature. In fact, the walls of the capillaries are formed from a single, thin layer of cells (called **endothelium**). It is the thinness and permeability of the capillary wall that permits the rapid exchange of materials between the blood passing through the capillaries and the surrounding tissue cells.

Blood is collected from the capillary networks by **venules**. A number of venules converge and form a **vein**, which then directs the blood out of the organ. The major veins return the blood to the heart (indeed, veins are defined as major vessels conducting blood *toward the heart*; again, composition of the blood is ignored in the definition). Like the arteries and arterioles, the venules and veins have muscular walls to regulate blood flow and blood pressure. The larger venules and the veins contain **valves**; these act like one-way gates that open only in the direction that leads to the heart. The action of the valves assures that the blood does not back up in the circulatory system.

Also considered part of the circulatory system are the **lymphatic vessels**. These vessels arise *blindly* within a tissue (fig. 7.1) and act to conduct fluid toward the heart. The fluid carried by the lymphatics is called **lymph** (not blood). Like the veins and larger venules, large lymphatics contain valves. At intervals along the lengths of the larger lymphatics are **lymph nodes** that act to filter the lymph (see later). Ultimately, the lymph drained from the body's tissues is emptied into major veins and the *mixture* of blood and lymph (referred to as *blood*) enters the heart to be recirculated.

THE PULMONARY AND SYSTEMIC CIRCULATIONS

The human body contains two separate and distinct circulations of the type described in figure 7.1. The two circulations are called the **pulmonary circulation** and the **systemic circulation** (fig. 7.2). Each circulation has its own pump (i.e., heart) and its own set of arteries, arterioles, capillaries, venules, and veins. Only one organ of the body is served by the pulmonary circulation and that is the *lungs*. The sole purpose of this circulation is

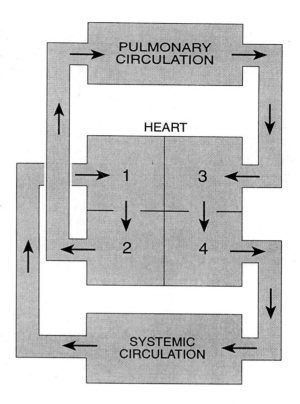

Figure 7.2
There are two blood circulations in the body: pulmonary and systemic. The pulmonary circulation is fed by the right side of the heart (sometimes called the "right heart") and directs blood through the lungs. The systemic circulation is fed by the left side of the heart (sometimes called the "left heart") and directs blood through all of the body's organs and tissues.

to enrich the blood's oxygen content and to diminish its content of carbon dioxide. This testifies to the critical role of oxygen in the body's metabolism and the continuous need to eliminate the waste carbon dioxide that is produced during metabolism.

The systemic circulation carries blood to *all* of the body's organs (muscles, brain, liver, stomach, kidneys, etc.). This implies that the lungs are *also* served by the systemic

163

circulation; that is, there are *two* circulations of blood in the lungs. The pulmonary circulation carries blood through the lungs in order to oxygenate the blood and eliminate carbon dioxide, while the systemic circulation carries blood through the lungs in order to nourish the lungs and carries away the lungs' wastes. The two bloods (pulmonary and systemic) do not mix in the lungs.

7-1

CIRCULATION OF BLOOD THROUGH THE HEART

The pumps that drive blood through the pulmonary and systemic circulations comprise the right and left hearts, usually referred to as the *right and left sides of the heart*. We'll begin our consideration of the circulatory system by examining the organization and functions of the different parts of this most important organ.

The major anatomical parts of the heart are shown in figure 7.3. Blood returning to the heart from the *systemic* circulation enters the *right* side of the heart through two major veins: the **superior vena cava** (structure *1* in figure 7.3) and the **inferior vena cava** (structure *2* in figure 7.3). The superior vena cava returns blood to the heart from the upper parts of the body (head, neck, arms, upper chest, etc.), whereas the inferior vena cava returns blood to the heart from the lower parts of the body (abdomen, legs, etc.). The terms "superior" and "inferior" have nothing at all to do with the relative importance of the two veins; rather, the terms refer only to their anatomical positions (the superior vena cava entering the heart from above, and the inferior vena cava entering the heart from below). The heart chamber into which the blood is emptied is the **right atrium** (structure *3* in figure 7.3).

A valve called the **tricuspid valve** (structure *4* in figure 7.3) separates the **right atrium** from the **right ventricle** (structure *5* in figure 7.3). This valve is open during atrial filling, so that much of the blood entering the right atrium immediately descends through the valve and enters the right ventricle. The right ventricle is already about 80% filled when a small volume of additional blood is pushed into the right ventricle by contraction of the right atrium's walls. The walls of the right ventricle are quite muscular. This is because the contractions of the right ventricle must push the blood through the entire *pulmonary* circulation. Contraction of the right ventricle exerts considerable force on the blood and pushes it up and out of the heart through the **pulmonary artery** (structure *8* in figure 7.3). In order to enter the pulmonary artery, the blood must push open the second valve of the right side of the heart, namely the **pulmonary semilunar valve** (structure *7* in figure 7.3). The pulmonary artery divides to produce two branches that direct blood into the right and left lungs.

The contraction of the right ventricle is quite vigorous and were it not for the actions of the **papillary muscles** (structure *6* in figure 7.3), the upward push of the blood might cause regurgitation into the right atrium. This is prevented by the contractions of the papillary muscles, which are attached to the walls of the right (and left) ventricle and which pull downward on the cusps of the tricuspid valve. A strong enough downward pull exactly balances the upward push by the blood and keeps the cusps of the valve in the horizontal (closed) position. As a result, the blood of the right ventricle is pushed into the pulmonary artery.

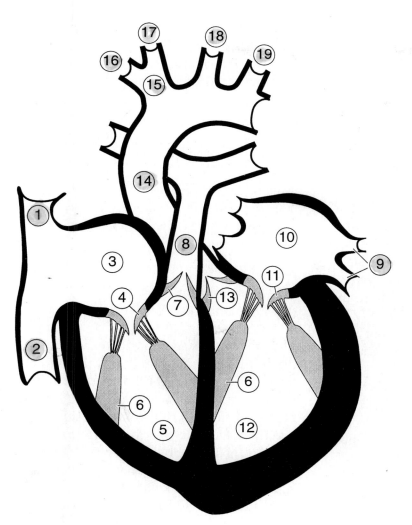

Figure 7.3
Anatomy of the heart as seen from the front. *1*, superior vena cava; *2*, inferior vena cava; *3*, right atrium; *4*, tricuspid valve; *5*, right ventricle; *6*, papillary muscle; *7*, pulmonary semilunar valve; *8*, pulmonary artery; *9*, pulmonary veins (2 at each side of the left atrium); *10*, left atrium; *11*, bicuspid (mitral) valve; *12*, left ventricle; *13*, aortic semilunar valve; *14*, aortic arch; *15*, brachiocephalic (innominate) artery; *16*, right subclavian artery; *17*, right common carotid artery; *18*, left common carotid artery; *19*, left subclavian artery.

Blood returns from the pulmonary circulation to the left side of the heart through the four **pulmonary veins** (structures *9* in figure 7.3), two veins carrying blood from the right lung and two from the left lung. This blood enters the **left atrium** (structure *10* in figure 7.3) and descends through the open **bicuspid** (or **mitral**) **valve** (structure *11* in figure 7.3) into the **left ventricle** (structure *12* in figure 7.3). The left ventricle is already about 80% filled when a small volume of additional blood is pushed into the left ventricle by contraction of the left atrium's walls. The walls of the left ventricle are very muscular. This is because the contractions of the left ventricle must push the blood through the entire *systemic* circulation. Contraction of the left ventricle exerts considerable force on the blood and pushes it up and out of the heart through the **aortic arch** (structure *14* in figure 7.3). In order to enter the aortic arch, the blood pushes open the **aortic semilunar valve** (structure *13* in figure 7.3). At the same time that

7-2

the left ventricle contracts, so do the papillary muscles attached to the walls of the left ventricle (structure *6* in figure 7.3). These muscles pull down on the bicuspid valve, thereby opposing the upward push exerted by the blood. As a result, the bicuspid valve remains closed and no blood is regurgitated into the left atrium. Branches of the aorta direct the blood to all organs of the body. As the aorta arches over the heart, it gives rise to several major arteries (fig. 7.3). The first of these is the **brachiocephalic** (or **innominate**) artery (structure *15* in figure 7.3). This is a very short artery that almost immediately branches into the **right subclavian artery** (structure *16* in figure 7.3) and the right **common carotid artery** (structure *17* in figure 7.3). The right subclavian artery directs blood into the right upper chest and the right arm, whereas the right common carotid artery directs blood into the right sides of the neck and head. The second major branch of the aortic arch is the **left common carotid artery** (structure *18* in figure 7.3), which directs blood into the left sides of the neck and head. The **left subclavian artery** (structure *19* in figure 7.3) branches directly off the aortic arch and directs blood to the left upper chest and the left arm.

| 7-3 |

Not seen in figure 7.3 are the right and left **coronary arteries**. These small vessels branch off the aortic arch just beyond the aortic semilunar valve. The coronary arteries carry blood into (and nourish) the muscular walls of the heart itself (i.e., the heart is not nourished directly by the blood that fills its chambers).

Arteries and Veins

It is worth pointing out once again that the terms *artery* and *vein* identify blood vessels on the basis of the *direction* in which the vessel conducts blood through the circulatory system; that is to say, arteries conduct blood *away from the heart*, whereas veins conduct blood *back to the heart*. Note that blood vessels are *not* identified as arteries or veins on the basis of whether the blood that they contain is "rich" or "poor" in oxygen. Consequently, it should not be a source of confusion for you that the pulmonary artery is poor in oxygen, whereas the pulmonary veins are rich in oxygen.

The Cardiac Cycle (Systole and Diastole)

The previous discussion of the anatomy of the heart noted the order in which blood travels through the right and left sides of the heart. For convenience, the right side was described first, and then the left side was described. However, it is important to recognize that, in fact, the right and left sides of the heart act synchronously. That is, (1) both the right and left atria fill with blood at the same time; (2) the right and left ventricles are filled with blood at the same time; and (3) the right and left ventricles contract to expel blood into the pulmonary artery and the aortic arch at the same time.

The sequential actions of the heart's chambers constitute what is called the **cardiac cycle**. In a heart that is beating 75 times per minute (i.e., a common resting rate), each cycle lasts about 0.8 seconds (fig. 7.4). The contraction of a heart chamber is called **systole**, and the relaxation that allows the chamber to fill is called **diastole**. Thus, right and left atrial systole occur at the same time and are accompanied by right and left ventricular diastole; atrial systole lasts

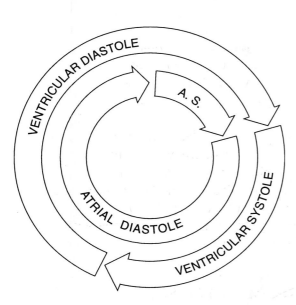

Figure 7.4
The cardiac cycle. The relative time intervals of atrial/ventricular systole/diastole for a heart that is beating at 75 cycles per minute (resting rate). A. S. = atrial systole.

0.1-0.2 seconds. Following the end of atrial systole, the right and left ventricles enter their systolic phase, directing blood into the pulmonary artery and aortic arch. Ventricular systole lasts about 0.3 seconds. The onset of ventricular systole is accompanied by atrial diastole. The end of ventricular systole is followed by a pause lasting about 0.4 seconds. During this portion of the cardiac cycle, *both* the atria and ventricles are in diastole (see figure 7.4).

The entry of blood into the atria is not an entirely passive phenomenon. Rather, the elastic recoil of the atrial walls at the end of systole acts to suck blood into the two chambers from the superior and inferior vena cavas and from the pulmonary veins. This action is similar to what happens when you squeeze and then release the rubber bulb of an eyedropper. As the bulb returns to its normal shape, the eyedropper fluid is drawn up toward the bulb.

The Frank-Starling Law of the Heart

Systole of the ventricles does not eject all of the blood that these chambers contain. For example, in a heart that is beating at rest, the volume of blood in the ventricles just prior to the onset of their systole is about 130 c.c. (i.e., the left and right ventricles *each* contain 130 c.c.); this volume of blood is known as the **end-diastolic volume**. Systole of the ventricles ejects 50-60% of the end-diastolic volume, which corresponds to about 70 c.c. of blood. The volume of blood remaining in the ventricles at the end of their systole is called the **end-systolic volume** and is about 60 c.c. (i.e., 130 c.c. minus 70 c.c.).

As noted in chapter 5, heart rate is controlled by the sympathetic and parasympathetic divisions of the autonomic nervous system. Sympathetic stimulation of the heart not only increases the rate at which the heart beats but also increases the percentage of the end-diastolic volume that is ejected by ventricular systole. During such stimulation, the end-systolic volume may be reduced to about 30 c.c.

The extent to which the walls of the ventricles are stretched as the ventricles are filled with blood influences the force with which the ventricular walls contract during ventricular systole. That is, the greater the volume of blood entering the ventricles from the atria (and the greater the resulting distention of the ventricles' walls), the greater will be the force with which the ventricles contract. This innate ability of the ventricles to adjust their force of contraction to match the incoming volume of atrial blood is known as the **Frank–Starling law of the heart**. Consequently, during exercise, when the contractions of skeletal muscle push greater volumes of blood back to the heart, the chambers fill with more blood; this is followed by a corresponding increase in the force of contraction to ensure that the extra volume of blood received by the ventricles is being expelled.

Stroke Volume

The volume of blood expelled from the left ventricle with each systole is called the **stroke volume**. As already noted, for a heart that is beating at rest, this amounts to about 70 c.c. of blood (remember that an additional 60 c.c. of blood remains in the left ventricle at the end of systole [i.e., the *end-systolic volume*]). Because equal volumes of blood are pumped into both the pulmonary and systemic circulations, the resting stroke volume of the right ventricle also is 70 c.c. During exercise, as the rate and force with which the heart beats increase, there is an increase in stroke volume (and a decrease in the end-systolic volume).

Cardiac Output (Minute Volume)

In a person at rest, the heart goes through its cycle about 75 times per minute. If the stroke volume is 70 c.c., then in one minute, 5,250 c.c. of blood (i.e., 70 x 75) are pumped from the left ventricle. The volume of blood pumped from the left ventricle in one minute is called the **cardiac output** or **minute volume**. If, during exercise, the stroke volume increases to 100 c.c. and the heart rate increases to 120 cycles per minute, then the cardiac output would be increased to 12,000 c.c. (i.e., 120 x 100). It is possible to estimate the total volume of blood in the body; in the average adult male, it is about 5,000 c.c. (i.e., 5 liters). Therefore, if the cardiac output at rest is 5,250 c.c., then in one minute's time, a volume of blood slightly greater than the total volume of blood in the body is pumped from the heart. This also implies that the *average* drop of blood takes about one minute to circulate through the body.

The Heart Sounds

In a quiet room, if you place your ear against the chest of another person, you will hear sounds emanating from the heart; these are called the **heart sounds**. (The heart sounds may be heard even more effectively using an instrument known as a **stethoscope**.) Each cardiac cycle produces two heart sounds; the first sound is called the **"lub"** sound and the second sound is called the **"dupp"** sound. These names were chosen because the words sound very much like the heart sounds themselves. The sounds are produced by the closures of the four heart valves. The lub sound is produced by the closing of the tricuspid and bicuspid (i.e., mitral) valves and occurs at the onset of ventricular systole. The dupp sound is produced by the closing of the pulmonary and aortic semilunar valves and occurs at the end of ventricular systole. It is important to note that neither sound is produced by the "beating" action of the heart *per se* (i.e., you are not hearing the contraction of muscle). Defects in the actions of the heart valves (e.g., leakage) are usually accompanied by abnormal heart sounds called **murmurs**.

7-4

BLOOD PRESSURE

Units of Hydrostatic Pressure

As the blood travels through the body it exerts pressure against the walls of the vessels through which it is flowing. The pressure is called **blood pressure**. Blood pressure is a special form of a more general pressure called **hydrostatic pressure**, which is the pressure exerted by *any* fluid on the walls of the channels or vessels through which the fluid travels.

How pressure changes as the fluid travels forward can be illustrated by considering water from an open faucet traveling through a garden hose that is lying on the ground. If tiny holes are made in the hose along its length, water will not only come out of the end of the hose, but will also spray out of these holes. If all of the holes face upwards, the water will shoot upward to a height that is proportional to the hydrostatic pressure in the hose at the position of the tiny hole. The closer the hole is to the faucet, the greater will be the height reached by the spray.

If narrow glass tubes (columns) are inserted into the tiny openings in the hose (so that the water cannot spray out onto the ground (fig. 7.5) something different is observed. At each opening, the water rises up in the glass tube until a specific height is reached. At that height, the weight of the column of water inside the glass tube is great enough to resist the upward pressure from below. For holes that are further and further away from the faucet, the height reached by the rising water column is proportionately lower.

The amount of hydrostatic pressure present at each point along the hose can be quantitated by measuring the height of the column of water supported in each vertical column (i.e., h_1, h_2, h_3, etc.). For example, if the height of water attained in the second tube from the faucet (i.e., h_2) is 18 inches, then the hydrostatic pressure at that point along the hose would be stated as "18 inches of water." If the height attained in the fifth tube is seven inches, then the hydrostatic pressure at that point along the hose would be stated as "7 inches of water."

Usually, hydrostatic pressure values are *not* expressed in "inches of water." Instead, the liquid *mercury* (atomic symbol *Hg*) is used. This is because mercury is much heavier (denser) than water, and so the heights of the columns of mercury sustained by hydrostatic pressure are considerably lower (making measurements more convenient). Since mercury is 13.6 times as dense as water, a

Figure 7.5
Relationship between the distance from the source of hydrostatic pressure and the magnitude of the pressure at that distance. The hydrostatic pressure is equal to the height (h_1, h_2, etc.) that the water reaches in each narrow tube.

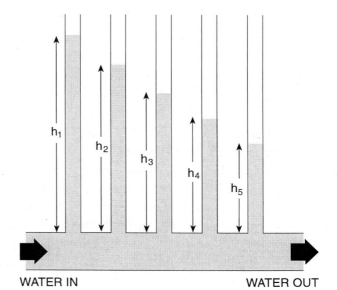

WATER IN   WATER OUT

169

hydrostatic pressure of 18 inches of water is the same as a pressure of 1.32 inches of mercury (i.e., 18/13.6). This pressure would be written as *1.32 in. Hg*. (During a television weather report, atmospheric pressure is usually given as about 30 inches of mercury. What this means is that the pressure of the atmosphere pushing down on the surface of the earth is equivalent to that of a column of mercury that is 30 inches high.)

Physiologists employ the metric system, and in the case of pressure measurements, millimeters (i.e., mm) are used instead of inches. Since one inch equals 25.4 mm, a hydrostatic pressure of *1.32 in. Hg* is the same as *33.5 mm Hg* (i.e., 1.32 x 25.4).

Measuring Blood Pressure

In a manner similar to the garden hose analogy given above, the pressure exerted by blood on the walls of the blood vessels progressively diminishes the further the blood moves from the heart (i.e., the blood vessels are analogous to the hose, and the heart is analogous to the faucet). The decrease in blood pressure that occurs between the aorta and the superior and inferior vena cava is the result of the resistance presented by the intervening blood vessels. This resistance, called the **peripheral resistance**, is due to friction *within* the blood itself and *between* the blood and vessel walls. The magnitude of the

Figure 7.6
Changes in blood pressure and blood velocity as blood circulates from the left ventricle, through the systemic circulation, to the right atrium. The pressure curve is for a person who is lying down; if standing up, gravity affects the venous pressure, lowering the pressure in veins above the heart and raising the pressure in veins below the heart.

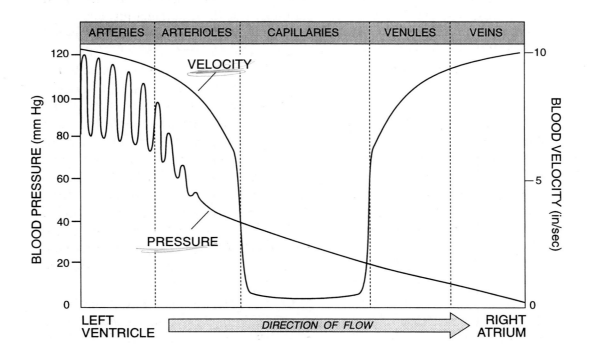

changes in blood pressure that take place as the blood travels through the systemic circulation is shown in figure 7.6.

Although there is a continuous fall in pressure from the aorta to the veins, the fall is not uniform (whereas in a garden hose, the fall in hydrostatic pressure would be uniform with increasing distance from the faucet). In a person with normal blood pressure, blood expelled from the left ventricle into the aortic arch raises the pressure in that artery to about 120 mm Hg at the peak of ventricular systole; this pressure is called the **systolic blood pressure**. Then, as the left ventricle relaxes and fills with blood, the pressure in the aortic arch falls to about 80 mm Hg; this pressure is called the **diastolic blood pressure**. Thus, the pressure in the aortic arch fluctuates between 120 and 80 mm Hg, depending on whether the left ventricle is in systole or diastole (fig. 7.6).

Although blood is pumped into the aortic arch in cycles, it moves forward continuously (not intermittently) in the blood vessels. In the arteries, the flow takes the form of pulsations. Continuous flow is made possible because the walls of the aortic arch are elastic. During systole, the arch is distended, but during diastole the arch recoils, thereby continuing to push the blood forward (the blood cannot back up because the aortic semilunar valve is slammed shut).

By the time that the blood is well into the arteriolar branches of the body's arteries, the fluctuations in pressure between ventricular systole and diastole are hardly discernible. As seen in figure 7.6, the blood pressure continues to fall across the capillaries, venules, and veins. In a person who is reclining, blood pressure in the veins near the right atrium is nearly 0 mm Hg. (In a person standing upright, gravity has an effect on venous blood pressure, lowering it several mm Hg in veins above the heart and raising it several mm Hg in veins below the heart.)

Measuring Blood Pressure Using a Sphygmomanometer. Certainly, it is not feasible to measure blood pressure in a person in the same way as hydrostatic pressure was measured in the garden hose example. That is, one cannot insert tall glass tubes into blood vessels and measure the height of the column of blood supported by the pressure in the blood vessel. (This can be done in the laboratory under experimental conditions and is known as the *direct method* of measuring blood pressure.) Instead, during a physical exam, blood pressure is measured *indirectly* using an instrument known as a **sphygmomanometer** (fig. 7.7).

The sphygmomanometer is used to measure the variation in pressure in the **brachial artery** of the upper arm as blood is pumped through this vessel. The procedure takes the following form. The inflatable sleeve or **cuff** of the sphygmomanometer is wrapped around the upper arm of the subject just above the elbow. The bell of a **stethoscope** is placed on the arm just below the sleeve. With no air in the sleeve and no pressure applied to the upper arm, the only sounds heard (if any) will be the smooth flow of blood through the artery. The instrument's **rubber bulb** is then repeatedly squeezed in order to pump air into the cuff. As the cuff inflates, it applies pressure to the upper arm, further and further squeezing the walls of the brachial artery. The amount of pressure that is being applied to the arm can be read on the scale of the **mercury manometer** which is also connected to the air lines leading to the inflatable sleeve.

Typically, air is pumped into the cuff until the mercury level in the manometer reaches about 180 mm (i.e., the pressure in the cuff

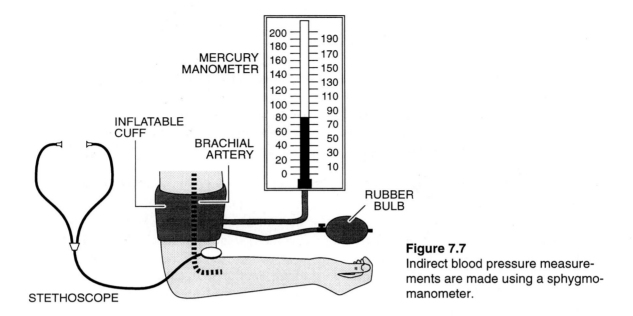

Figure 7.7
Indirect blood pressure measurements are made using a sphygmomanometer.

at that point is 180 mm Hg). At 180 mm Hg, the pressure in the cuff is great enough to squeeze the brachial artery of the upper arm shut, thereby stopping the flow of blood into the lower arm. With the flow of blood in the brachial artery halted, no sounds are heard with the stethoscope (fig. 7.8). While continuing to listen to the brachial artery, air is slowly released from the cuff. As this is done, the level of mercury in the manometer starts to fall. In a normal person, faint thumping sounds, called **Korotkoff sounds**, are heard when the mercury level falls to about 120 mm (above 120 mm Hg, no sounds are heard). The faint thumping sounds that are heard at 120 mm Hg are produced by small quantities of blood being pushed through the partially obstructed brachial artery *at the peak of ventricular systole* (fig. 7.8). The pressure at which the first faint sounds are heard is called the **systolic pressure**. As more air leaves the cuff and the pressure falls further, the thumping sounds get more intense. This is because with each ventricular

systole, the volume of blood getting through the partially blocked brachial artery increases. Eventually, the pressure in the cuff reaches a low enough value that blood again flows smoothly through the brachial artery (fig. 7.8). The pressure at which the loud thumping sounds give way to the quieter sounds of even flow is called the **diastolic pressure**. In a normal person, this pressure is about 80 mm Hg.

7-5

Usually, the systolic and diastolic pressure values are represented as a ratio. Thus, in the above example, the blood pressure would be represented as *120/80*. The numerical difference between the systolic pressure and the diastolic pressure is called the **pulse pressure**. In a normal person, the pulse pressure is 40 mm Hg (i.e., 120 mm Hg minus 80 mm Hg). The pulse pressure represents the pressure variation to which the major arteries are subjected during the cardiac cycle.

The sphygmomanometer makes it possible

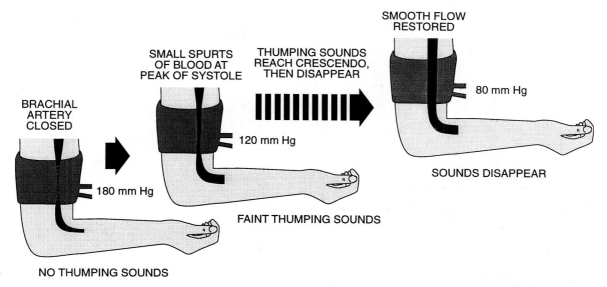

Figure 7.8
The appearance and disappearance of the Korotkoff sounds indicate the systolic and diastolic blood pressure values.

to obtain an indirect measurement of the fluctuation in blood pressure in a major artery. In stricter terms, the systolic and diastolic blood pressures are defined as follows. The systolic pressure is *the maximum pressure that the blood exerts on the walls of the blood vessel through which the blood flows*, whereas the diastolic blood pressure is *the minimum pressure exerted by the blood on the walls of the blood vessel through which the blood flows*.

Venous Return

As seen in figure 7.6, the return of systemic blood to the heart through the body's major veins occurs under very little pressure. Indeed, were it not for several accessory mechanisms, the return of blood to the right atrium would be quite inefficient. Venous return is assisted by the presence of *valves*

that are positioned every few inches along the lengths of the veins. These valves open in one direction only—namely, the direction that leads the blood back to the right atrium. As a result, blood cannot "back-up" in the veins (even in the leg veins where gravity pulls the blood downward, away from the right atrium).

A second important mechanism that assists in the return of blood to the right atrium is the **muscle pump**. The term "muscle pump" refers to the effect that contraction of any of the musculature in the body has on the flow of blood in the veins. The effect of this muscle contraction is to "pump" blood toward the heart. The mechanism is illustrated in figure 7.9 and may be explained as follows. Many of the body's veins course between muscles, and when this musculature contracts, it squeezes the veins. Compression of a short length of a vein forces blood from that length. Although blood is forced in both the

173

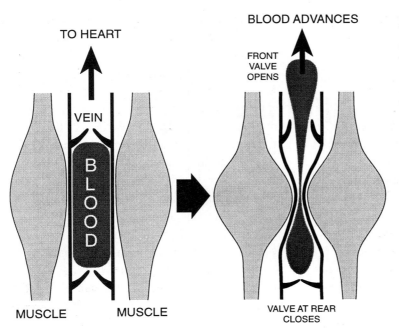

TO HEART

VEIN

B
L
O
O
D

MUSCLE   MUSCLE

BLOOD ADVANCES

FRONT
VALVE
OPENS

VALVE AT REAR
CLOSES

Figure 7.9
The muscle pump. Contraction of the body's muscles squeezes the veins, thereby displacing the blood. The blood in the veins cannot back up because the valves to the rear are forced shut. Instead, the blood is pushed forward (toward the right atrium).

forward (i.e., *toward* the heart) and backward (i.e., *away from* the heart) directions, the valves of the veins prevent any net backward flow of the blood. Consequently, the blood can only be displaced in the direction that leads to the right atrium.

So important is the influence of the muscle pump on venous return that in a person who stands motionless, blood has great difficulty returning to the right atrium. Pooling of blood in the lower half of the body due to gravity so deprives the brain of blood (and oxygen) that the person may loose consciousness and collapse.

Velocity of the Blood

In addition to describing blood pressure changes during circulation, figure 7.6 shows how the *velocity* (i.e., speed) of the blood changes. Blood pumped into the aortic arch from the left ventricle travels through this

major artery at about 10 inches per second. However, as the blood is distributed into the smaller arteries (and then into the arterioles), the velocity of the blood progressively decreases. By the time the blood reaches and flows in the body's capillary networks, its velocity is reduced to about 1/100th of an inch per second. In the venous half of the circulation, there is a progressive increase in blood velocity as the blood passes from the capillaries into the venules and from the venules into the veins.

Changes in blood velocity are analogous to changes in the velocity of any liquid flowing through a closed network of large and small tubes and may be explained as follows (see figure 7.10). When a liquid flowing at a given velocity in one tube enters a single, narrower tube, the flow velocity increases (fig. 7.10, *top*). Indeed, the respective rates of flow are *inversely* related to the *cross-sectional areas* of the two tubes. Since a tube's cross-sectional area is a function of the square of the tube's radius, the rate of flow of a liquid increases four-fold when the liquid passes from one tube into another tube that has half the diameter.

In the case of the circulatory system, however, large arteries branch into a *number* of narrower arteries, and each of these arteries branches into a *number* of still narrower arterioles. Unlike the conditions described in the previous paragraph, in the circulatory system a single large tube gives rise to *many* smaller ones (fig. 7.10, *bottom*). To understand how the velocity of the blood changes as it passes from a vessel into the vessel's many branches, we must consider the *sum of the cross-sectional areas of all of the branches*. In the body, the sum of the cross-sectional areas of the branches of a vessel is a value that is greater than the cross-sectional area of the original vessel. Therefore, as blood flows from a major artery into the artery's branches, the blood *slows down*. As the blood flows from an artery into the many arterioles formed by that artery, the blood again slows down. And, as blood flows from

an arteriole into the many capillaries that the arteriole forms, the blood's velocity slows still more.

When blood is drained from a capillary network by a venule, the blood's velocity increases again. This is because the cross-sectional area of the venule is less than the sum of the cross-sectional areas of the capillaries that merge to form that venule. The blood velocity again increases as blood passes from a number of venules into the vein that drains them. Finally, as blood enters the superior and inferior vena cava from the body's smaller veins, it again speeds up, with the result that blood enters the right side of the heart at nearly the same velocity that it left the left side of the heart. | 7-6 |

THE SINOATRIAL AND ATRIOVENTRICULAR NODES

As noted in chapter 4, cardiac muscle is **myogenic**. That is, heart muscle tissue does not rely on the nervous system for stimulatory impulses. Instead, the electro-chemical events that cause the heart muscle to contract arise within the heart itself. When a small piece of heart muscle tissue is carefully removed from the wall of a heart and is pushed through a steel sieve that separates the tissue into individual cardiac cells, each cell can be seen to go through a rhythmic cycle of contraction and relaxation Interestingly, when a number of cells are pushed together so that they touch one another, the entire group takes on the rhythm of the fastest cycling individual cell.

In the intact body, the pace of the heart is set by a small mass of tissue buried in the wall of the right atrium near the superior

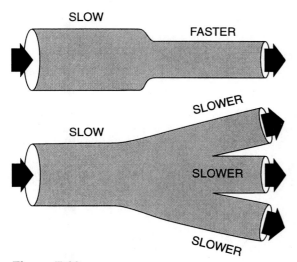

Figure 7.10
Changes in the velocity of the blood .

vena cava's entry point. This small mass of specialized cardiac tissue is called the **sinoatrial node** (usually abbreviated **SAN** and also called the heart's **pacemaker**; figure 7.11). Large numbers of postganglionic fibers of the sympathetic and parasympathetic divisions of the autonomic nervous system terminate in the SAN and were it not for their influence on the SAN, the SAN would set a resting pace of about 100 beats per minute. However, responding to the release of nor-

epinephrine from the postganglionic sympathetic fibers and acetylcholine from the postganglionic parasympathetic fibers, the resting pace is set to 70 to 75 beats per minute. The dominance of the parasympathetic pathway (which reaches the SAN via the vagus nerve [cranial nerve X]) is referred to as **vagal inhibition**.

Approximately once each second, the SAN initiates a wave of depolarization that spreads across the atria. The spread of depolarization

Figure 7.11

The sinoatrial node (SAN), atrioventricular node (AVN), bundle of His, bundle branches, and Purkinje fibers of the heart. In each cardiac cycle, a wave of depolarization spreads across the atria from the SAN. This wave, which is followed by atrial systole, activates the AVN. The AVN initiates a second wave that reaches the ventricular musculature via the bundle of His, bundle branches, and Purkinje fibers, and is followed by ventricular systole.

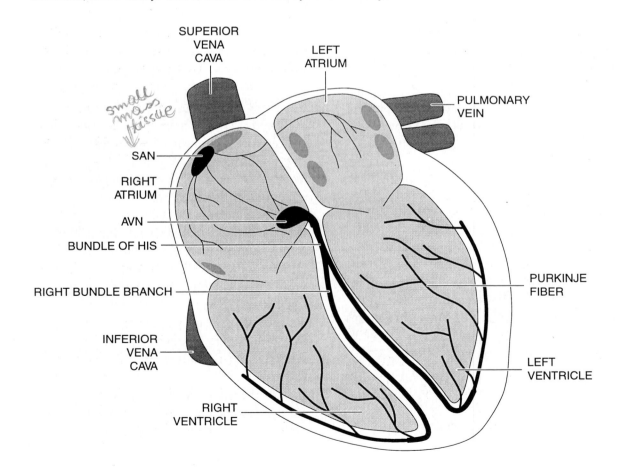

through the atria occurs in two ways: (1) it passes from one cardiac cell to another through the gap junctions that link their adjacent plasma membranes (see chapter 4), and (2) it is conducted by three specialized tracts of cardiac fibers that radiate through the walls of the right and left atria (fig. 7.11). The wave of depolarization spreads across the atria at 800-1,000 cm per second, with the result that both atria are depolarized in a small fraction of a second. The depolarization wave (which is an electrochemical event) is immediately followed by a wave of contraction (which is a mechanical event) spreading through the atrial musculature in the same direction. The depolarization wave does not cross the wall or **septum** that separates the two atria from the two ventricles (the septum contains connective tissue that acts to insulate the atria from the ventricles).

Because the SAN lies in the upper right corner of the wall of the right atrium, the wave of depolarization moves *across* and *downward* as it spreads through the heart tissue. Since the depolarization wave has direction to it, so does the contraction wave that follows. The effect is to drive any blood that remains in the atria downward, through the open tricuspid and bicuspid valves and into the two ventricles. Because the SAN is in the wall of the right atrium, the right atrium *begins* to contract a tiny fraction of a second ahead of the left atrium. This difference is hardly discernible in a heart beating at the normal rate but can be seen in a slow motion picture of the normal heart in action.

At the base of the atria and at the heart's midline is the **atrioventricular node** (AVN; fig. 7.11). This node triggers the contractions of the two ventricles. When atrial depolarization reaches the AVN, the AVN initiates a wave of depolarization that travels down through the atrioventricular septum into the **bundle of His** (pronounced "hiss"). The speed with which the wave of depolarization spreads through the AVN is quite slow (30-50 cm per second) in comparison with atrial depolarization (and ventricular depolarization). This slowness provides an important delay in which atrial systole is completed before the onset of ventricular activation.

| 7-7 |

The bundle of His divides into the left and right **bundle branches** which carry the wave of depolarization down toward the heart's apex along the interventricular septum. At the apex of the heart, the bundle branches give rise to individual **Purkinje fibers** that radiate upward along the inside walls of the ventricles, terminating in the various regions of the ventricular musculature. When the wave of depolarization carried by the bundle of His, bundle branches, and Purkinje fibers reaches the ends of the Purkinje fibers, the membranes of adjacent cardiac muscle cells are depolarized and the cells contract. Depolarization of the bundle of His, bundle branches and Purkinje fibers proceeds at about 5,000 cm per second. Because of this very fast conduction speed, the ventricular musculature is activated quickly.

| 7-8 |

Just as in the case of the atria, contraction of the ventricles has direction. The depolarization wave first reaches the ventricular muscle cells near the heart's apex because the Purkinje fibers terminating in this region are the shortest (i.e., the depolarization wave travels the length of a short fiber in less time than it travels the length of a long fiber). The last ventricular cells to be depolarized are those near the atrioventricular septum, because this area is reached by the longest Purkinje fibers. Thus, contraction of the ventricles begins at the apex of the heart and spreads upward toward the atria, pushing the blood ahead of it. Blood cannot be regurgitated into the two atria because the tricuspid and bicuspid valves are kept closed. Therefore, the blood passes into the pulmonary artery and aortic arch, pushing the pulmonary and aortic semilunar valves open as the blood advances.

As noted earlier, the AVN (and the bundle of His) lie at the heart's midline; additionally, there is symmetry in the organization of the left and right bundle branches and Purkinje fibers of the two ventricles. Consequently the contractions of each ventricle begin (and end) at the same time. As already noted, this is not so in the case of the atria. Because the SAN is located in the wall of the right atrium, right atrial systole begins (and ends) slightly ahead of left atrial systole.

THE ELECTROCARDIOGRAPH

The changes in electrical activity of the heart that precede and follow atrial and ventricular systole are conducted through other tissues of the body to the body's surface and may be detected and recorded using an instrument known as an **electrocardiograph**. The recording that is obtained is called an **electrocardiogram** (usually abbreviated **ECG** or **EKG**).

An ECG of the heart's activity is obtained by attaching the sensing electrodes of the electrocardiograph to the skin of the arms, legs, and/or chest. The so-called "standard leads" are depicted in figure 7.12 and involve electrode attachment to the wrists and the left leg. A permanent ECG recording may be obtained using graph paper, or the tracing may be displayed transiently on an

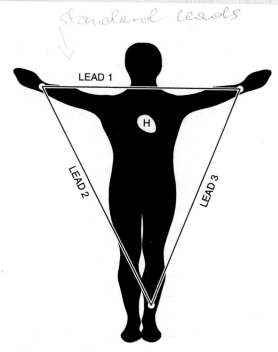

standard leads

Figure 7.12
Placement of the standard electrode leads when obtaining an electrocardiogram. Lead 1 measures the electrical changes detected at the left and right wrists; lead 2 measures the changes at the right wrist and left leg; and lead 3 measures the changes at the left wrist and left leg. ("H" shows the position of the heart.)

oscilloscope or television monitor. In either case, what is seen is the relationship between elapsing time (usually measured in seconds) and the magnitude of the electrical changes taking place in the heart tissues (usually measured in millivolts [i.e., thousandths of a volt]). A typical tracing for a normal heart is shown in figure 7.13.

The electrocardiogram shows three major "waves" (rises and falls) for each cycle of heart activity. The first wave is the **P-wave**. This wave is a measure of the electrical changes in the heart that accompany the *depolarization* of the sinoatrial node and the atrial musculature; the trailing edge of the P-wave also includes electrical changes that

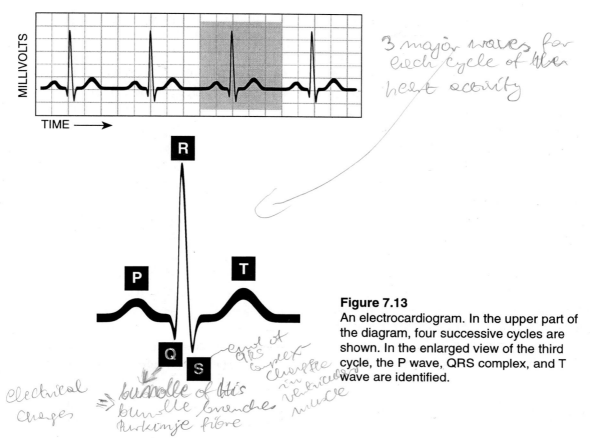

3 major waves for each cycle of the heart activity

Figure 7.13
An electrocardiogram. In the upper part of the diagram, four successive cycles are shown. In the enlarged view of the third cycle, the P wave, QRS complex, and T wave are identified.

electrical changes → *bundle of His bundle branches Purkinje fibre*

end of QRS complex Changes in ventricular muscle

occur in the AVN. It is important to remember that electrical depolarization *precedes* the contractions of the two atria (i.e., the P-wave is not a recording of atrial systole).

The P-wave is followed by the **QRS-complex**, which really is comprised of one major wave (R) and two much smaller ones (Q and S). Wave Q is a small downward deflection of the electrocardiogram at the beginning of the QRS-complex that corresponds to electrical changes occurring in the bundle of His, bundle branches, and Purkinje fiber. Wave R represents the large electrical change associated with the depolarization of the extensive ventricular musculature (especially the musculature of the left ventricle). Wave S reflects the electrical changes occurring in the ventricular muscle at the end of

the QRS-complex. Notice how much larger the QRS-complex is than the P-wave. The QRS-complex is greater because so much more tissue is undergoing depolarization. The QRS-complex is followed by ventricular systole.

The third (and last) wave of the electrocardiogram is the **T-wave**. This wave represents the *repolarization* of the ventricular tissues. To be sure, the atrial tissues are also repolarized before the next cardiac cycle begins; however, no distinct wave is produced during atrial repolarization. This is because the repolarization of the atria takes place at the same time as the depolarization of the ventricles. Since the depolarization of the ventricles represents such a dramatic electrical change, it

7-9

masks the effects of atrial repolarization.

It is worth repeating that the ECG tracing is a measure of electrical changes in the heart tissues and not mechanical changes. That is, the EKG is not a measure of the timing or force of heart muscle contraction. Rather, it is a measure of the electrical events in the *heart that precede muscle contraction.*

CONTROL OF HEART ACTIVITY

Although heart tissue is myogenic, the rate and force with which the heart beats can be altered by actions of the nervous system. Regulation of the heart is effected through sympathetic and parasympathetic pathways that link the brain and spinal cord with the SAN (and AVN). Increases in the flow of nerve impulses to the heart over sympathetic pathways serve to increase heart rate; in contrast, increases in the flow of nerve impulses to the heart over parasympathetic pathways serve to decrease heart rate. Usually, increased activity of one pathway is accompanied by decreased activity of the other pathway.

The parasympathetic pathway begins in a center in the medulla of the brain stem called the **cardioinhibitory center** (CIC). Motor impulses are conducted from the CIC to the heart over preganglionic fibers that are a part of the Vagus nerve (i.e., cranial nerve X). In the wall of the right atrium is the terminal ganglion in which the preganglionic fibers synapse with short postganglionic fibers. The postganglionic fibers terminate in the SAN.

The center for the sympathetic pathway also resides in the medulla and is called the **cardioaccelerator center** (CAC). Impulses originating in this center are conducted down the spinal cord, where they are transmitted to preganglionic fibers exiting the spinal cord at the level of the heart. In a ganglion that is part of the sympathetic chain, the impulses are transmitted to postganglionic fibers that conduct these signals to the SAN. The effect of this pathway is to increase the heart rate.

Under resting conditions, the heart rate is determined primarily by the inhibitory influence of the parasympathetic pathway (i.e., "vagal inhibition"). Further slowing of the heart rate is brought about by an increase in parasympathetic discharge and a decrease in sympathetic discharge. The heart rate is increased by increasing the sympathetic impulses and decreasing the parasympathetic impulses.

The respective activities of the CIC and CAC are determined by a number of factors including (1) the nature of impulses reaching these centers from higher regions of the brain (e.g., the cerebrum), (2) impulses reaching these centers from interoceptors in the aorta and carotid arteries, and (3) the chemical composition of the blood flowing through the medulla itself.

The effects of the cerebrum on heart rate should already be quite familiar to you. Watching a frightening movie or the exciting finish to a sports event can cause an increase in heart rate and an increase in blood pressure. These effects are due to the influence of higher brain centers of the cerebrum upon the cardiovascular centers of the brain stem.

Changes in heart rate and blood pressure can also be induced reflexively through the actions of interoceptors located in the walls of blood vessels near the heart. As shown in figure 7.14, blood entering the aortic arch from the left ventricle is apportioned among a number of arterial branches. The first

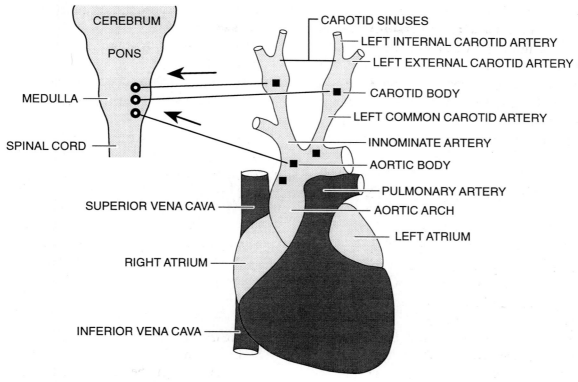

Figure 7.14
Major arteries and veins near the heart. Some of these vessels contain interoceptors that sense blood pressure and the levels of oxygen and carbon dioxide dissolved in the blood. Sensory nerve impulses initiated by these interoceptors serve to reflexively regulate heart rate and blood pressure.

branches (not seen in figure 7.14) are the two **coronary arteries**, which exit the aortic arch just beyond the aortic semilunar valve. The coronary arteries carry blood into the walls of the heart. (The heart itself is, therefore, the first organ whose tissues receive fresh, oxygen-rich systemic blood). The second branch of the aorta is the **innominate artery** (also known as the **brachiocephalic artery**). This is a short artery that soon divides into two branches. One of these branches is the **right subclavian artery**, which directs blood to the chest and shoulder regions on the right side of the body and also into the

right arm. The other branch is the **right common carotid artery**, which directs blood up through the neck and into the head. The third branch of the aortic arch is the **left common carotid artery**. (Note that unlike its right counterpart, the left common carotid artery branches directly from the aortic arch; there is no innominate [or brachiocephalic] artery on the left side of the body.) Like the right common carotid artery, the left common carotid artery courses upward through the neck. The fourth branch of the aortic arch is the **left subclavian artery** (not seen in figure 7.14). Unlike its right counterpart, the left subclavian artery branches directly from the aortic arch. The left subclavian artery directs blood to the chest and shoulder regions on the left side of the body and also into the left arm.

Especially important insofar as the regulation of heart rate and blood pressure are concerned are the left and right common carotid arteries of the neck and their respective external and internal branches. Each common carotid artery gives rise to two branches. One of these branches, the **external carotid artery**, carries blood upward from the neck into the superficial regions of the head (facial muscles, scalp, etc.), whereas the other branch, the **internal carotid artery**, delivers blood to the deeper tissues of the neck and head, including the brain. At the branch points, there is a dilation of each internal carotid artery; these dilations are the **carotid sinuses** (fig. 7.14). Among the cells that make up each carotid sinus are *pressure receptors* that respond to changing blood pressure. Just below the carotid sinuses are the **carotid bodies**; these are clusters of *chemical receptors* that sense the concentrations of oxygen and carbon dioxide dissolved in the blood. Similar pressure and chemical receptors are located in the walls of the aortic arch (the **aortic bodies**).

| 7-10 |

The pressure receptors of the carotid sinuses and aortic arch are associated with sensory nerve fibers that carry nerve impulses to the cardiovascular centers in the medulla. Suppose that something you see or hear causes a sudden increase in heart rate. When this occurs, there is a parallel increase in blood pressure in the aortic arch and carotid arteries. This, in turn, causes the frequency of discharge from the receptors to increase. The increased frequency of sensory impulses reaching the medulla reflexively causes a widespread dilation (relaxation) of the walls of arterioles, bringing about an immediate reduction in blood pressure. Usually, this is quickly followed by a decrease in sympathetic and an increase in parasympathetic impulses sent to the SAN, thereby decreasing the heart rate. A fall in blood pressure in the aortic arch and carotid arteries has the opposite effects.

The chemical receptors of the carotid bodies sense the levels of oxygen and carbon dioxide dissolved in the blood. These receptors are associated with sensory nerve fibers that carry a continuous flow of impulses to the medulla. When the oxygen level falls and/or the carbon dioxide level rises, there is an increase in the frequency of these impulses and this triggers an increase in heart rate so that larger amounts of blood are sent through the pulmonary circulation. An increase in the oxygen content of the blood (or a decrease in the blood's carbon dioxide) has the opposite effect.

The amounts of oxygen and carbon dioxide dissolved in the blood that flows through the medulla also act to influence heart rate. The quantities of these two gases in the blood vary inversely; that is, a decrease in blood oxygen (usually due to increased body activity) is accompanied by an increase in carbon dioxide, and vice versa. The CAC and CIC of the medulla sense changes in the levels of these dissolved gases and trigger the necessary changes in heart rate.

It is interesting that the cardiovascular centers of the brain are much more sensitive to carbon dioxide than they are to oxygen. That is to say, a much earlier and rapid response is triggered by a rise in the level of carbon dioxide in the blood than is unleashed by a fall in oxygen. (This is why during cardiopulmonary resuscitation [i.e., CPR], you are asked to blow your own "used" air into the mouth of the victim.)

Finally, changes in heart rate can also be induced chemically through substances that act directly on the SAN. For example, epinephrine (a hormone secreted by the adrenal glands) stimulates the SAN, causing it to discharge with greater frequency. The result is an increase in heart rate.

THE LYMPHATIC CIRCULATION

As noted at the beginning of this chapter, in addition to blood, a second fluid circulates in the body; this fluid is **lymph** and the vessels through which it circulates comprise the **lymphatic circulation**.

Unlike the blood, which travels in a *circular* route beginning and ending at the heart, the lymph originates in various tissues of the body and travels *toward* the heart only. The process that gives rise to lymph is illustrated in figure 7.15. Blood entering the capillary beds of a tissue is the source of nutrients, raw materials, oxygen, and other life-sustaining substances. Quantities of these materials, together with water, are filtered from the bloodstream at the arterial ends of the capillaries and enter the **tissue space** that surrounds the tissue cells The fluid that fills this

Figure 7.15
Nutrients, raw materials, and oxygen pass from the bloodstream into the tissue space as blood circulates through the body's capillary beds. Much of the waste produced by tissue metabolism and many of the tissue's secretions enter the bloodstream near the capillary bed's venous end and are carried back to the heart. Additional waste and other tissue secretions pass from the tissue space into blind-ended lymphatic vessels, thereby forming lymph. The lymph slowly progresses toward, and eventually enters, the blood of major veins near the heart. (Arrows indicate the direction of flow.)

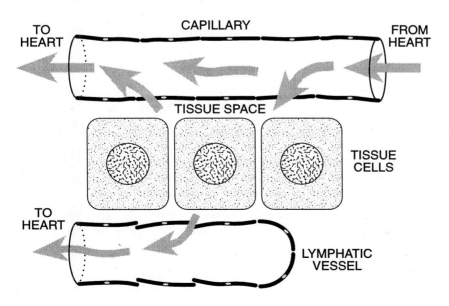

space and bathes the tissue cells is called **tissue fluid**. No red blood cells are filtered from the blood, so that tissue fluid is colorless. Cells draw upon the tissue fluid for their metabolic needs and also empty their wastes and secretions into this fluid. (The forces that cause materials to leave the bloodstream and enter the tissue space at the arterial end of a capillary bed, and to enter the bloodstream at the capillary bed's venous end, are considered in chapter 8).

Much of the material emptied into the tissue spaces from the tissue cells enters the blood near the venous ends of the capillaries. However, some of the tissue fluid enters more porous vessels that arise blindly within the tissue; these blind-ended vessels are the **lymphatics**. Once it is within the lymphatic vessels, the fluid is now called lymph.

Lymph is drained from a tissue by successively larger lymphatics that ultimately convey and empty the lymph into specific veins. The unidirectional flow of lymph within the lymphatics, which is maintained despite the absence of a pump, relies on the contractions of visceral and skeletal muscles. These contractions squeeze upon the lymphatics that course through the tissue.

The lymphatics contain a rich supply of valves that open only in the direction that leads to the heart.

All lymphatics ultimately converge upon two major lymphatic vessels that empty the lymph into the blood of the **subclavian veins** (fig. 7.16). These vessels are the **right lymphatic duct** and the **thoracic duct** (the latter is also known as the left lymphatic duct). Of the two lymphatic ducts, it is the thoracic duct that is responsible for carrying by far the largest amount of lymph into the bloodstream (about 75%). (This is because many more lymphatic vessels converge upon the thoracic duct than converge upon the right lymphatic duct). The areas of the body served by the right lymphatic duct and the thoracic duct are shown in figure 7.17. Once the lymph enters the subclavian veins, it is rapidly mixed with the blood. The subclavian veins merge with the **internal jugular veins** to form the left and right **brachiocephalic veins**; the latter then merge to form the **superior vena cava**, which empties the mixture of blood and lymph into the heart's right atrium.

It is important to keep in perspective the relative amounts of lymph and blood being carried to the heart through the body's veins

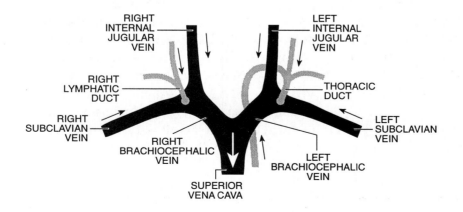

RIGHT INTERNAL JUGULAR VEIN

LEFT INTERNAL JUGULAR VEIN

RIGHT LYMPHATIC DUCT

THORACIC DUCT

RIGHT SUBCLAVIAN VEIN

LEFT SUBCLAVIAN VEIN

RIGHT BRACHIOCEPHALIC VEIN

LEFT BRACHIOCEPHALIC VEIN

SUPERIOR VENA CAVA

Figure 7.16
Lymph produced throughout the body is eventually collected by the right lymphatic duct and the thoracic duct. These two major lymphatic vessels empty their lymph into the right and left subclavian veins.

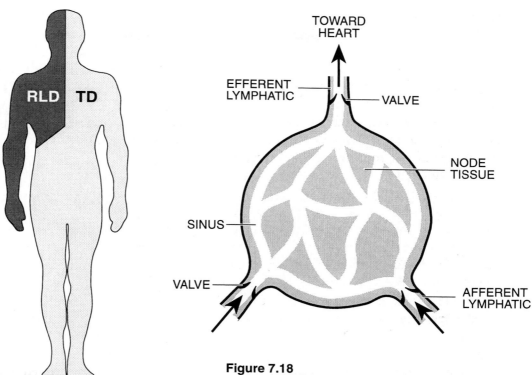

Figure 7.17
The body areas drained by the right lymphatic duct (RLD, dark grey) and the thoracic duct (TD, light grey). The right lymphatic duct serves the right side of the head and chest and the right arm; the thoracic duct serves the remainder of the body.

Figure 7.18
Organization of a lymph node. Lymph flows into the node through one or more afferent lymphatic vessels and leaves the node through an efferent lymphatic vessel. The vessels leading into and out of the node contain valves that ensure one-way flow through the node (shown by arrows). Within the node, the lymph flows through narrow sinuses and directly contacts the nodal tissue cells.

and lymphatic vessels. In an average adult at rest, approximately five liters of blood flows into the right atrium through the superior and inferior vena cava each minute. This is to be compared with about two liters of lymph entering the subclavian veins *in 24 hours* (which is less than 0.03% of the volume of blood).

The lymphatic circulation is characterized by large numbers of **lymph nodes** (fig. 7.18)

that interrupt the flow of lymph. One or more lymphatic vessels (called **afferent lymphatics**) may enter a single node, but the node is usually drained by a single **efferent lymphatic**. The nodes are large masses of lymphoid tissue that are perforated by **sinuses**. The lymph is filtered as it flows through these sinuses, foreign matter and particulate debris being removed and degraded by the nodal tissue.

SELF TEST[*]

True/False Questions

1. Arteries are defined as large blood vessels that carry oxygen-rich blood.

2. The circulatory system acts to distribute heat through the body so that all of the body's parts are at the same (or nearly the same) temperature.

3. The lungs receive blood from both the pulmonary circulation and the systemic circulation.

4. The pulmonary semilunar valve is on the same side of the heart as the tricuspid valve.

5. The oxygen content of the pulmonary arteries is higher than the oxygen content of the pulmonary veins.

6. Systole of the ventricles ejects 50-60% of the end-diastolic volume.

7. Movements of the legs and arms help to move blood through the veins and back into the right side of the heart.

8. For someone who is resting (not exercising), the stroke volume of the heart would be about 500 c.c.

9. Blood entering the heart's right atrium is a recent mixture of lymph and blood.

10. Like blood, lymph is driven forward in the lymphatics by the actions of the heart.

Multiple Choice Questions

1. At rest, a volume of blood equal to that of the entire body is pumped from the left ventricle (A) every 10 seconds, (B) every one minute, (C) every 10 minutes, (D) every hour, (E) every 24 hours.

2. The papillary muscles of the right ventricle (A) pull the bicuspid valve open, (B) pull the tricuspid valve open, (C) keep the bicuspid valve closed, (D) keep the tricuspid valve closed, (E) push blood into the aortic arch.

3. Which one of the following properly lists the sequence of heart structures through which the blood flows? (A) left atrium, left ventricle, right atrium, right ventricle, (B) right atrium, left atrium, right ventricle, left ventricle, (C) right atrium, right ventricle, left atrium, left ventricle, (D) right ventricle, right atrium, left ventricle, left atrium.

4. The heart sound called "lub" marks (A) the start of ventricular systole, (B) the start of ventricular diastole, (C) the start of atrial systole, (D) the end of ventricular systole.

5. The sounds heard when using a sphygmomanometer are known as Korotkoff sounds. The first, faint Korotkoff sounds correspond to (A) the closure of the bicuspid and tricuspid valves, (B) the closure of the aortic and pulmonary semilunar valves, (C) the systolic blood pressure, (D) the diastolic blood pressure, (E) the movements of the stapes against the oval window.

6. Which one of the following is closest to the "normal" blood pressure for a healthy adult? (A) 180/140. (B) 150/100, (C) 120/80, (D) 100/60.

---

[*] *The answers to these test questions are found in Appendix III at the back of the book.*

7. With regard to the passage of blood from an artery into its many arteriole branches, (A) the velocity of the blood and the blood pressure rise, (B) the velocity of the blood and the blood pressure fall, (C) the velocity of the blood falls but the pressure rises, (D) the velocity of the blood rises but the pressure falls.

8. Blood travels most slowly in the (A) arteries, (B) arterioles, (C) capillaries, (D) venules, (E) veins.

9. Which one of the following is also known as the heart's "pacemaker?" (A) the sino- atrial node, (B) the atrioventricular node, (C) the aortic node, (D) the caval node, (E) none of these choices is correct.

10. The QRS complex of an electrocardio- gram represents (A) depolarization of the atria, (B) repolarization of the atria, (C) depolarization of the ventricles, (D) repolarization of the ventricles.

11. Most of the body's lymph returns to the bloodstream through the (A) aortic duct, (B) right lymphatic duct, (C) inferior vena cava, (D) superior vena cava, (E) thoracic duct.

THE BLOOD

In this chapter, we will be concerned with the composition and physiological functions of the blood. Although we will consider each function in some detail, the various functions of the blood may be summarized briefly as:

1. the transport of oxygen, nutrients, and raw materials to the tissues of the body,

2. the removal of carbon dioxide and metabolic wastes from the body's tissues,

3. the transport of hormones and other secretory substances from one tissue (or organ) to another, and

4. the distribution of the body's heat of metabolism.

COMPOSITION OF THE BLOOD

Since, under normal conditions, a sample of blood removed from the body undergoes a transformation from a liquid state to a solid state (i.e., the blood **coagulates**), the only simple way to study the composition of the blood is to collect a sample into a container that has been coated with a chemical substance that prevents coagulation (i.e., an **anticoagulant**). If this is done and the "whole blood" is allowed to remain undisturbed for several hours, the blood will separate into two major phases (fig. 8.1). The lower phase is intensely red in color and is called the **formed element** phase (as we will see later, this phase consists principally of *red blood cells* or *erythrocytes*). The upper phase is a clear, yellow fluid and represents the medium in which the formed elements were previously suspended; this phase is called the **plasma**.

A number of substances can act as anticoagulants. The body's natural anticoagulant is **heparin**; its function is to prevent blood from coagulating during its circulation through the body. Other common anticoagulants include *sodium citrate*, *sodium oxalate*, and *ethylenediaminetetraacetic acid* (usually abbreviated *EDTA*).

In a normal, adult male the formed elements represent about 46% of the total volume of a sample of whole blood; the other 54% is represented by the plasma (see figure 8.1). This ratio (i.e., 46% or 0.46) is called the **hematocrit** and is a convenient, quickly measured, and simple diagnostic indicator of

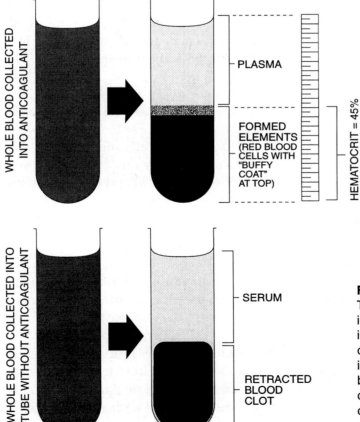

Figure 8.1
Top: When whole blood is collected into anticoagulant, it soon separates into the formed element phase and the overlying plasma. Bottom: If the blood is collected without anticoagulant, the blood clots and retracts, the retracted clot settling to the bottom of the container. The overlying liquid is called serum.

the condition of the blood. For example, if a person's hematocrit were 30% (or 0.30), this would be an indication of the existence of an **anemia**, there being too few (red) blood cells in the blood. In normal, adult females the hematocrit is about 42%.

When fresh, whole blood is collected into a container that lacks anticoagulant, the blood quickly undergoes a transformation from a liquid to a gelatinous solid (fig. 8.1). The solid is called a **blood clot**. If the clot is allowed to stand for several hours, it shrinks or **retracts**, exuding a clear fluid and slowly forming a smaller and more solid mass. Eventually, the retracted clot settles to the bottom of the container, leaving the clear fluid above. The clear fluid exuded from the retracted clot looks very much like plasma but it is called **serum**. There is an important difference between the compositions of blood plasma and blood serum; namely, plasma contains the chemicals needed in order to cause the coagulation of blood, whereas serum lacks these substances because they were consumed during the formation of the clot.

| 8-1 |

THE FORMED ELEMENTS

The formed element phase of a blood sample represents all of the solids that settle to the bottom of the container when whole blood is collected into anticoagulant. These solids include a number of different cells and particles. The predominant constituents of the formed element phase are the **red blood cells** or **erythrocytes**. These contain the red protein **hemoglobin**, which transports oxygen in the blood and also gives the blood its red color. In normal adult males, there are about 5,100,000,000 –5,800,000,000 (that's 5.1 to 5.8 billion) erythrocytes in each c.c. (cubic

centimeter) of whole blood. In normal adult females, there are 4.3 to 5.2 billion red blood cells per c.c. of blood. (It is common to express blood cell counts as cells per mm^3 [where 1 c.c. contains 1,000 mm^3]. Using this standard, the red blood cell counts in adult males and females are 5.1–5.6 and 4.3–5.2 million/mm^3 of blood, respectively.)

Also present among the formed elements but much fewer in number are the white blood cells or **leukocytes**. In normal adults, there are 5,000,000-9,000,000 leukocytes in each c.c. of whole blood (or 5,000-9,000/mm^3). As we will see later, the leukocytes play a variety of roles in the body. Although most leukocytes are larger than erythrocytes, they are considerably less dense. Therefore, being lighter, they settle from whole blood more slowly than the erythrocytes and form a thin layer called the **buffy coat** at the top of the formed element phase (see figure 8.1).

Finally, the formed element phase also includes a large number (i.e., 130,000,000–360,000,000/c.c.) of noncellular particles called **platelets** or **thrombocytes**. These play an important role in the coagulation of blood that occurs following injury. The platelets are much smaller than the red and white blood cells and settle very slowly onto the top of the buffy coat.

THE BLOOD PLASMA

The blood plasma is the fluid in which the formed elements are suspended. However, plasma should not be thought of as an inert liquid playing a passive physiological role. Despite being mainly water, the plasma is rich in dissolved proteins that serve a variety of important physiological roles in the body and is the medium in which such vital substances as vitamins, sugars, amino acids, and hormones are transported from one tissue site to another. We will look more closely at the various functions of the plasma later in the chapter, but at this point let's examine the red blood cells, white blood cells, and platelets in greater detail.

RED BLOOD CELLS (ERYTHROCYTES)

Shape and Number of Red Cells

The red blood cells (erythrocytes) are the most abundant of the formed elements (more than five billion per /c.c. of whole blood) and are also among the simplest and smallest of the cells in the body (fig. 8.2). Red blood cells have no nucleus and also lack most of the other intracellular components that characterize typical human cells (e.g., mitochondria, Golgi bodies, endoplasmic reticulum, ribosomes, etc.). The interior of the cells is packed with molecules of the conjugated protein **hemoglobin**, which serves to transport oxygen from the lungs to the body's tissues. Indeed, the amount of hemoglobin present in red cells is so great that it is necessarily arranged in an orderly manner with the result that it appears almost crystalline (i.e., "para-crystalline"). The para-crystalline organization of hemoglobin in red blood cells is believed to contribute to the peculiar shape of the cell which is that of a **biconcave disk** (fig. 8.2). The biconcavity creates a pale zone near the center of the cell, which many years ago was mistaken for a nucleus. Because erythrocytes lack a cell nucleus and other components, they are capable of only limited metabolism and are certainly unable to grow and divide. Red blood cells have the rather limited life span

of about 120 days.

Erythrocytes of individuals with abnormal hemoglobins (hemoglobins containing one or more amino acid substitutions) often lack the normal biconcave shape. The most notorious example is the "sickle" or crescent shape of erythrocytes in people with **sickle-cell anemia**. In this genetically determined disease, a single amino acid substitution occurs in the beta globin polypeptide chains of the hemoglobin molecule (see later). Under conditions of oxygen deprivation or shortage, the hemoglobin molecules aggregate to form long fibers and these progressively group together into bundles. The bundles of hemoglobin molecules make the cells less flexible and also deform the cells, causing their sickling.

The oxygen transported by erythrocytes enters and leaves the cells by diffusion, and the biconcave shape of the cell facilitates oxygen movement by increasing the surface

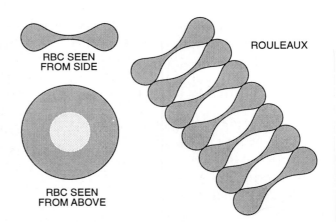

Figure 8.2
Red blood cells lack nuclei and other intracellular components and have the shape of a biconcave disk. This shape lends itself to an orderly stacking of cells, thereby forming rouleaux. A rouleaux may consist of dozens of cells, not just the few that are shown here.

area-to-volume ratio of the cell. The disk shape of erythrocytes also allows the formation of **rouleaux** or long stacks of cells. Rouleaux formation (see figure 8.2) appears to be related to the concentration of certain plasma proteins (see later) and materially increases the blood's viscosity.

Hemoglobin

Blood is red because of the **hemoglobin** that fills each of the red blood cells (each red blood cell contains about 280,000,000 molecules of hemoglobin). Hemoglobin is a *conjugated* protein; that is, it is a protein that contains some nonprotein portions. The function of hemoglobin is to transport oxygen. Therefore, hemoglobin molecules can exist in either of two states: *oxygenated* and *unoxygenated* (fig. 8.3). In its oxygenated state, a hemoglobin molecule carries 4 molecules of oxygen (i.e., 8 atoms of oxygen).

The most common type of human hemoglobin is called hemoglobin **A** (A = "adult") and is usually abbreviated **HbA**. HbA is a relatively small conjugated protein and has a molecular weight of 64,500 daltons (fig. 8.4). The protein portion of the molecule, called **globin**, contains 574 amino acids distributed among four polypeptide chains.

| In the lungs: | $Hb + 4\,O_2 \longrightarrow Hb(O_2)_4$ |
| In other tissues: | $Hb(O_2)_4 \longrightarrow Hb + 4\,O_2$ |

Figure 8.3
The reactions of hemoglobin with oxygen. In the lungs, one molecule of hemoglobin combines with four molecules (8 atoms) of oxygen. In other tissues (e.g., muscles, brain, etc.), hemoglobin releases its oxygen.

The four **globin chains** consist of two pairs: two identical **alpha** (α) chains (141 amino acids each) and two identical **beta** (β) chains (146 amino acids each). Each of the four polypeptide chains of hemoglobin is twisted and folded to form a more compact three-dimensional shape (as shown for an alpha chain in fig. 8.4). The four globin chains neatly fit against one another to form an $\alpha_2\beta_2$ **tetramer** that has an overall shape much like that of a sphere.

The nonprotein portions of each hemoglobin molecule are four **heme groups**, one heme group associated with each of the four globin chains. The atoms of each heme group all lie in one spatial plane and the group as a whole is partially embedded in a fold in the globin chain with which the heme is associated (fig. 8.4). At the center of each heme group is an all-important atom of **iron**. It is the iron atoms of the heme groups with which molecular oxygen (i.e., O_2) combines in the lungs to form oxygenated hemoglobin and from which oxygen is released in other tissues of the body (fig. 8.3).

In adults, HbA (i.e., $\alpha_2\beta_2$) is the most abundant hemoglobin and accounts for 98% of all hemoglobin present. A second adult hemoglobin, called **HbA$_2$**, accounts for the remaining 2%. The difference between HbA and HbA$_2$ is that a pair of **delta** (δ) chains replaces the beta globin chains (i.e., HbA$_2$ is represented as $\alpha_2\delta_2$). Still other kinds of hemoglobin molecules are present in the blood at earlier stages of human development. In the very early stages of embryonic development, a hemoglobin called **Gower-1** appears (Gower-1 is $\zeta_2\epsilon_2$– a pair of **zeta** (ζ) chains and a pair of **epsilon** (ϵ) chains). The next hemoglobin to appear is called **Gower-2** ($\alpha_2\epsilon_2$– a pair of alpha chains and a pair of epsilon chains) and this is quickly followed by the

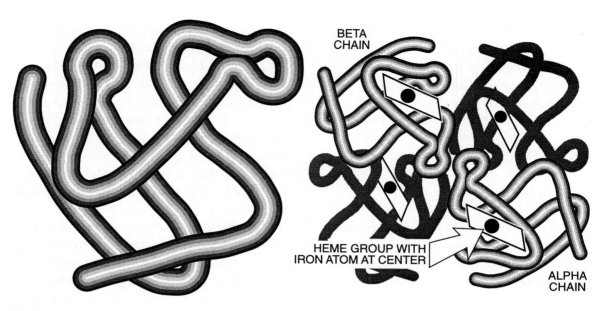

BETA
CHAIN

HEME GROUP WITH
IRON ATOM AT CENTER

ALPHA
CHAIN

Figure 8.4
Molecular structure of human hemoglobin A. Shown on the left is the folded and twisted shape taken by an alpha globin chain. All human globin chains are similarly shaped. The 4-chain tetramer consisting of two alpha and two beta chains is seen on the right. (The two uniformly shaded chains are at the rear.) Also shown are the flat heme groups, each group inserted in a fold in the surface of its associated globin chain). At the center of each heme group is the iron atom (black circle) that binds molecular oxygen.

appearance of **Hb Portland** ($\zeta_2\gamma_2$– a pair of zeta chains and a pair of **gamma** chains). Both Gower-2 and Portland are soon replaced by **HbF** or **fetal hemoglobin** ($\alpha_2\gamma_2$, consisting of a pair of alpha chains and a pair of gamma chains). HbF predominates in the blood through the remainder of fetal development. Beginning soon after the twelfth week of fetal development, HbA appears and progressively increases in relative abundance. Shortly before birth, HbA_2 appears. By about six months after birth, little if any of the HbF can be found in the blood. At this time, about 98% of the hemoglobin is HbA and the remaining 2% is HbA_2, and in a normal person, this ratio persists through adult life. The specialized embryonic and fetal hemoglo-bins have somewhat different affinities for molecular oxygen, properties that no doubt are important if oxygen bound to hemoglobin A (and A_2) in the mother's blood is to be transferred to the hemoglobin in the red blood cells of the developing fetus.

8-2

Erythropoiesis

The production of blood cells begins in the **yolk sac** of the early embryo, then shifts to the **liver** and **spleen**, and in the latter stages of fetal development moves to the **bone marrow**. In adults, most blood cells are produced in the marrow occupying the epiphyses of the body's long bones (e.g., arm and leg bones), the vertebrae, ribs, sternum, and skull. Blood flows into the soft marrow tissue via small arterioles that enter the bone through tiny surface perforations. The vessels terminate within the marrow, with the result that the blood is free to percolate through the marrow tissue. The narrow spaces through which the blood passes are called the **marrow sinuses**. Having traveled through the sinuses, the blood is then collected by small venules that direct the blood back out of the bones. While percolating through the marrow sinuses, new red and white blood cells are added to the blood. The process by which the blood cells are created in the bone marrow is called **hemopoiesis** (or **hematopoiesis**).

The particular development of red blood cells in the bone marrow is called **erythropoiesis**, and it is one of the most striking examples of cell specialization occurring in the body. Erythropoiesis begins with *pluripotent* **stem cells** that give rise in 4-7 days to mature, hemoglobin-filled erythrocytes. While the specialization of the red cell is accompanied by the production of large amounts of hemoglobin, it is also accompanied by the loss of nearly all internal cell structures and many physiological properties. Mature red blood cells lack nuclei, mitochondria, endoplasmic reticulum, Golgi bodies, ribosomes, and most other typical cell organelles.

As noted earlier in the chapter, the mature erythrocyte is a relatively simple cell, bounded at its surface by a plasma membrane and containing internally a highly concentrated, para-crystalline array of hemoglobin molecules that are used for oxygen transport. This highly specialized state is attained during the course of a number of developmental stages which are depicted in figure 8.5. Hemopoiesis begins with pluripotent stem cells capable of producing white blood cells as well as red blood cells. When pluripotent stem cells undergo division, one of the two daughter cells remains undifferentiated (i.e., "uncommitted") and pluripotent, so that depletion of the marrow's pluripotent stem cells does not take place (in figure 8.5 [and figure 8.9], this is what is meant by "self renewal"). If the *other* daughter cell produced by division of a pluripotent stem cell is to give rise to erythrocytes (or certain leukocytes), then that daughter cell is called a **myeloid stem cell** ("myeloid" implies development in the marrow). The alternate possibility is that the cell will give rise to various kinds of **lymphocytes** (lymphocytes are discussed later in this chapter and also in chapter 9); in the latter case, the daughter is called a **lymphoid stem cell** (fig. 8.5).

Division of myeloid stem cells produces either **proerythroblasts** (forerunners of red blood cells) or **proleukoblasts** (forerunners of leukocytes; figure 8.5). In proerythroblasts, there is a modest onset of hemoglobin synthesis.

With very little intervening cell growth, the proerythroblast soon divides into two smaller cells called **erythroblasts**. Once this stage is reached, hemoglobin synthesis becomes the dominant aspect of cell metabolism. Again, little cell growth precedes the division of erythroblasts into **normoblasts**. In normoblasts, hemoglobin synthesis reaches its peak. Shortly before the completion of hemoglobin synthesis, each normoblast begins to loose its intracellular structures, most notably the nucleus, and the cell is transformed into a **reticulocyte**. The term reticulocyte refers to the reticulum (i.e., lacy network) that can be seen in this cell when it

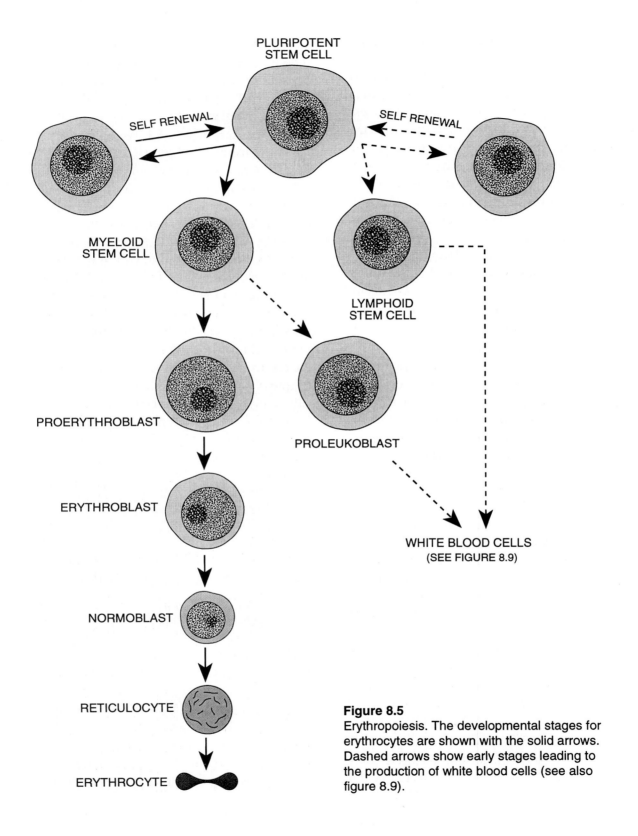

Figure 8.5
Erythropoiesis. The developmental stages for erythrocytes are shown with the solid arrows. Dashed arrows show early stages leading to the production of white blood cells (see also figure 8.9).

is stained for microscopy. The reticulum represents a precipitate of the vestiges of intracellular structures. Reticulocytes continue to synthesize hemoglobin for several hours, but afterwards all protein synthesis comes to a permanent halt. It is at the reticulocyte stage that the maturing red blood cell is released from the bone marrow into the circulating blood. Within a few hours, the reticulocyte loses its reticulum and takes on the characteristic biconcave shape of the mature erythrocyte.

Although all protein synthesis comes to a halt during reticulocyte maturation, a red cell retains a limited metabolic capacity after release from the bone marrow. What limited metabolism is retained by the mature cell serves to sustain it during its 120-day and 700-mile journey through the circulatory system. We will return to the subject of erythropoiesis later in the book in connection with our discussion of hormones, for erythropoiesis is regulated by the secretion of a hormone (called **erythropoietin**) by the kidneys.

8-3

Rate of Red Blood Cell Production

Since in the average, normal adult male there are about 5.4 billion red blood cells in each c.c. of blood, and these cells have an average life span of 120 days, a few simple calculations quickly reveal the rate at which erythrocytes are released from the bone marrow into the circulating blood. About 8% of the body weight is represented by blood; therefore, a person weighing 78 kilograms or 172 lb has 6.24 kilograms or about 6.2 liters (i.e., 6,200 c.c.) of blood in his circulation. If each c.c. of blood contains 5.4 billion red cells,

then altogether there would be 33.5×10^{12} (i.e., 33,500,000,000,000) circulating erythrocytes. Since all of these are replaced during an interval of 120 days or 10.4×10^6 seconds, this means that 3.22×10^6 red blood cells reach full maturity and are released into the bloodstream each second. In other words, in a typical adult the differentiation and maturation of 3 million erythrocytes is completed *each second*.

Obviously, an appreciable amount of the body's metabolic energy, as well as chemical resources, are continuously consumed to support erythropoiesis. This is in stark contrast with other highly differentiated cells, such as those of muscle and nerve, whose proliferation ceases shortly after birth.

Elimination of Old Red Blood Cells

If more than 3,000,000 new red blood cells are released into the circulating blood each second, then it stands to reason that a corresponding number of aged red cells must be lost each second. Old red cells are lost from the circulating blood in two ways: (1) by **intravascular hemolysis**, and (2) **sequestration** by the **reticuloendothelial system**.

Intravascular Hemolysis. *Intravascular hemolysis* means the breakage of a red blood cell while it is circulating in the bloodstream. About 5-10% of all red blood cells lost from the circulation are lost in this way. Although this is a small percentage, it is nonetheless a large number of cells (e.g., 5% of 3,000,000 per second is 150,000 cells per second). When a red blood cell lyses in circulation, its hemoglobin is "spilled" into the blood plasma. Since hemoglobin is a fairly small protein, it is capable of filtration in the kidneys, where, under the acid conditions that exist in that organ, the hemoglobin would form a precipitate. Such filtration and precipitation of free hemoglobin could do serious injury to the kidney tissue. This is prevented in the following way. Circulating in the blood plasma is another protein called **haptoglobin**, which combines with hemoglobin to form a nonfilterable and non-precipitating complex that is later removed from the blood by the reticuloendothelial tissues (see below).

Sequestration by the Reticuloendothelial System.

The reticuloendothelial system (or **RES**) is a collection of tissues that seek out (i.e., "sequester") aged red cells and remove them from the circulating blood. Tissues of the RES are found in the **bone marrow**, **liver**, and **spleen** and function similarly in all

of these organs. The principal mechanism for sequestering aged red cells is shown in figure 8.6.

RES tissue forms **sinuses** (narrow channels) through which the peripheral blood flows. Within these sinuses the blood directly contacts the surrounding tissue cells (i.e., there are no capillary walls separating the blood from the surrounding tissue). Unlike young red cells, which readily pass through the sinuses, aged red cells encountering the sinuses' walls become attached to the surfaces of the cells lining the sinus (figure 8.6, stages 1 and 2). The cells lining the sinuses are **macrophages** (derived from blood monocytes, see later), and having adsorbed an aged red cell, they begin to engulf the cell in a surface pocket (figure 8.6, stage 3). The process by which this engulfing occurs is a common phenomenon in the

Figure 8.6
Sequestration of aged red blood cells by macrophages of the reticuloendothelial system (RES). Numbers 1 through 6 show successive stages of phagocytosis in which the aged red cell is engulfed and digested by the macrophage. Young red blood cells pass uninterrupted through the sinuses of the RES.

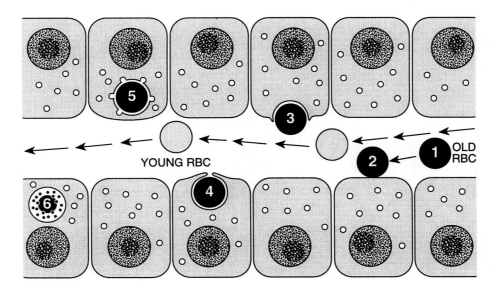

body and is called **phagocytosis** (i.e., "cell eating"). Eventually, the red blood cell is completely enveloped in a small chamber or **vacuole** within the macrophage (figure 8.6, stage 4). Small cytoplasmic vesicles (called **lysosomes**), containing a rich complement of digestive enzymes, fuse with the vacuole, emptying their contents into it (figure 8.6, stage 5). The digestive enzymes carry out the complete degradation of the entrapped red cell (figure 8.6, stage 6).

The body expends considerable effort in the production and destruction of red blood cells. The degradation of red blood cells produces a variety of end-products, many of which are reused or "recycled" by the body (fig. 8.7). Excluding water, the bulk of a red cell is represented by hemoglobin, and we may therefore limit our consideration of red cell destruction to the fate of this material.

The initial stage of hemoglobin breakdown separates the molecule into its protein component (i.e., *globin*) and the iron-containing *heme* (fig. 8.7). The globin is broken down to individual amino acids that are reused by the body for the synthesis of new proteins. Heme is separated into two components: iron and **biliverdin** (a greenish pigment). The iron released from heme is reused by the body. Since most of the body's iron is present in hemoglobin, the reutilization of iron implies that most of it will once again be incorporated into the hemoglobin of newly developing red blood cells. Iron released during the breakdown of hemoglobin is either stored in the liver, spleen, or bone marrow as an iron-rich conjugated protein called **ferritin** or it is transported in the blood (bound to a plasma protein called **transferrin**) to the erythropoietic tissues for

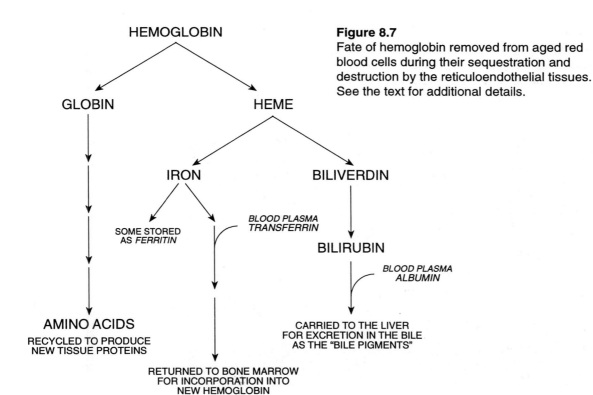

Figure 8.7
Fate of hemoglobin removed from aged red blood cells during their sequestration and destruction by the reticuloendothelial tissues. See the text for additional details.

incorporation into new hemoglobin.

Of all of the breakdown products of hemoglobin, only the biliverdin is destined for excretion from the body. First, most of the biliverdin is converted to **bilirubin** (a reddish pigment). The bilirubin is then coupled to the plasma protein **albumin** and carried from the spleen and bone marrow to the liver. In the liver, biliverdin and bilirubin are added to the bile, which is secreted from the liver into the small intestine (biliverdin and bilirubin are known as the **bile pigments**).

Strangely enough, biliverdin and bilirubin are not the only chemical forms in which the breakdown products of heme finally leave the body. This is because bacteria living in the small intestine take up much of the bile pigments and convert them to **urobilinogen**. Some of the urobilinogen, together with residual bilirubin and biliverdin leave the body with the feces (the body's solid waste). Indeed, these pigments are responsible for the color of the fecal material. However, much of the urobilinogen produced by intestinal bacteria is absorbed into the bloodstream. Eventually, this urobilinogen is removed from the blood by the kidneys, which then excrete the urobilinogen in the urine (where it contributes to the color of the urine).

8-4

WHITE BLOOD CELLS (LEUKOCYTES)

The white blood cells or leukocytes differ from the red blood cells in their abundance, appearance, and physiological functions. In the average adult male, one c.c. of blood contains between five and nine million leukocytes. Leukocytes are subdivided into two major groups: *granular* leukocytes or granulocytes and *agranular* leukocytes or **agranulocytes** (fig. 8.8).

Granulocytes

Granulocytes are white blood cells whose cytoplasm is characterized by the presence of large numbers of small granules. There are three kinds of granulocytes, named according to the properties of the histological stains that are used to visualize and characterize the cells during microscopy. Granulocytes that stain with alkaline (or basic) stains are called **basophils** (i.e., "alkali-attracting"). Granulocytes that stain with neutral stains are called **neutrophils** (i.e., "neutral [stain]-attracting"), and granulocytes that stain with acidic stains are called **acidophils** (i.e., "acid-attracting"). Because the acidic dye *eosin* is frequently used as a stain for acidophils, acidophils are also commonly known as **eosinophils** (i.e., "eosin-attracting").

Among the granulocytes, the neutrophils are the most abundant. Indeed, neutrophils are the most common of all leukocytes, accounting for 55-75 percent of the white blood cells present in normal blood (fig. 8.8). Acidophils are relatively rare and usually account for 2-4 percent of all white cells. Rarest of all are the basophils, which account for 0.5-1 percent of all white cells. All three types of granulocytes are produced in the bone marrow by the division and differentiation of stem cells (a process called **leukopoiesis**, see below). Because the nuclei of granulocytes typically have an unusual and varied shape, granulocytes are also known as **polymorphonuclear leukocytes**.

Figure 8.8 *(opposite)*
The blood's formed elements (red blood cells, white blood cells, and platelets).

	NAME	PERCENT	ORIGIN	FUNCTION
RED BLOOD CELLS (ERYTHROCYTES)	ERYTHROCYTE	100	BONE MARROW	OXYGEN TRANSPORT
WHITE BLOOD CELLS (LEUKOCYTES) — GRANULOCYTES	NEUTROPHIL	55-75	BONE MARROW	PHAGOCYTOSIS
	ACIDOPHIL (EOSINOPHIL)	2-4	BONE MARROW	PHAGOCYTOSIS
	BASOPHIL	0.5-1	BONE MARROW	HISTAMINE RELEASE
AGRANULOCYTES	MONOCYTE	3-8	BONE MARROW	PHAGOCYTOSIS
	LYMPHOCYTE	20-40	BONE MARROW LYMPH NODES THYMUS SPLEEN	IMMUNE RESPONSE
PLATELETS	PLATELETS	100	BONE MARROW	BLOOD COAGULATION

Neutrophils and acidophils are involved in protecting the body against infection. For example, these cells scavenge the blood and tissues, seeking out and phagocytosing bacteria and other microorganisms (neutrophils and acidophils are also referred to as "microphages," a term that implies their capacity for phagocytosing small particles). Neutrophils are usually the first white blood cells to arrive at a site of infection, where they specialize in attacking bacteria. Neutrophils and acidophils die within a few hours after carrying out the phagocytosis and digestion of foreign matter. For this reason, these granulocytes have a variable but rather brief life span (estimated at about 12-24 hours).

In addition to their role as scavengers, acidophils are believed to release enzymes that break up blood clots in the circulation. One of the distinct structural characteristics of the neutrophils and acidophils is the multilobed nature of the cell nucleus (fig. 8.8).

Basophils are believed to play a variety of roles including the release of heparin (the body's natural anticoagulant) and the secretion of histamines during an allergic reaction. In structure and action, the basophils resemble **mast cells** of connective tissue (which also contain histamine and heparin). All three types of granulocytes are produced in the bone marrow (fig. 8.9).

Agranulocytes

The agranulocytes are white blood cells whose cytoplasm lacks distinct granules (at least, granules are not visible when the cells are examined by conventional light microscopy). There are two kinds of agranulocytes: **monocytes** and **lymphocytes** (fig. 8.8). The monocytes, which represent 3-8 percent of all leukocytes are the largest of the white blood cells (fig. 8.8). Like the neutrophils and acidophils, the monocytes act as scavengers of foreign organisms and foreign particles by carrying out phagocytosis. After circulating in the blood for a short time, monocytes migrate into the surrounding body tissues, where they increase in size and are converted to macrophages. These macrophages wander through the tissues, continuing to scavenge foreign particles as well as dying or malfunctioning tissue cells. Some macrophages take up permanent residence in the lungs where they scavenge air-borne particles that enter the delicate lung tissue. Other macrophages take up residence in the liver and spleen where they function as reticuloendothelial tissue (see figure 8.6). Like the granulocytes, monocytes are produced in the bone marrow (fig. 8.9).

The lymphocytes are the second most abundant leukocytes, typically accounting for about 20-40 percent of all white cells. They are smaller than the granulocytes and monocytes and have a large, spherical nucleus with scant cytoplasm (fig. 8.8). All lymphocytes are initially derived from the early fetal bone marrow. However, during fetal development, some of these lymphocytes migrate to and take up residence in the **thymus gland**, where (under the influence of the hormone **thymosin**) they undergo important physiological changes. Many of these lymphocytes subsequently leave the thymus and migrate to (and populate) the body's lymphoid tissues (e.g., **lymph nodes** and **spleen**); some return to the bone marrow.

A consequence of all this is that in adults there are two functionally different types of lymphocytes: **B-lymphocytes** and **T-lymphocytes** (discussed in detail in chapter 9, which deals with the body's immune system).

B-lymphocytes are lymphocytes that originate in and continue to be produced by the bone marrow, whereas the T-lymphocytes are those that initially migrated to the thymus gland and were altered there. Whereas the thymus gland is an organ of substantial size in children (it lies in the upper part of the chest, just under the sternum or breastbone), by the age of puberty the thymus shrinks and begins to atrophy and degenerate. By the time of adulthood, most T-lymphocytes are produced by the lymph nodes, spleen, and bone marrow.

Leukopoiesis

The stages of development of white blood cells, a process that is called **leukopoiesis**, are summarized in figure 8.9. The production of white blood cells begins in the bone marrow with the division of pluripotent stem cells into either **myeloid stem cells** or **lymphoid stem cells**. Certain lymphoid stem cells remain in the bone marrow, ultimately giving rise to B-lymphocytes. Other lymphoid stem cells leave the bone marrow during embryonic and fetal development, migrating first to the thymus gland and then to the spleen and lymph nodes. In these organs, the lymphoid stem cells give rise to T-lymphocytes. As noted above, the thymus gland is a major site of T-lymphocyte production in a child but by puberty shrinks to a considerably smaller size. Production of T-lymphocytes in the lymph nodes, spleen, and bone marrow continues throughout a person's life.

The myeloid stem cells of the bone marrow are the source of both red blood cells (see figure 8.5) and white blood cells. During leukopoiesis, the myeloid stem cells produce **proleukoblasts** which, in turn, give rise to **monoblasts** and **progranulocytes**. The monoblasts are the source of monocytes (which subsequently become tissue macrophages), whereas the progranulocytes are the source of neutrophils, acidophils and basophils. The functions of these white blood cells were described earlier.

8-5

PLATELETS (THROMBOCYTES)

The last of the formed elements of the blood are the **platelets**. Strictly speaking, platelets are not cells; rather, they are small *fragments* of cells that are produced in the bone marrow and then released into the peripheral blood. In a normal person, the platelets number about 130 to 360 million per c.c. of whole blood.

Like the erythrocytes, granulocytes, and monocytes, platelets originate in the bone marrow's myeloid stem cells (fig. 8.10). The myeloid stem cells produce **megakaryoblasts**, which then give rise to **megakaryocytes**–the direct source of platelets. Megakaryocytes are bone marrow cells characterized by finger-like extensions of their cytoplasm. These extensions fragment (or bud) yielding small pieces of membrane-enclosed cytoplasm, which are then released into the bloodstream as the platelets (fig. 8.10).

Whereas in humans (and other mammals) platelets are not true cells, in other vertebrate animals platelets are whole cells. In these animals, the platelets are more properly identified as **thrombocytes**–a term that suggests their physiological role (a **thrombus** is a blood clot). The platelets are rich in secretory granules whose contents help to trigger the coagulation of blood (see later in the chapter).

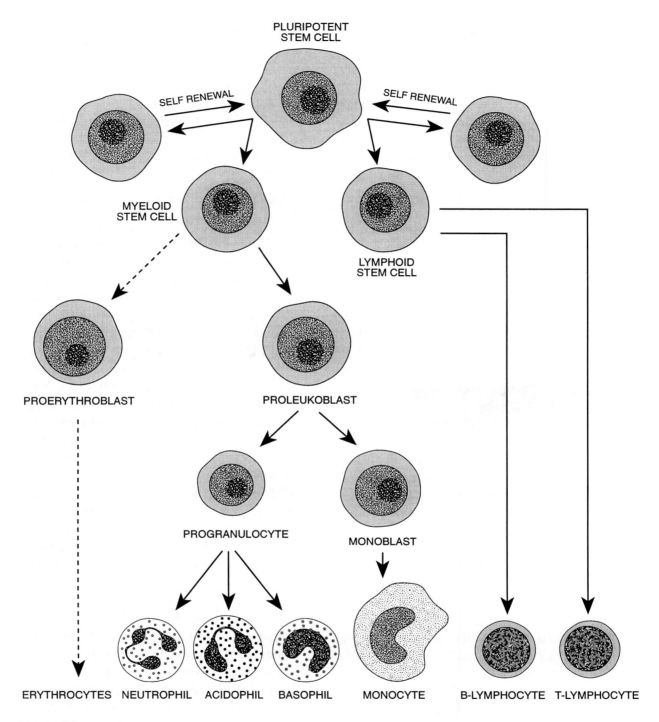

Figure 8.9
Leukopoiesis. The developmental stages of the various types of white blood cells are shown using solid arrows (dashed arrows show stages leading to production of red blood cells; see also figure 8.5).

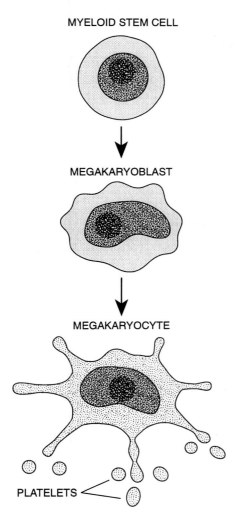

MYELOID STEM CELL

MEGAKARYOBLAST

MEGAKARYOCYTE

PLATELETS

Figure 8.10
Platelets are produced in the bone marrow by
the fragmentation of megakaryocytes.

BLOOD PLASMA

The formed elements of the blood are suspended in a liquid called **plasma**, a fluid that
also has a number of important physiological
functions of its own. Let's begin our consideration of the plasma by identifying this fluid's
principal chemical constituents.

Chemical Composition

Plasma is about 90 percent water, and in this
water are dissolved a variety of inorganic
and organic substances that make up the
remainder of the plasma's volume (table
8.1). Most of the inorganic constituents of
the plasma are salt ions, such as the ionic
forms of sodium (Na^+), potassium (K^+),
calcium (Ca^{++}), chlorine (Cl^-), and phosphate ($PO4^{---}$). Also present are varying
amounts of small organic compounds such
as sugars (primarily glucose), amino acids,
urea, and uric acid. These represent substances in transit between one organ and
another. For example, glucose dissolved in
the plasma may be in transit from the liver
(where it is stored as glycogen) to tissues that
will then withdraw the glucose from the
blood in order to fuel their metabolism. Urea
and uric acid are metabolic wastes in transit
from the tissues in which they were produced
to the organs that will excrete them from the
body (i.e., the kidneys). Also included among
the small organic constituents of the plasma
are a number of hormones (not listed in table
8.1). The hormones of the plasma are chemical messages released into the blood by one
organ (i.e., an *endocrine* gland) and carried
in the plasma to "target" organs where the
hormone causes a modification in the behavior of the target.

The largest of the organic substances dissolved in the plasma are proteins (including
some that are also hormones). The **plasma
proteins** typically account for about 8 percent of the plasma's volume and are of three
types: **albumins**, **globulins**, and **fibrinogen**.
The albumins, which represent about 60% of
the plasma proteins, are secreted into the
blood plasma by the liver. These proteins are
highly soluble and are responsible for the

205

TABLE 8.1   COMPOSITION OF THE BLOOD PLASMA

SUBSTANCE	AMOUNT*	COMMENT/FUNCTION
Water	90-92 g	The plasma's solvent and suspending medium
Nutrients		
Glucose	70-100 mg	In transport to tissues
Amino acids	35-65 mg	In transport to tissues
Wastes		
Urea	20-30 mg	In transport to kidneys
Uric acid	2-6 mg	In transport to kidneys
Creatinine	1-2 mg	In transport to kidneys
Lactic acid	8-15 mg	In transport to liver
Salt Ions		Maintain electrolyte balance
Chloride	369 mg	
Sodium	325 mg	
Potassium	16 mg	
Calcium	9.8 mg	
Sulfate	3.7mg	
Phosphate	3.5 mg	
Magnesium	2.1 mg	
Proteins		
Albumins	3.5-5.0 g	Lipid transport; source of colloid osmotic pressure
Globulins		
Alpha	0.7-1.5 g	Lipid transport; hormone transport
Beta	0.6-1.1 g	Lipid transport; iron transport
Gamma	0.7-1.5 g	Antibodies (contribute to immunity)
Fibrinogen	0.2-0.4 g	Blood coagulation

* Values are given per 100 c.c. of blood plasma.

plasma's *colloid osmotic pressure* (see below). The albumins also serve in the transport of certain lipids (e.g., fatty acids) and bile pigments in the bloodstream.

The globulins, which account for about 35% of the plasma proteins, are chemically much more diverse than the albumins; they also play a greater variety of physiological roles. Three classes of globulins are identified and are termed **alpha, beta**, and **gamma globulins** (i.e., α-globulins, β-globulins, and γ-globulins). Like the albumins, the alpha and beta globulins are produced by the liver and secreted into the blood. Included among the alpha and beta globulins are those that bind and shuttle lipids, vitamins, metal ions, and other small molecules from one tissue to another; these alpha and beta globulins are often referred to as *transport proteins*. Included among these globulins are **transferrin** (the protein that transports iron in the bloodstream), **ceruloplasmin** (the protein that transports copper in the bloodstream), **haptoglobin** (the protein that binds hemoglobin released during intravascular hemolysis, see earlier), and those that transport fat-soluble vitamins and lipids.

Because of their low solubility in water, the major plasma lipids (triglycerides, phospholipids, and cholesterol) are carried in the bloodstream bound to plasma proteins. The union between lipid and protein is called a **lipoprotein**. Lipids have a very low density (less than the density of water); therefore, lipoproteins are less dense than unconjugated proteins. Accordingly, lipoproteins are divided into three groups on the basis of their lipid content; these are (1) **very low density lipoproteins** (called **VLDL**s and which are rich in lipid, especially triglycerides), (2) **low density lipoproteins** (called **LDL**s and which have less lipid but in which the principal lipid present is cholesterol), and (3) **high density lipoproteins** (called **HDL**s and which contain only small amounts of lipid).

The last of the three classes of plasma globulins are the gamma globulins. Gamma globulins are *not* produced by the liver; rather, they are produced by a special class of B-lymphocytes called **plasma cells** and are then secreted into the blood. As is discussed at length in the next chapter, the gamma globulins act to protect you against infection; gamma globulins are also known as **antibodies** or **immunoglobulins**.

Fibrinogen, which is neither an albumin nor a globulin, accounts for about 5% of the protein present in blood plasma. Like the albumins and the alpha and beta globulins, fibrinogen is synthesized in the liver and then secreted into the bloodstream. Fibrinogen's physiological role is in blood coagulation (discussed at length later in the chapter), where it forms the fibrous matrix in which blood cells become entrapped. As you might expect, whereas fibrinogen is present in blood plasma, it is absent from blood serum.

Transcapillary Exchange

The exchange of nutrients, wastes, and other substances between the bloodstream and the tissue fluid that bathes the cells of the body occurs at the level of the **capillary beds** (see figure 7.15). The walls of capillaries are extremely thin, being formed by a single layer of flattened cells called the capillary **endothelium**. Although there is some variation from one capillary bed to another, the capillary endothelium is generally permeable to most substances having molecular weights less than about 10,000 daltons. This implies that the plasma proteins (nearly all of which have molecular weights greater than 10,000 daltons) cannot readily cross the capillary wall and enter the surrounding tissue space. In contrast, small molecules like O_2, CO_2, glucose, and urea are freely permeable. Needless to say, the blood cells are far too large to permeate the capillary walls.

The exchange of materials between the blood and tissue fluid is called **transcapillary exchange** and involves two major processes: **diffusion** and **filtration**. As discussed in chapter 3, the driving force behind diffusion is a **concentration gradient**; that is, substances will cross a permeable barrier from the side on which they are at higher concentration to the side on which they are at lower concentration until a concentration equilibrium is established. Diffusion accounts for the transcapillary exchange of many small molecules, including O_2, CO_2, and glucose. For example, the concentration of oxygen dissolved in the systemic blood of the body's capillaries is greater than in the tissue fluid. Consequently, oxygen diffuses from the blood into the tissue fluid as blood passes through these capillary networks.

Unlike diffusion, which is based on concentration gradients, filtration between the bloodstream and the tissue fluid is based on an interaction between **blood pressure** (or **hydrostatic pressure**) and **colloid osmotic pressure** (also called **oncotic pressure**). This interaction may be explained in the following way.

Whenever a fluid rich in dissolved protein (e.g., the plasma) is separated from a fluid that contains little dissolved protein (e.g., the tissue fluid) by a barrier (e.g., the capillary wall) that is permeable only to small molecules, an *osmotic* pressure gradient is created across the barrier. This pressure acts to draw water and small molecules through the barrier from the fluid that contains little protein into the fluid that is rich in protein. In the case of the capillary beds, the pressure gradient is created by the plasma proteins (mainly albumin) and acts to draw tissue fluid into the blood flowing through the capillaries. The pressure causing the liquid flow is called *colloid osmotic pressure* (a "colloid" is any substance that will not cross a biological membrane and usually implies that the substance in question is protein). Like hydrostatic pressure, colloid osmotic pressure is measured and expressed in the units *mm Hg.*

Colloid osmotic pressure is not the only pressure at play across the walls of capillaries. There is also (1) *blood pressure* (the hydrostatic pressure exerted against the walls of the blood vessels as the blood is driven through the vessel by the actions of the heart), (2) *tissue fluid osmotic pressure* (i.e. the osmotic pressure created by the small amount of protein and other large molecules present in tissue fluid), and (3) *tissue fluid hydrostatic pressure* (i.e., the pressure created in the tissue spaces by the tissue fluid).

The effects of all of these pressures are illustrated in figure 8.11. In this figure the pressure values given are those for a generalized capillary bed. As you learned in chapter 7, blood pressure falls as the blood crosses the capillaries from arterioles to venules. At the arterial end of a capillary bed, the blood pressure (i.e., P_B) is about 35 mm Hg but drops to about 15 mm Hg by the time the blood reaches the bed's venous end. The colloid osmotic pressure of the blood (i.e., COP_B) is about 25 mm Hg, and since no plasma proteins cross the capillary wall into the tissue fluid, this osmotic pressure is maintained throughout the capillary bed. (Keep in mind that the blood pressure acts to push materials out of the bloodstream by driving fluid through the capillary walls, whereas colloid osmotic pressure acts to draw fluid into the bloodstream.)

The tissue fluid also exerts hydrostatic and colloid osmotic pressure, although the values are considerably lower than those of the blood. Although there is some disagreement, it is generally recognized that the hydrostatic pressure of the tissue fluid is extremely low (perhaps even negative). For illustrative purposes, we will take the tissue fluid hydrostatic pressure (i.e., P_T) to be close to zero

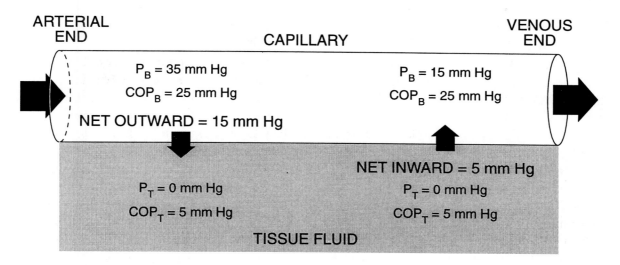

ARTERIAL END — CAPILLARY — VENOUS END

$P_B = 35$ mm Hg
$COP_B = 25$ mm Hg
NET OUTWARD = 15 mm Hg

$P_B = 15$ mm Hg
$COP_B = 25$ mm Hg

NET INWARD = 5 mm Hg

$P_T = 0$ mm Hg
$COP_T = 5$ mm Hg

$P_T = 0$ mm Hg
$COP_T = 5$ mm Hg

TISSUE FLUID

Figure 8.11
Transcapillary filtration. At the arterial end of a capillary bed, the pressures acting to drive fluid out of the capillary (i.e., the blood pressure [P_B] plus tissue fluid colloid osmotic pressure [COP_T]) are *greater* than the pressures resisting this action (i.e., the blood's colloid osmotic pressure [COP_B] plus the tissue fluid's hydrostatic pressure [P_T]). As a result, there is net outward filtration. In contrast, at the venous end of the capillary bed, P_B plus COP_T are *less* than COP_B plus P_T, so that there is net inward filtration. Therefore, a microcirculation is created in which fluid leaves the blood at the arterial end of a capillary bed and reenters the blood at the venous end.

mm Hg (fig. 8.11). What small amount of protein is dissolved in the tissue fluid creates a modest colloid osmotic pressure (i.e., COP_T) of about 5 mm Hg. (Again, it is important to note that the tissue fluid's hydrostatic and osmotic pressure act in directions that are *opposite* to the corresponding pressures of the blood.)

As seen in figure 8.11, at the arterial end of a capillary bed, the pressures acting to drive fluid *out of* the capillary (i.e., the blood pressure [P_B] plus tissue fluid colloid osmotic pressure [COP_T]) are greater than the pressures resisting this action (i.e., the blood's colloid osmotic pressure [COP_B] plus the tissue fluid's hydrostatic pressure [P_T]). Indeed, the pressure difference is 15 mm Hg (i.e., {[35 + 5] - [25 + 0]}. As a result, there

is net outward filtration (i.e., water and small molecules leave the blood and enter the tissue fluid). In contrast, at the venous end of the capillary bed, P_B plus COP_T are less than COP_B plus P_T. Indeed, the difference is -5 mm Hg (i.e., {[15 + 5] - [25 + 0]}). Consequently, there is net inward filtration (i.e., water and small molecules leave the tissue fluid and enter the bloodstream). Thus, a *microcirculation* is created in which fluid leaves the blood at the arterial end of a capillary bed and reenters the blood at the venous end.

It should be apparent from the preceding discussion that the tissue fluid bathing the body's cells undergoes continuous turnover. The tissue space receives a continuous infusion of new materials at the arterial end of a capillary bed, where blood pressure dominates colloid osmotic pressure. At the venous end of the capillary bed, where the blood's colloid osmotic pressure dominates blood pressure, tissue fluid is drawn back into the bloodstream. Along with the reabsorbed fluid go the tissue secretions and cellular wastes. As a result, the composition of the fluid bathing the body's cells remains fairly constant. Nutrients that are taken up by the cells are replaced with fresh nutrients from the arterial blood. Secretions and wastes emptied into the fluid by the tissue cells do not accumulate because they are taken up in the venous blood. The relative constancy of the tissue fluid that results is known as **homeostasis**.

HEMOSTASIS

Whenever there is an injury to a blood vessel (e.g., suppose that you accidentally cut your finger), the body executes several mechanisms in order to stem the loss of blood through the wound. The body's effort to minimize the amount of blood that is lost is called **hemostasis**.

Reflexive Vasoconstriction

One action that is taken immediately upon injury to a blood vessel is **vasoconstriction**. That is, smooth muscle tissue in the walls of the blood vessel contracts bringing about a constriction of the vessel. The constriction reduces the flow of blood through the vessel and brings about a corresponding reduction in blood loss. Vasoconstriction is triggered reflexively, presumably as a result of the stimulation of receptor cells at or near the site of the injury.

Platelet Plug Formation

When blood circulates through uninjured vessels, the platelets are repelled by one another and are also repelled by the lining or **endothelium** of the blood vessels. However, when there is injury to a blood vessel wall, **collagen fibers** in the underlying connective tissue become directly exposed to the blood. Platelets adhere to these collagen fibers. At the same time, damaged subendothelial cells release a protein called **von Willebrand factor** that coats the injured surfaces. Platelets also stick to surfaces coated by von Willebrand factor. Platelets adhering to and accumulating at the site of injury then release **secretory granules**.

The secretory granules of platelets contain **adenosine diphosphate (ADP), thrombaxane A2**, and **serotonin**. Thrombaxane A2 and serotonin are *vasoconstrictors* that act to reduce the flow of blood into the injury site. ADP acts on the surfaces of platelets making them sticky, thereby promoting the accumulation of still greater numbers of platelets in the injury site. As more and more platelets adhere to the injured tissue, a **platelet plug** is formed that acts to seal the wound temporarily. Platelets also release a phospholipid called **PF3** that plays a role in several steps of the coagulation process (table 8.2).

Blood Coagulation

The most important of the three hemostatic mechanisms is **blood coagulation**. Indeed, reflexive vasoconstriction and formation of a platelet plug should be thought of as preludes to the coagulation process. By coagulation is meant the blood's capacity to undergo a change from fluid to solid, forming a resilient long-lasting seal for the wound.

TABLE 8.2 SUBSTANCES IMPORTANT IN HEMOSTASIS

SUBSTANCE		PATHWAY / COMMENT / FUNCTION
von Willebrand factor		Released by injured tissue; binds platelets
Collagen fibers		Exposed at injury site, binds platelets
Thrombaxane A$_2$		Released by platelets; strong vasoconstrictor
Serotonin		Released by platelets; strong vasoconstrictor
PF3		Released by platelets; phospholipid factor
Adenosine diphosphate		Released by platelets; increases stickiness of platelet surfaces
Plasma Clotting Factors		
I	Fibrinogen	Common; source of fibrin threads
II	Prothrombin	Common; active form acts on fibrinogen
III	Thromboplastin	Extrinsic; contributes to factor VII complex
IV	Calcium ions	Intrinsic, extrinsic, and common
V	Proaccelerin	Common; helps activate thrombin
VII	Proconvertin	Extrinsic; contributes to factor VII complex
VIII	Antihemophilic factor	Intrinsic; lack of this factor is cause of Hemophilia A
IX	Christmas factor	Intrinsic; lack of this factor is cause of Hemophilia B
X	Stuart-Prower factor	Common; helps activate thrombin
XI	Plasma thromboplastin antecedent	Intrinsic; activates factor IX
XII	Hageman factor	Intrinsic; activates factor XI
XIII	Fibrin-stabilizing factor	Common; acts to polymerize fibrin monomers

As important as coagulation is, it is also important that blood *not* coagulate while it circulates in the body. Whereas small internal clots do not pose a serious health threat and are quickly removed, a large internal **thrombus** could obstruct a blood vessel, thereby causing serious harm to the tissues normally fed by that vessel. One of the body's natural anticoagulants is **heparin**, which is produced by lung cells and mast cells. Small amounts of heparin are also carried in the blood by basophils and may act to prevent intravascular coagulation. Purified heparin is used to prevent donated blood (or blood drawn from the body for a blood test) from coagulating. Since purified heparin is rather expensive, other chemicals possessing anticoagulant properties are more commonly used. Among these anticoagulants are citrates, oxalates, and EDTA (ethylenediaminetetraacetic acid). The actions of heparin and other anticoagulants are described later.

Formation of the Fibrin Network. The blood clot that seals a wound consists of more than a platelet plug. Also present is a network of protein fibers, called **fibrin threads**, that support the platelet plug (acting much like scaffolding). The network of fibrin threads also acts to trap red and white blood cells, thereby increasing the size of the clot (fig. 8.12). The formation of the fibrin thread network is at the heart of the coagulation process.

The fibrin threads of a blood clot are formed from a plasma protein called **fibrinogen** that is manufactured in the liver and secreted into the bloodstream. The chemical conversion of fibrinogen to fibrin occurs as a result of a series of chemical reactions that can be initiated either intrinsically or extrinsically (fig. 8.13).

The Intrinsic Pathway. The intrinsic pathway is initiated when (1) blood contacts a foreign surface (e.g., when blood contacts the surface of a glass test tube that has not been coated with anticoagulant) or (2) blood is exposed to the collagen proteins that are present in the subendothelial tissue of a blood vessel (e.g., as occurs when the wall of a blood vessel is injured). Such contact causes the activation of a soluble plasma protein called **factor XII** (also known as **Hageman factor**; see figure 8.13).

Active factor XII then activates a second plasma protein called **factor XI** (also known as **plasma thromboplastin antecedent**). Activated factor XI then activates a third plasma protein called **factor IX** (or **Christmas factor**). Active factor IX, a fourth plasma protein called **factor VIII** (or **anti-**

Figure 8.12
Coagulated blood. Blood coagulates when fibrinogen dissolved in the plasma is converted into insoluble fibrin threads. These threads create a meshwork that entraps blood cells (and platelets), thereby forming a resilient plug that seals the wound.

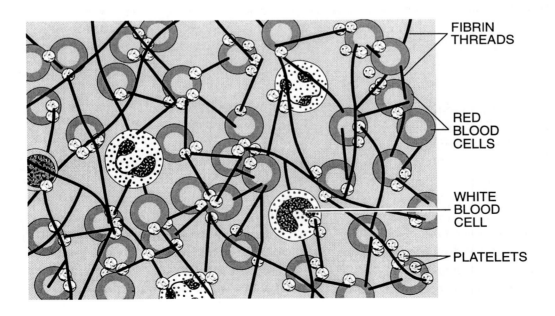

FIBRIN
THREADS

RED
BLOOD
CELLS

WHITE
BLOOD
CELL

PLATELETS

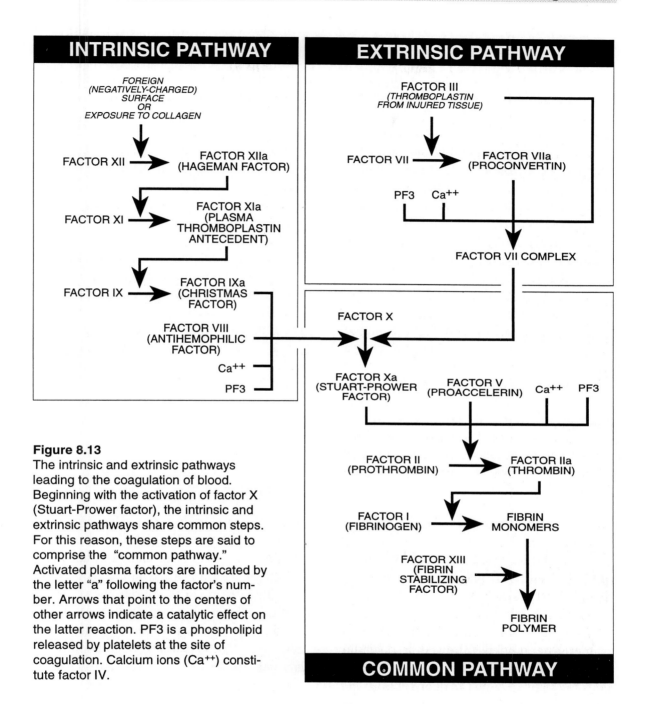

Figure 8.13
The intrinsic and extrinsic pathways leading to the coagulation of blood. Beginning with the activation of factor X (Stuart-Prower factor), the intrinsic and extrinsic pathways share common steps. For this reason, these steps are said to comprise the "common pathway." Activated plasma factors are indicated by the letter "a" following the factor's number. Arrows that point to the centers of other arrows indicate a catalytic effect on the latter reaction. PF3 is a phospholipid released by platelets at the site of coagulation. Calcium ions (Ca⁺⁺) constitute factor IV.

hemophilic factor), calcium ions (**factor IV**), and PF3 released from platelets then combine to form a complex that activates a fifth plasma protein called **factor X** (or

Stuart-Prower factor).

Active factor X, together with a sixth plasma protein called **factor V** (or **proaccelerin**) and additional calcium ions and

PF3, then forms a complex that converts **prothrombin** (also known as **factor II**) into **thrombin**. Thrombin is a protein-digesting enzyme that strips away portions of the plasma's **fibrinogen** (**factor I**) molecules, converting them into **fibrin monomers**. In a final step, the fibrin monomers are caused to *polymerize* (i.e., to combine with one- another into long chains), thereby forming the fibrin threads that are the basis of the fibrin network. Polymerization of the fibrin monomers is catalyzed by the last of the plasma clotting factors, namely **factor XIII** (or **fibrin stabilizing factor**).

The series of reactions just described may be triggered by exposure of plasma to collagen lying just under the endothelial lining of a blood vessel (e.g., as would occur when a blood vessel is injured). Because all of the elements involved in the process are native to the blood vessel and its contents, the reactions are said to be *intrinsic*. In the *extrinsic* pathway to be described, the reactions are not wholly intravascular but are initiated by catalyst released from the surrounding injured tissue.

The Extrinsic Pathway. The reactions of the extrinsic pathway (fig 8.13) are initiated when damaged tissue releases **thromboplastin** (**factor III**) into the injury site. Although most tissues contain thromboplastin, thromboplastin is especially abundant in the skin and in the brain. Injury in these tissues is, therefore, quickly followed by coagulation of the blood. Thromboplastin activates **factor VII** (also called **proconvertin**), and then combines with proconvertin, calcium ions, and PF3 to form a large molecular complex (*factor VII complex*); this complex then activates factor X. From this point on, the sequence of reactions is identical to that described earlier (i.e., the reactions of the

common pathway depicted in figure 8.13).

As noted earlier in the chapter, the coagulation of blood may be prevented using anticoagulants such as heparin (the body's natural anticoagulant) and certain salts, such as citrates, oxalates, and EDTA. Heparin prevents blood from coagulating by indirectly inactivating thrombin. Citrates, oxalates, and EDTA stop coagulation by combining with calcium ions (a process called *chelation*).

The fully developed blood clot serves to stem the loss of blood while the injured blood vessel (and surrounding tissue) begins to repair itself by new growth. Once healing is well underway, the blood clot is removed. The removal of the blood clot involves the action of a plasma enzyme called **kallikrein**, which converts another plasma protein, **plasminogen**, into the fibrin-digesting enzyme **plasmin**. Plasmin digests fibrin polymers, thereby eliminating the scaffolding on which the clot was organized.

Hemophilia

A number of abnormalities exist among the human population in which one or another clotting factor is not produced in adequate quantities, resulting in the failure of the blood to coagulate properly. Among these abnormalities, the most common is the disease **hemophilia** (the disease can be identified in about 1 in 10,000 people). In hemophiliacs, an excessive loss of blood from the body may occur even when the injury to the body is minor.

There are two major forms of hemophilia called **hemophilia A** (about 85% of all hemophilias) and **hemophilia B** (about 15% of all hemophilias). In people afflicted with hemophilia A, the problem is an inability to produce properly functioning factor VIII (antihemophilic factor). In hemophilia B (also called **Christmas disease**), there is an inability to produce proper levels of factor IX (Christmas factor). Both hemophilias are inherited diseases. The diseases are more common in males than in females because the genes controlling the production of clotting factors VIII and IX are carried on the X-chromosomes of cells (i.e., the *sex chromosomes*). The cells of females posses two X-chromosomes, so that if one of them carries the defective clotting factor gene, the potentially harm will be counteracted by the normal gene that is present on the other X-chromosome. In the cells of males, however, there is only one X-chromosome. As a result, one defective gene is all that is necessary for the disease to occur. Because of their differential occurrences in males and females, diseases like hemophilia are called "sex-linked" diseases. Sex-linked diseases are considered in chapter 15, which deals with the subject of heredity.

SELF TEST*

True/False Questions

1. The blood's "formed elements" include red blood cells, white blood cells, and small particles called platelets.

2. The hematocrit of a normal male's blood is about 25%.

3. Because they lack a cell nucleus and are extremely fragile, human red blood cells rarely survive in circulation for more than a month after their release into the bloodstream.

4. As a person sleeps, his (or her) RES removes millions of erythrocytes per second from the bloodstream and destroys them.

5. Alpha and beta globulins present in blood plasma are the breakdown products of molecules of hemoglobin released from aged red blood cells.

6. Proper coagulation of the blood requires the participation of more than ten so-called "clotting factors" from the blood plasma.

Multiple Choice Questions

1. Suppose that a blood sample is collected from a normal male into a test tube that contains no anticoagulant. If the blood is allowed to coagulate and then fully retract, it will exude (release) a clear fluid called (A) lymph, (B) tissue fluid, (C) serum, (D) coagulin, (E) plasma.

2. The "buffy coat" contains (A) myelin, (B) actin and myosin, (C) the sarcolemma of striated muscle cells, (D) leukocytes and platelets, (E) none of these choices is correct.

---

* *The answers to these test questions are found in Appendix III at the back of the book.*

215

3. Which of the following are classified as agranulocytes? (A) erythroblasts and normoblasts, (B) neutrophils, basophils, and acidophils, (C) monocytes and lymphocytes, (D) thrombocytes and macrophages, (E) muscle cells and nerve cells.

4. Most of the hemoglobin synthesized by a red blood cell as it matures in the bone marrow is produced during the (A) erythroblast, (B) normoblast, (C) reticulocyte, (D) monocyte, (E) erythrocyte stage of development.

5. Which of the following are not derived from the bone marrow? (A) eosinophils, (B) erythrocytes, (C) monocytes, (D) B-lymphocytes, (E) all of these are derived from the bone marrow.

6. Aged red blood cells are removed from the circulation (A) only in the bone marrow, (B) only in the liver, (C) only in the spleen, (D) in the bone marrow, liver, and spleen, (E) none of these choices is correct.

7. In the abnormality called "Christmas disease," (A) the hematocrit is too low, (B) too few red blood cells are produced by the bone marrow, (C) there are too many acidophils in circulation, (D) the blood fails to coagulate properly.

CHAPTER 9

THE IMMUNE SYSTEM

In this chapter, we will consider the organization and actions of the body's **immune system**. At the same time, we will take up several closely related subjects including **blood typing**, **blood transfusion**, and **cancer**.

It might be well to ask at the outset what

the subjects of immunity and cancer have in common. Immunity is a state attainable in humans (as well as other higher vertebrates) in which the body is protected against the harmful effects of certain disease-causing (i.e., pathogenic) agents such as parasites, bacteria, and viruses. Cancer, on the other hand, is a condition characterized by the abnormal and rapid growth of tissue cells to form tumors that, in many instances, interfere with normal tissue functions and often lead to death. In recent years, it has become clear that the body's system of defense against parasites, bacteria, viruses, and other infectious agents also serves as a line of defense against cancer. Moreover, it appears that some cancers are intimately associated with viral infections; that is, certain viruses are **oncogenic** (cancer-creating). Later in the chapter, when we explore the causes of and defenses against cancer, other relationships between the immune system and cancer will become apparent.

NONSPECIFIC (FIXED) AND SPECIFIC (ADAPTIVE) DEFENSE SYSTEMS

The body employs two major systems of defense against infection; these are called *nonspecific* (or *fixed*) and *specific* (or *adaptive*) systems. A nonspecific system of defense is one that protects the body against infection in a general way and includes the protection that is provided by *physical barriers* (such as the skin), *phagocytic blood and tissue cells*, *natural killer cells*, and localized *inflammatory responses*. A nonspecific defense system does not discriminate among different bacterial species or different parasites. In contrast, a specific system of defense is one that is mounted against a particular intruder (e.g., against a particular

bacterial species or against a particular type of virus) and is based on the detection of the intruder as the result of a "foreign" substance that the intruder carries or secretes. As you will learn, the foreign substance that is at the heart of the adaptive response is what is known as an **antigen**.

Nonspecific Defense Systems

Physical Barriers. In order to present a real threat, toxins (poisons) and foreign agents (such as bacteria or viruses) must get *into* the body's tissues or bloodstream. Several physical barriers act to prevent this. For example, the **skin** serves as a nonspecific barrier against the penetration of different toxic chemicals and most microorganisms. Though somewhat more delicate, epithelial linings such as those that cover the surfaces of the respiratory, excretory, and reproductive tracts also act to bar entry to the body's tissues. The strong acid that is secreted by the gastric pits of the stomach is a major impediment to infectious agents that enter the digestive tract.

Phagocytic Blood and Tissue Cells. As discussed in chapter 8, the blood contains cells capable of phagocytosing and destroying other cells. Phagocytic blood cells include **neutrophils**, **acidophils** (or eosinophils), and **monocytes**–cells that act in a nonspecific way to phagocytose and destroy foreign cells that have gained access to the bloodstream through a wound. Neutrophils, acidophils, and monocytes can also leave the bloodstream and wander through the tissues where they pursue their phagocytic activity, destroying foreign invaders and dying, deformed, or poorly functioning tissue cells. Such neutrophils and acidophils are called tissue **microphages**. Monocytes that wander into the tissues and provide the same sort of protection develop into **macrophages**.

Tissue macrophages can be *mobile* or *fixed*. Mobile macrophages continue to move about within a tissue or organ after having arrived there following their exit from the bloodstream. Included among these macrophages are the *alveolar* macrophages that scour the air sacs of the lungs (see chapter 10). Fixed macrophages are macrophages that take up permanent residence at specific loci within a tissue. Included among these cells are the *reticuloendothelial* macrophages (see figure 8.6), *microglia* of the central nervous system, and *Kupffer* cells of the liver.

Although the actions taken by neutrophils, acidophils, and monocytes are nonspecific (i.e., monocytes and neutrophils scavenge any of a variety of foreign agents of disease), their interactions with products of the immune system (see below) increase their specificity and efficiency.

Natural Killer Cells. As you will learn in the sections that follow, nearly all of the body's lymphocytes are components of the specific system of defense. These lymphocytes "sense" the presence (and cause the eradication) of specific foreign agents. The process is initiated by interactions that occur between the foreign agent's antigens and receptor molecules that project from the lymphocyte's surface. However, there is a small percentage of lymphocytes that carry no antigen-specific surface receptors; these are called *null cells*. Among the null cells, there is a population that nonspecifically interacts with and destroys foreign cells (i.e., these null cells attack any foreign cell and even destroy abnormal human cells such as those that may produce a cancerous growth

or cells that are infected by virus). These "antigen-independent" lymphocytes are called **natural killer** (or **NK**) cells. The lethal effects of NK cells are the result of "cytotoxin" (i.e., cell poison) molecules that they transfer to their prey and to their secretion of granules that contain pore-forming proteins (i.e., *perforin*). These proteins perforate the membranes of the target cell, leading to the cell's osmotic lysis. Although NK cells account for as few as one in ten lymphocytes, they are an important line of defense.

Inflammatory Responses. When the physical barriers of the skin or epithelial lining of the body are crossed by a foreign agent, what soon follows is **inflammation** (a swelling and reddening of the affected area). Inflammation results from the local release of certain chemicals such as **histamine**. Histamine released by injured tissue (and by mast cells) acts to make blood vessel walls more penetrable by the blood's phagocytic leukocytes (neutrophils, acidophils, and monocytes), as they now squeeze their way through the narrow spaces that separate adjacent endothelial cells (a process called **diapedesis**). The capillary walls also become more permeable to water and dissolved substances. The influx of water into the tissue space is the principal source of the swelling that accompanies inflammation. Neutrophils, acidophils, and monocytes reaching the inflamed area engage in phagocytosis of the invading agents and also release chemicals that act to stimulate the growth of scar tissue at the site of injury.

Specific Defense Systems

In contrast to nonspecific defense systems, the body's immune system is an adaptive system and is highly specific. Through what is called an **immune response**, agents of infection are recognized by the antigens with which they are associated and are destroyed by the direct and indirect actions of **B-** and **T-lymphocytes**. Additionally, the immune system exhibits long-term memory of its encounters with agents of infection. This **immunologic memory** (described later) permits more rapid response to a second, third, or subsequent attack by a given agent. The rest of this chapter is devoted to an up-close examination of this adaptive defense system.

DUALITY OF THE IMMUNE SYSTEM

There are two ways in which the immune system responds to infection. In the **cellular immune response**, there is a direct interaction between lymphocytes and infected tissue cells. In contrast, the **humoral immune response** acts principally through the secretion of proteins called **antibodies** or **immunoglobulins** into the blood plasma, lymph, and tissue fluid. As we shall see, these antibodies combine with the antigens that are carried in the surface of or are released by the pathogen. The combination of antibody with antigen then initiates a response leading to the elimination of the antigen and, more importantly, its source.

Classes of Lymphocytes

There are two major classes and one minor class of lymphocytes. These are called **B-lymphocytes** (or **B-cells**), **T-lymphocytes** (or **T-cells**) and **natural killer** cells (or **NK-cells**). B- and T-lymphocytes are components of the body's specific (adaptive) de-

fense system, whereas NK-cells (described briefly above) belong to the body's nonspecific (fixed) defense system.

B-lymphocytes and natural killer cells are initially derived from stem cells in the bone marrow and liver of the developing embryo and fetus. Some of these cells migrate from the bone marrow to the spleen and lymph nodes, where they continue to proliferate. Most, however, remain in the bone marrow. Therefore, in children and adults, B-cells and NK-cells are acquired from the bone marrow, spleen and lymph nodes (fig. 9.1).

T-lymphocytes are derived from embryonic stem cells that migrate from the bone marrow to the thymus gland, where they mature under the influence of the hormone **thymosin**. From the thymus gland, some T-cells migrate to and take up residence in the spleen, lymph nodes, and bone marrow (fig. 9.1). Whereas a child's thymus gland is a principal source of T-cells, this organ gradually degenerates, so that in an adult, T-cells are derived principally from the spleen, lymph nodes, and bone marrow. They continue to be called T-cells ("T" for "thymus") because of the specializations that they underwent during their time spent in the thymus gland. The distribution of the three classes of lymphocytes in a sample of normal blood is about 75% T-cells, 15% B-cells and 10% NK-cells.

The activation of T- and B-lymphocytes by the presence of foreign agents in the body leads to the production of families of cells that combat the infection. The T-lymphocytes give rise to **cytotoxic**, **helper**, **and suppressor** T-cells, whereas the B-lymphocytes give rise to antibody-secreting **plasma cells**. Both T- and B-lymphocytes also give rise to **memory cells**–cells that are responsible for immunologic memory.

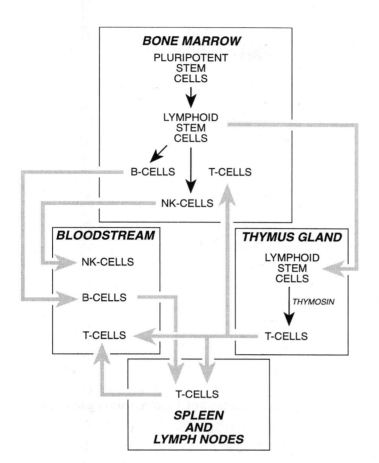

Figure 9.1
Origins and distribution of the three major forms of lymphocytes. Grey arrows indicate the migration of cells from one tissue or organ to another. Black arrows indicate cellular developmental changes.

ANTIGENS, HAPTENS, ANTIBODIES, AND T-CELL RECEPTORS

Antigens and Haptens

An **antigen** is a molecule (or part of a molecule) that acts to stimulate the body's immune system and trigger an **immune response**. In nearly every instance, the antigen is "foreign" to the person whose immune system is stimulated (i.e., it is chemically different from any of the molecules normally present in the tissues of the person). Although antigens may have diverse chemical character, most are proteins, carbohydrates, glycolipids, or nucleic acids.

Related to antigens are **haptens**. These are small molecules that *alone* cannot stimulate the immune system (i.e., in a solitary state they are not *antigenic*). However, when haptens are combined with another (usually larger) carrier molecule, they can trigger an immune response. Haptens are important because they are so widespread; for example, haptens occur in certain pharmaceuticals (e.g., penicillin can act as a hapten), dust particles, and animal danders.

Antibodies

Antigens and haptens are "recognized" by two families of proteins peculiar to the immune system: **antibodies** and **T-cell**

receptors. Antibodies are found in the plasma membranes of B-cells, where the antibody molecules project from the cell surface and act as antigen receptors. Antibodies are also secreted by B-lymphocytes during an immune response. **T-cell receptor** proteins are similar to antibodies but project from the plasma membranes of T-cells. However, unlike the antibodies of B-cells, T-cell receptor proteins are *not* secreted by T-cells. T-cell receptors provide T-lymphocytes with the ability to attach directly to foreign cells that have entered the body and also to the body's own cells when these tissue cells have been infected by viruses (see later).

The antibodies (also called **immunoglobulins**) that circulate in the bloodstream make up the "gamma globulin" fraction of blood plasma (indeed, antibodies account for about 20% of the protein present in blood plasma). The production of antibodies is stimulated by antigens that are present in or released from pathogens that have entered the body; typically, the antigens are constituents of the membranes of bacteria or the coats of viruses and have been recognized by the immune system as being foreign or alien to the body.

The antibodies produced during an immune response have the capacity to bind to the type of antigen that triggered the response. The reaction between the antibody and antigen is highly specific; each type of antibody reacts with a particular antigen and no other. It is estimated that the human body is capable of producing between ten million and one billion ($10^7 - 10^9$) different kinds of antibody molecules, each antibody capable of reacting with a different antigen.

Although millions of different antibodies can be produced by the human immune system, all can be assigned to one of five antibody classes on the basis of chemical and functional similarities. These classes are called *IgA*, *IgD*, *IgE*, *IgG*, and *IgM* ("Ig" stands for "immunoglobulin," a term used interchangeably with the term "antibody;" "A", "D", "E", "G", and "M" stand for *alpha*, *delta*, *epsilon*, *gamma*, and *mu* [letters of the Greek alphabet]). The most abundant antibodies are those that belong to the IgG class (about 80% of all antibodies), and the organization of the typical IgG molecule is shown in figure 9.2. Each antibody is composed of four polypeptide chains of two different sizes; there is a pair of identical high molecular weight chains called **heavy chains** and a pair of identical low molecular weight chains called **light chains**. (The heavy chains contain about 450 amino acids, whereas the light chains consist of about 214 amino acids.) Each light chain is linked to a heavy chain, and the two light chain/heavy chain pairs are linked together. The bonds that link chains together are formed between the sulfur atoms of nearby cysteine amino acids (see figure 2.17); these bonds are called *disulfide bridges*.

The chemical reaction that takes place between antigen and antibody occurs at the ends of the paired heavy and light chains of the antibody molecule (see figure 9.2). It is in this region that the numbers, kinds, and sequences of amino acids that comprise the heavy and light polypeptide chains differ from one antibody molecule to another. Much of the remaining chemical structure of the antibody molecule is the same in all members of an antibody class. The chemical variability that is possible at the antigen-binding sites of antibodies is almost limitless, so that antibodies capable of combining with any antigen can be produced. The complementarity of

9-1

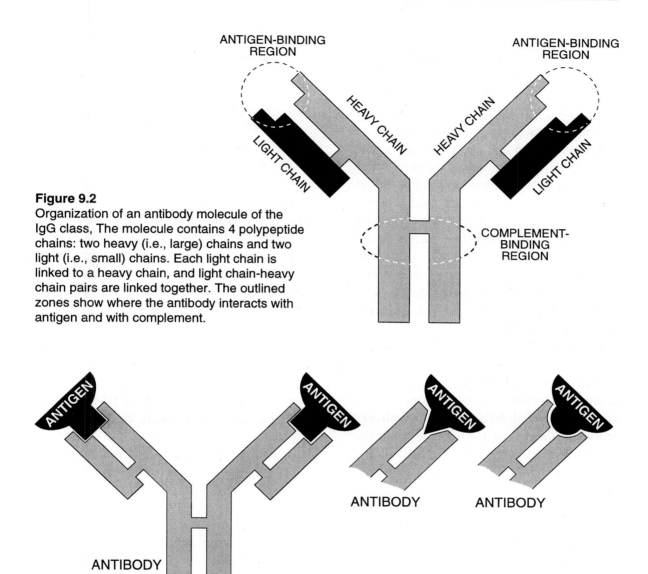

Figure 9.2
Organization of an antibody molecule of the IgG class, The molecule contains 4 polypeptide chains: two heavy (i.e., large) chains and two light (i.e., small) chains. Each light chain is linked to a heavy chain, and light chain-heavy chain pairs are linked together. The outlined zones show where the antibody interacts with antigen and with complement.

Figure 9.3
The interaction between antibody and antigen at the antigen-binding site involves only a portion of the antigen molecule, namely that portion of the antigen that is complementary to (and chemically reactive with) the paired ends of the antibody's light and heavy chains. Three different antigen-binding sites and three different antigens are depicted here.

the shapes of antigens and antigen-binding sites is illustrated in figure 9.3. Note that the interaction at the binding site involves only the complementary portion of the antigen molecule (not the entire antigen molecule). The left half of figure 9.4 depicts a B-cell that has antibody molecules projecting from its surface.

T-Cell Receptors

Like the plasma membranes of B-cells, T-cell surfaces also contain antigen-binding receptors. Although **T-cell receptors** have many properties in common with antibodies, they are a distinct class of proteins and are not secreted from the cells into the bloodstream. A T-cell receptor is about two-thirds the size of an antibody and consists of only two polypeptides (called an "alpha" chain and a "beta" chain) interconnected by a single disulfide bridge (fig. 9.5).

One end of each of the two polypeptides that comprise a T-cell receptor is anchored to the surface of the T-cell; the other end of each polypeptide projects away from the cell surface. Together, the unanchored ends form the foreign antigen binding sites. The interaction between a T-cell receptor and antigen involves the entire T-cell. As a result, T-cells are able to attach to the surface of the antigen-bearing particle (such as a virus-infected human tissue cell).

CLONAL SELECTION THEORY

The **clonal selection theory** explains the mechanism by which the body's immune system responds to the appearance of antigens that are present in (or released from) an agent of infection (e.g., bacteria). At birth, the body's immune system includes many millions of different families (or **clones**) of B-lymphocytes. Each one of these clones consists of a population of cells whose surface antibodies are identical to one another, with each antibody able to recognize and combine with a single, distinct antigen. The members of each one of these clones of B-

Figure 9.5
Organization of a T-cell receptor. The receptor is comprised of two interconnected polypeptide chains: an alpha chain and a beta chain. One end of the receptor is anchored in the surface of the T-cell (see figure 9.4); the other end forms a site that binds foreign and MHC antigens.

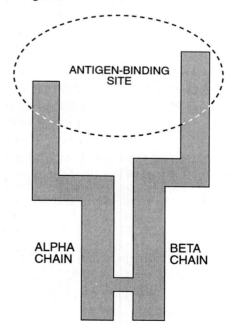

Figure 9.4
B- and T-lymphocytes, showing the antibodies that project from the surface of a B-cell and the receptors that project from the surface of a T-cell.

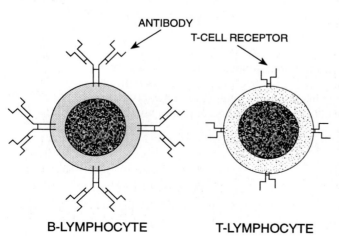

lymphocytes are derived from a single precursor cell and the origin of the clone is independent of whether or not the body has previously been exposed to the antigen with which the B-cell surface antibodies react. Thus, the body's immune system is primed at birth with B-cells responsive to any of millions of different antigens and is continuously "on alert" in anticipation of the appearance of these antigens.

A B-cell immune response is triggered when antigens of an agent of infection become bound to antibody molecules in the plasma membranes of a particular B-cell clone. Because the interaction between antigen and antibody is so specific, the B-cell clone that reacts with the antigen is said to have been "selected" by the antigen (this is why the phenomenon is called "clonal selection"). The interaction between a particular antigen and the surface antibodies of a preexisting clone of B-cells is illustrated in figure 9.6.

Binding of antigen to the surface antibodies of a particular B-cell clone leads (via a mechanism detailed later in the chapter) to a succession of rounds of cell division that greatly increases the number of cells belonging to this B-cell clone (the clone is said to expand or proliferate; fig. 9.6). Some of the newly produced B-cells become **plasma cells**, which now synthesize large quantities of antibody but which secrete the molecules instead of inserting them in their cell surfaces. The copious amounts of antibody that are secreted are available to react with additional antigen (fig. 9.6). Other cells (not shown in figure 9.6) produced by these divisions become **memory B-cells;** these cells are kept in reserve and respond with antibody secretion only if there is a subsequent exposure to the same antigen (discussed later under "immunologic memory"). A similar form of clonal selection leads to the activation and proliferation of specific clones of T-lymphocytes (e.g., *cytotoxic* T-cells and *memory* T-cells).

MAJOR HISTOCOMPATIBILITY COMPLEX PROTEINS

As already noted, the surfaces of B-cells contain antibodies (and the surfaces of T-cells contain receptors) that react with specific antigens. In addition to these "markers," the surfaces of lymphocytes–indeed the surfaces of *all nucleated human cells*–contain protein markers that identify the cell as human and as belonging to a particular individual (i.e., belonging to a particular member of the human species). Although they are present in nearly all cells of the body, having been discovered first in leukocytes, these markers are called the *human leukocyte antigens* (*HLAs*). Because of their importance in tissue transplantation and because they are encoded by a small cluster of genes on human chromosome number 6, these protein markers are also known as the **major histocompatibility complex** proteins (or **MHC** proteins). For consistency, only the latter terminology will be used here.

The more closely two people are related, the more alike are their MHC proteins; in identical twins, the MHC proteins are the same. Unrelated individuals have different sets of MHC proteins; thus, like the antibodies and T-cell receptors, the MHC proteins are incredibly diverse. However, unlike antibodies and T-cell receptors, which differ among the millions of different clones of B-cells and T-cells of an individual, the MHC proteins differ only among individuals; that

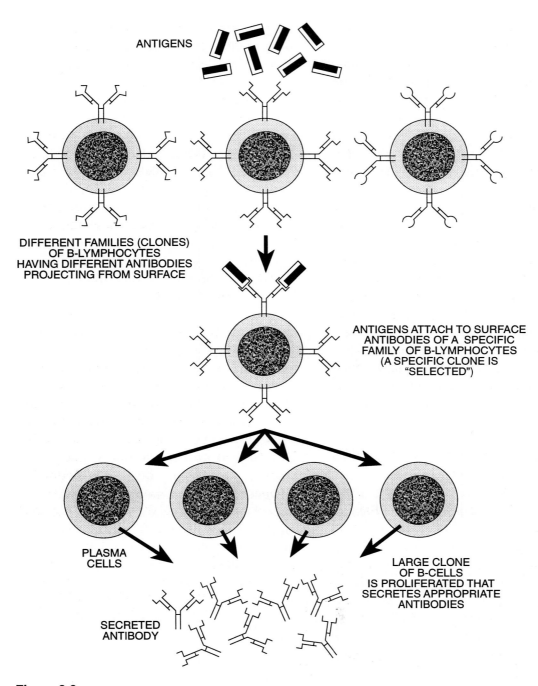

Figure 9.6
Clonal selection. Antigen binding to surface antibodies of one of millions of different clones of B-lymphocytes (only three shown here) is part of the mechanism that triggers a B-cell immune response. In the response, the selected clone is expanded to form very large numbers of identical B-cells. Some of these cells (called plasma cells) secrete antibodies that will react with additional antigen. This leads to the destruction of the antigen and its source.

is, all cells of a single individual carry the same MHC proteins.

MHC proteins are divided into two major classes: **class I MHC proteins** and **class II MHC proteins**. Class I MHC proteins are found in the surfaces of nearly all cells, whereas class II MHC proteins are limited to cells that play a role in the immune response. For example, class II MHC proteins are found in B-cells, T-cells, and "antigen-presenting" macrophages (see later). Class I MHC proteins consist of one large and one small polypeptide chain, whereas class II MHC proteins consist of two polypeptide chains of equal size. (Like T-cell receptors, the paired chains are called "alpha" and "beta" chains; however, unlike T-cell receptors, the chains are not linked by disulfide bridges.) Class I and class II MHC proteins are depicted in figure 9.7. The distributions of antibodies, T-cell receptors, and MHC proteins in human cells are summarized in figure 9.8.

Interactions Between MHC Proteins and Antigenic Determinants

After antigens combine with the antibodies

that project from the surfaces of B-cells, the antigen-antibody complex is internalized by the cell. Once inside the cell, the antigen is partially degraded. One of the antigen fragments produced during the degradation of the antigen is combined with an MHC protein and is then displayed at the cell's surface. The display of the antigen fragment is one of the essential steps in an immune response. Antigen fragments that are displayed at the cell surface in combination with MHC proteins are called **antigenic determinants**.

PARTICULARS OF THE B-CELL IMMUNE RESPONSE

To understand how B-lymphocytes are caused to secrete antibodies, let's consider a case in which a person acquires a bacterial infection (fig. 9.9). Antigens from the surfaces of some of the bacteria become bound to antibodies in the plasma membranes of one or more of the millions of clones of B-lymphocytes that are present in the body (figure 9.9, step 1). The antigen-antibody complexes are then internalized by the B-cell. After chemical alteration within the

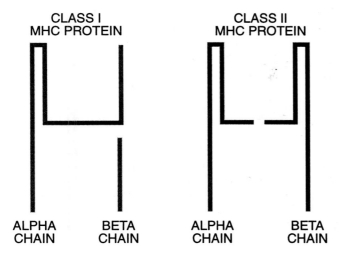

Figure 9.7
Organization of MHC class I and class II proteins. Each protein consists of two polypeptide chains: an alpha chain and a beta chain.

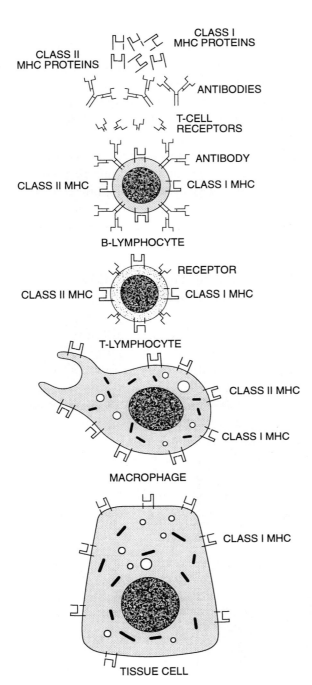

Figure 9.8
Distributions of class I and II MHC proteins, antibodies, and T-cell receptors in the surfaces of human cells. Class I MHC proteins are in the surfaces of all nucleated cells. Class II MHC proteins are in the surfaces of lymphocytes and phagocytic cells.

B-cell in which the antigen is partially degraded, antigenic determinants are combined with class II MHC proteins and displayed at the B-cell's surface (figure 9.9, step 2).

The acts of binding antigen, internalizing it, digesting it, and displaying antigenic determinants in combination with class II MHC proteins are not sufficient for activation of the B-cell clone. Rather, bacterial antigens must also be taken up during nonspecific phagocytosis of some of the bacteria by phagocytic cells such as neutrophils, acidophils, and monocytes that act as scavengers in the bloodstream and in the body's tissues (figure 9.9, step 3).

The bacteria phagocytosed by these scavenging blood cells are degraded, and antigenic determinants are then displayed at the cell surface in combination with class II MHC proteins (figure 9.9, step 4). Phagocytic cells that carry out this process are referred to as **antigen-presenting cells**. The antigenic determinants are recognized by T-cells possessing receptors for the displayed determinants (figure 9.9, step 5). T-cells that interact with and are activated by antigen-presenting macrophages are called **helper T-cells**. Activated helper T-cells then interact with the B-lymphocytes displaying the same antigenic determinants (figure 9.9, step 5). The interaction between helper T-cells and B-lymphocytes serves to activate the B-lym-

Figure 9.9 *(opposite)*
Major events during the activation and proliferation of antibody-secreting lymphocytes. Note that the interactions that take place between helper T-cells, antigen-presenting macrophages, and B-lymphocytes not only involve the association of bacterial antigens with surface antibodies and T-cell receptors, but also involves the cells' respective MHC proteins.

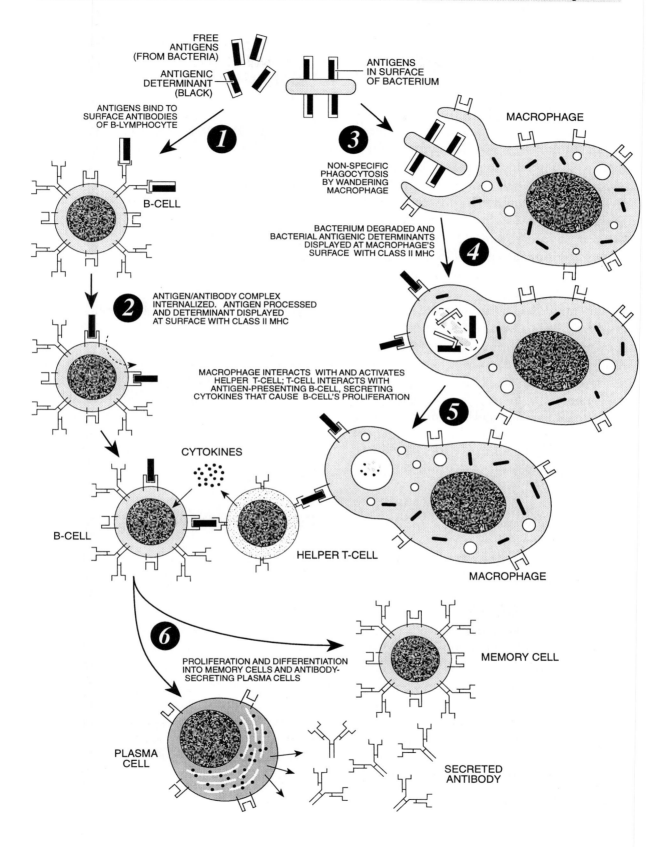

FREE
ANTIGENS
(FROM BACTERIA)

ANTIGENIC
DETERMINANT
(BLACK)

ANTIGENS
IN SURFACE
OF BACTERIUM

ANTIGENS BIND TO
SURFACE ANTIBODIES
OF B-LYMPHOCYTE

1

3

MACROPHAGE

NON-SPECIFIC
PHAGOCYTOSIS
BY WANDERING
MACROPHAGE

B-CELL

BACTERIUM DEGRADED AND
BACTERIAL ANTIGENIC DETERMINANTS
DISPLAYED AT MACROPHAGE'S
SURFACE WITH CLASS II MHC

4

2 ANTIGEN/ANTIBODY COMPLEX
INTERNALIZED. ANTIGEN PROCESSED
AND DETERMINANT DISPLAYED
AT SURFACE WITH CLASS II MHC

MACROPHAGE INTERACTS WITH AND ACTIVATES
HELPER T-CELL; T-CELL INTERACTS WITH
ANTIGEN-PRESENTING B-CELL, SECRETING
CYTOKINES THAT CAUSE B-CELL'S PROLIFERATION

5

CYTOKINES

B-CELL

HELPER T-CELL

MACROPHAGE

6

PROLIFERATION AND DIFFERENTIATION
INTO MEMORY CELLS AND ANTIBODY-
SECRETING PLASMA CELLS

MEMORY CELL

PLASMA
CELL

SECRETED
ANTIBODY

phocytes causing the rapid multiplication of members of the clone, thereby yielding large numbers of identical **plasma cells** and **memory cells** (figure 9.9, step 6). Activation (and resulting cloning) of the B-lymphocytes is believed to be mediated by substances called **cytokines** (also called **lymphokines**) that are secreted by helper T-cells. Only the plasma cells that are derived from activated B-lymphocytes produce and secrete antibodies (figure 9.9, step 6). The memory cells are kept in reserve and will be called on to respond during a second (or subsequent) infection by the same antigen-bearing pathogen (see later).

The antibodies that are secreted by plasma cells may have several different effects: (1) they may interact with antigens released by the pathogen, thereby neutralizing the antigens (rendering them harmless); (2) they may interact with surface antigens of the pathogen causing **agglutination** and leading to **opsonization**; and (3) they may interact with surface antigens of the pathogen thereby promoting **complement fixation**.

Neutralization of Antigenic Toxins

The sites on an antigen with which antibodies react and which initially act as a stimulus for the immune system are called **antigenic determinants**. Most of the antibodies secreted by plasma cells are **bivalent**, meaning that the antibody can combine simultaneously with two identical antigenic determinants. The kinds of complexes formed by interaction of antibody and antigen depend on the number of antigenic determinants that are present. If two or more antigenic determinants are present, *cross-linking* by the antibody can produce lattice-like complexes that are insoluble and form *pre-*

cipitates. These precipitates are eliminated by the phagocytic actions of scavenging macrophages.

Agglutination and Opsonization

Antibodies that interact with antigens present in the surfaces of invading microorganisms cause **agglutination** (fig. 9.10). During agglutination, microorganisms become cross-linked to form small masses. These masses are phagocytosed by macrophages (i.e., monocytes), which kill and digest the microorganisms. **Opsonization** refers to the mechanism by which the microorganisms coated with antibodies are "tagged" for phagocytosis. As illustrated in figure 9.10, the plasma membranes of macrophages possess receptors that bind the paired ends of the antibody heavy chains (i.e., they bind to the ends of the heavy chains that are opposite to the antigen-binding sites). The mechanisms of agglutination and opsonization are depicted in figure 9.10.

Complement Fixation

Complement fixation is yet another mechanism by which antibodies defend the body against invasion by pathogens. The term **complement** refers to a family of proteins that circulate in the blood plasma that are capable of forming **lytic complexes** on the surfaces of foreign cells. These lytic complexes create holes (pores) in the surfaces of the foreign cells, eventually leading to the death of these cells (i.e., cell *lysis*) as a result of the osmotic influx of water.

Unlike the immunoglobulins, the complement proteins are not synthesized in response

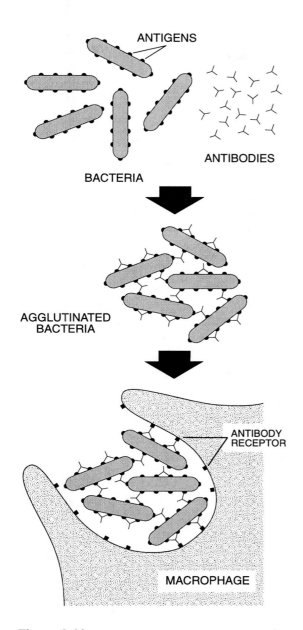

ANTIGENS

ANTIBODIES

BACTERIA

AGGLUTINATED
BACTERIA

ANTIBODY
RECEPTOR

MACROPHAGE

Figure 9.10
Opsonization and phagocytosis of agglutinated bacteria.

ment proteins have been identified and they are referred to as proteins C1, C2, C3, C4, C5, C6, C7, C8, and C9. Many of these proteins are **zymogens**–enzymes that are not yet in their active (i.e., catalytic) state. The complement zymogens are sequentially activated in a cascade of reactions that comprises complement fixation. Activation of these zymogens converts them to **proteolytic** enzymes–enzymes that can split protein molecules into smaller fragments.

As illustrated in figure 9.11, activation of the complement system begins when two or more antibodies attach to a cluster of antigenic sites on the surface of a pathogen such as a bacterium. Complement protein C1 then binds to the shared **complement binding site** (see figure 9.2) of the antibodies and is thereby activated. It is important to emphasize that the C1 binding site is created by two (or more) appropriately aligned antibody molecules bound to the bacterial surface.

Activated C1 enzymatically splits C4 into two protein fragments called C4a and C4b; the larger C4b fragment then attaches to a nearby site on the bacterium's surface (i.e., it chemically binds to the cell's plasma membrane; figure 9.12A). Activated C1 also splits C2 into two fragments (C2a and C2b; figure 9.12B); the C2b fragment then associates with the C4b fragment, which is already attached to the bacterial cell's surface (fig. 9.12B). The C4bC2b complex that is formed is a proteolytic enzyme that then splits C3 into fragments C3a and C3b, with C3b associating with the C4bC2b complex on the cell's surface (fig. 9.12C). Now, the C4bC2bC3b complex enzymatically splits C5 into C5a and C5b, with C5b joining the growing complement protein complex on the bacterium's surface (fig. 9.12 D). (The progressive attachment or binding of complement proteins to the cell surface is

to the appearance of an antigen; rather they are regular constituents of the bloodstream (regardless of the presence or absence of infectious agents in the body). Nine comple-

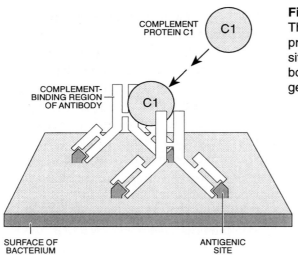

Figure 9.11
The complement cascade begins when complement protein C1 interacts with the complement-binding site formed by two or more antibodies that have bound to antigenic sites on the surface of a pathogen (such as a bacterial cell).

Figure 9.12
Complement fixation (see text for explanation).

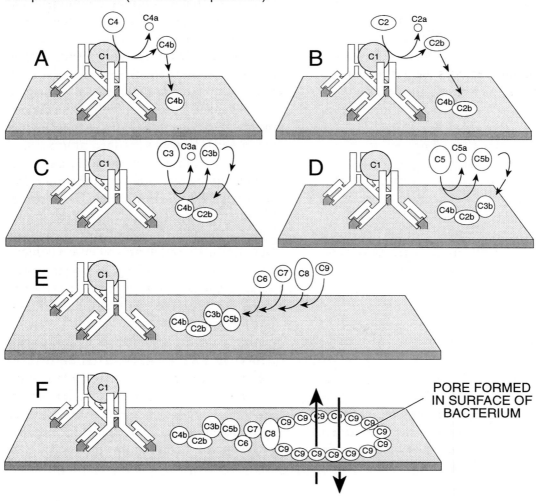

what is meant by *fixation*.) The splitting of C5 into C5a and C5b is the last *enzymatic* step in the cascade of reactions that characterizes complement fixation. The remaining steps involve the binding and polymerization of additional complement proteins.

Complement proteins C6, C7, and C8 now bind to the C4bC2bC3bC5b complex anchored to the bacterial cell's surface (figure 9.12E). Full lytic activity of the fixed complement is attained by the attachment of 12 to 15 C9 molecules (figure 9.12E and F) which complete the assembly of a **membrane attack complex** (MAC). In the MAC, the C9 molecules form a doughnut-shaped ring that buries itself in the cell's surface forming a pore that links the cell's interior with the outside medium (fig. 9.12F). The pore created by the MAC permits passage of water, ions, and small molecules through the plasma membrane but precludes the passage of cellular proteins. As a result, the cell's colloid osmotic pressure produces an influx of water that swells and lyses the cell.

The cascade of complement protein actions that has just been described constitutes what is known as the "classical pathway" of complement fixation. There is also an "alternative pathway." The two pathways share a number of steps, but the most distinguishing physiological feature of the alternative pathway is the lack of a requirement for the initial formation of antibody-antigen complexes on the foreign cell's surface. In other words, for *some* pathogens, MAC assembly can take place in the absence of initial antibody binding.

Complement fixation by antibody-coated bacteria (i.e., the classical pathway) and the lysis of the invading cells that follows is the most common defense mechanism attributable to B-cell-secreted antibodies. Inter-estingly, complement protein fragments that do not become part of the forming MAC (e.g., protein fragments C4a, C2a, and C5a) do play protective physiological roles. For example, these fragments appear to chemically attract microphages and macrophages to a site of infection and stimulate the release of histamines by basophils and mast cells. Histamines cause vasodilation (thereby allowing more blood to circulate through a site of infection) and also act to increase the permeability of capillary walls.

Immunologic Memory

Figure 9.13 compares the kinetics of plasma antibody production during a **primary immune response** and a **secondary immune response**. A primary immune response (the left half of the graph) occurs after the *first* exposure to a particular antigen. The response is characterized by a lag phase during which little or no antibody appears, followed by a rising antibody level in which the plateau that is reached amply deals with the antigen or antigen source (represented by the grey zone). The duration of the plateau phase varies according to the antigen responsible for triggering the immune response; the variable length of the plateau phase is indicated by the break in the graph's curve and the horizontal axis. During the plateau phase, a condition of **active immunity** exists. After some period of time, the antibody level begins to fall. The response to a second exposure to the same antigen (i.e., the secondary immune response shown in the right half of the graph) is more dramatic. The lag period is shorter, the response is more intense (greater quantities of antibody are produced), and the elevated antibody level is maintained

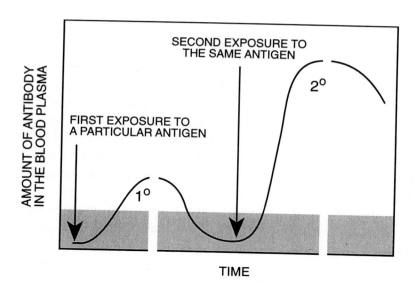

Figure 9.13
A comparison of primary (1°) and secondary (2°) immune responses. The grey zone represents the minimal level of antibody that must be attained to inactivate all of the antigen (or antigen source) that triggers the response. The two breaks in the curve represent time intervals of variable length.

for a longer period of time (i.e., there is a longer period of active immunity). Again, the variable length of the plateau phase during the secondary immune response is indicated by the break in the curve and the graph's horizontal axis. The difference between the two responses indicates that the body has "remembered" its earlier encounter with the antigen; this phenomenon is called **immunologic memory**.

Immunologic memory may be explained in the following way. The initial exposure to antigen causes differentiation of B-lymphocytes into memory cells as well as plasma cells. Whereas the plasma cells have a relatively short life span in which they are actively engaged in antibody secretion, memory cells do not secrete antibody and continue to circulate in the blood and lymph for months (or years). These memory cells are able to respond more quickly to the reappearance of the same antigen than undifferentiated B-lymphocytes. As we will see, memory cells are also produced by the multiplication and differentiation of T-lymphocytes.

Active vs. Passive Immunity

The antibodies produced during an immune response serve to destroy the pathogen whose antigens initially triggered the response. However, the unused antibody and the antibody that continues to be secreted by the plasma cells produced during the primary response provide continuing protection for some period of time. Moreover, a second (or subsequent) exposure to the antigen serves to activate memory cells, so that more antibody becomes available almost immediately. Consequently, whereas a person's initial exposure to an antigen may be accompanied by temporary illness, subsequent exposures to the same antigen are handled by the immune system with relative ease. This is what is meant by **active immunity**–the ability to easily fend off the harmful effects of a pathogen to which you have previously been exposed and to which you developed a primary response. For example, once you have had chicken pox, subsequent exposure to the chicken pox virus is rarely followed by any symptoms of the illness. For some pathogens (e.g., chicken pox virus and measles virus), active immunity lasts a lifetime.

Active immunity is to be distinguished from **passive immunity**. Passive immunity is conferred to people who have been exposed to extremely dangerous pathogens (e.g., snake venom, tetanus bacteria, etc.) and whose immune systems may not react quickly enough to the pathogen to prevent serious harm. The immunity is conferred by injecting **antiserum** containing antibodies against the pathogen into the bloodstream of the infected person. The antibody-rich antiserum is usually obtained by processing blood from an animal (e.g., horse, cow, pig, etc.) that was deliberately exposed to the antigen and which developed an active immune response. The antibodies injected into the bloodstream of the infected individual provide an immediate (although temporary) defense against the pathogen. Passive immunity is also referred to as "borrowed" immunity.

Autoimmune Diseases

One's immune system normally produces antibodies against foreign proteins but not against the native proteins of the body. That is, the immune system is able to distinguish between "self" and "nonself." Yet, one's own proteins may be regarded as antigens by the immune system of another person (or another animal). Thus, each person's tissues possess proteins (and other chemical substances) that are potential antigens. The ability to distinguish self from nonself develops very early in life.

In rare cases, individuals begin to produce antibodies or T-cells that are reactive against their own antigens. These antibodies are called **autoantibodies** and the diseases resulting from their presence are called **autoimmune diseases**. Among the autoimmune diseases are **insulin-dependent diabetes mellitus** (antibodies against insulin-producing cells of the pancreas), **paroxysmal cold hemoglobinuria** (antibodies against one's own red blood cells), **myasthenia gravis** (antibodies against one's own muscle cell acetylcholine receptors), **systemic lupus erythematosus** (antibodies against one's own DNA and nucleoproteins), and **Grave's disease** (antibodies against one's own thyroid tissue).

PARTICULARS OF THE T-CELL IMMUNE RESPONSE

As noted at the beginning of the chapter, there are three major types of T-cells; they are known as *helper* T-cells, *cytotoxic* T-cells, and *suppressor* T-cells. (It may be noted that there is still some disagreement regarding the existence of a discrete group of suppressor T-cells, many immunologists arguing that suppression of immune responses may occur as a result of actions taken by helper and cytotoxic T-cells.) Also, earlier in the chapter, you learned that one of the roles played by helper T-cells is the activation of B-cells during a humoral immune response. In this section, we will consider the cells that are involved and the events that take place during the mounting of a cell-mediated immune response.

Typically, cellular immune responses are mounted against infections by viruses and fungi, against cancer cells formed by the transformation of previously normal tissue cells, and against certain bacterial infections.

For purposes of illustration, let's consider the cellular immune response that is mounted against a viral infection.

The typical virus consists of a core of genetic material and enzymes enclosed within a shell (or capsid) comprised of proteins (see figure 3.9). During a viral infection, the nucleic acid-rich core of the virus (and sometimes the coat) enter the host cell, causing the infected cell to switch its metabolic machinery to the making of new virus coat proteins and new virus nucleic acids. These are then assembled into large numbers of new virus particles within the host. Eventually, the newly formed viruses exit the host cell (frequently killing the host cell in the process) and go on to infect additional host cells (see chapter 3 and particularly fig. 3.10). If not quickly checked, this cycle will cause major damage to the host's tissues.

The cellular immune response may be initiated in several ways but usually begins when wandering macrophages encounter and phagocytose free virus particles (figure 9.14, step 1). After partially degrading the phagocytosed viral material, portions of the virus coat proteins are displayed at the surface of the macrophage in conjunction with class II MHC proteins (figure 9.14, step 2). The displayed viral peptides are foreign to human tissues and therefore serve as antigens; at this point, the macrophage is known as an **antigen-presenting macrophage**. The presented antigens are now recognized by a member of a particular clone of helper T-cells. The helper T-cell engages the macrophage via its surface receptor molecules. Interaction between the macrophage and the helper T-cell serves to activate the T-cell, which now responds by secreting a variety of cytokines (figure 9.14, step 3). It is to be emphasized that helper T-cells interact only with cells that present a combination of class II MHC proteins and antigens at their surfaces.

The cytokines that are secreted by the helper T-cell play a variety of roles (figure 9.14, steps 4, 5, and 6). For example, these cytokines stimulate the growth and proliferation of the helper T-cells that initially secreted the cytokine, while also activating other helper T-cells of the clone (these two actions are respectively known as *autocrine* and *paracrine* actions). Additionally, the cytokines help to activate the cytotoxic T-cells that will have to deal with the host tissue cells infected by the virus (see below). Cytokines secreted by helper T-cells also attract monocytes to the area of infection and promote their development into actively phagocytic macrophages. Finally, cytokines act to prevent macrophages from migrating away from the infection site. The net effect of cytokine secretion by helper T-cells is the accumulation of macrophages and lymphocytes in the region of infection and is characterized by the inflammation that typically exists there.

The main features of cytotoxic T-cell action during viral infections are summarized in figure 9.15. At the same time that a virus-infected tissue cell produces new viral cores and coats, some of the newly produced viral proteins are combined with class I MHC proteins and displayed at the surface of the cell (figure 9.15, step 1). Cytotoxic T-cells belonging to the appropriate clone engage the infected cells; this engagement involves interactions between cytotoxic T-cell surface receptors and antigen-class I MHC protein complexes displayed at the surface of the infected cells (figure 9.15, step 2). The T-cells are now activated and secrete substances that will cause the death of the virus-infected

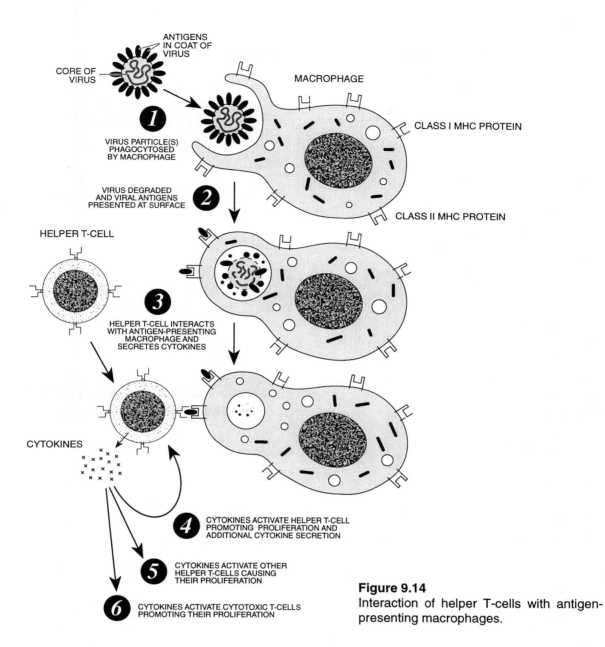

ANTIGENS IN COAT OF VIRUS

CORE OF VIRUS

MACROPHAGE

CLASS I MHC PROTEIN

CLASS II MHC PROTEIN

1 VIRUS PARTICLE(S) PHAGOCYTOSED BY MACROPHAGE

2 VIRUS DEGRADED AND VIRAL ANTIGENS PRESENTED AT SURFACE

HELPER T-CELL

3 HELPER T-CELL INTERACTS WITH ANTIGEN-PRESENTING MACROPHAGE AND SECRETES CYTOKINES

CYTOKINES

4 CYTOKINES ACTIVATE HELPER T-CELL PROMOTING PROLIFERATION AND ADDITIONAL CYTOKINE SECRETION

5 CYTOKINES ACTIVATE OTHER HELPER T-CELLS CAUSING THEIR PROLIFERATION

6 CYTOKINES ACTIVATE CYTOTOXIC T-CELLS PROMOTING THEIR PROLIFERATION

Figure 9.14
Interaction of helper T-cells with antigen-presenting macrophages.

cell (figure 9.15, step 3). These secretions include *cytotoxins* (cell poisons) that are taken up by the target cell and which soon kill it, and *pore-forming proteins* (e.g., **perforin**) that create pores in the target cell's surface. (The pore-forming proteins secreted by cytotoxic T-cells resemble protein C9 of the complement system discussed earlier.)

Passage of water and ions into the virus-infected cell through the newly created pores soon leads to the cell's lysis. The death of the infected tissue cell cancels the assembly of additional viruses by that cell and preempts the formation of any new viral constituents; although the tissue cell is sacrificed by this action, it stems the spread of the virus to

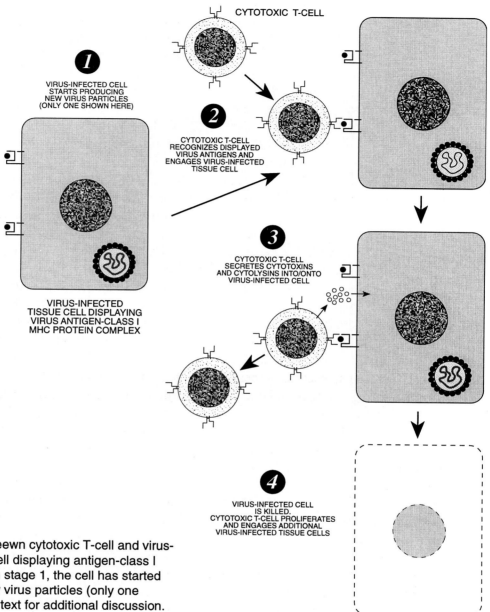

CYTOTOXIC T-CELL

❶
VIRUS-INFECTED CELL
STARTS PRODUCING
NEW VIRUS PARTICLES
(ONLY ONE SHOWN HERE)

❷
CYTOTOXIC T-CELL
RECOGNIZES DISPLAYED
VIRUS ANTIGENS AND
ENGAGES VIRUS-INFECTED
TISSUE CELL

VIRUS-INFECTED
TISSUE CELL DISPLAYING
VIRUS ANTIGEN-CLASS I
MHC PROTEIN COMPLEX

❸
CYTOTOXIC T-CELL
SECRETES CYTOTOXINS
AND CYTOLYSINS INTO/ONTO
VIRUS-INFECTED CELL

❹
VIRUS-INFECTED CELL
IS KILLED.
CYTOTOXIC T-CELL PROLIFERATES
AND ENGAGES ADDITIONAL
VIRUS-INFECTED TISSUE CELLS

Figure 9.15
Interaction betweewn cytotoxic T-cell and virus-
infected tissue cell displaying antigen-class I
MHC proteins. In stage 1, the cell has started
to assemble new virus particles (only one
shown). See the text for additional discussion.

neighboring healthy tissue. After releasing
cytotoxins (and pore-forming proteins), the
cytotoxic T-cell disengages its target and
seeks out other, similarly infected tissue cells
(figure 9.15, step 4).

The cytotoxic T-cell's encounter with its
target "activates" the T-cell. When it later
disengages its target, it may not only interact

directly with other target cells but may also
undergo a period of rapid growth and division, thereby producing additional members
of its T-cell clone. Keep in mind that (as
indicated earlier and depicted in figure 9.14)
additional cytotoxic T-cells are also acquired
as a result of the effects of cytokines released
by activated helper T-cells.

Not all of the progeny of activated helper and cytotoxic T-cells participate in fighting the infection responsible for their production. Rather, just like the B-cell story, some of these T-cells are kept in reserve and take action only when the body is again infected by the same foreign agent. These T-cells are called **memory cells** and are responsible for the more rapid and dramatic action taken by the immune system during secondary cell-mediated immune responses (i.e., immunologic memory).

The above discussion reveals that T-cells respond only to human cells carrying an MHC protein and an antigenic determinant from a foreign source (e.g., from viruses, intracellular bacteria, and fungi). Thus, two stimuli trigger the proliferation of the required T-cell clones. Cytotoxic T-cells respond to the combination of foreign antigen and class I MHC proteins, whereas helper T-cells respond to foreign antigen and class II MHC proteins. Note that the actions that are taken by T-cells are directed toward the body's own cells and not to free antigens. The rationale behind the cell-mediated immune response is that it is better to sacrifice (kill) even a large number of infected human cells than to allow the infection to spread to additional, healthy tissue.

It is generally agreed that the reduction in B- and T-cell proliferation that characterizes the final stages of an immune response is not simply due to the eradication of antigen. Furthermore, it is believed that T-cells are also the source of chemical agents that attenuate the immune response. It remains uncertain whether the factors that eventually suppress a humoral or cell-mediated immune response are produced by a special family of T-cells (which have been named suppressor T-cells) or whether they are the products of helper and cytotoxic T-cells. With the field

of immunology undergoing such intensive study and with progress in the field being so rapid, this is an argument that should soon be settled.

ACQUIRED IMMUNE DEFICIENCY SYNDROME (AIDS)

No disease in recent memory has attracted so much public concern and fear as **acquired immune deficiency syndrome** or **AIDS**. The disease was first described in the 1980s and is characterized by a drastic suppression of the body's immune system. This suppression ultimately leads to the occurrence of repeated opportunistic infections, the appearance of cancerous tumors, and the progressive deterioration of the central nervous system. In many instances, the infections that cause death would have caused little injury were the victim's immune system functioning normally. Unable to battle these infections in the normal manner, victims with a "full blown" case of AIDS eventually die.

AIDS is caused by a virus called **human immunodeficiency virus (HIV)**. HIV belongs to a family of viruses called **retroviruses**. What is special (and unusual) about retroviruses is that their genetic complement consists of RNA (not DNA) and that once they have infected a cell, they cause their host to use the viral RNA complement as a template for synthesizing viral DNA. (Usually, genetic information flows from DNA to RNA [to protein, see chapter 2], but in this case genetic information flows "backwards" from RNA to DNA (by a process called **reverse transcription**). This DNA is then used as a template to produce additional copies of viral RNA, which become incorporated into newly assembling virus particles.

At the time of this writing the HIV virus has infected about two million people in the United States and many more millions worldwide. Despite enormous research efforts by scientific communities worldwide (especially in the United States, Britain, and France) a cure for the disease has yet to be found. More than 40,000 Americans already have succumbed to the disease's ravages.

Biology of the HIV Virus

The structure of the AIDS virus is shown in figure 3.9. The nucleic acid core of the virus consists of two strands of RNA; also present in the core are a number of enzymes essential to the virus' proliferation within the host cell. The coat (or capsid) that encloses the virus consists of a lipid bilayer derived from its host's plasma membrane as it exits its host; also present in the coat are a number of virus-specific proteins.

Transmission of the virus occurs when free virus particles or virus-infected cells in the blood, semen, or other body fluids of one person are passed into the blood or tissues of another person. The three most common ways in which this occurs are (1) through sexual contact (homosexual and heterosexual), (2) injection of drugs using paraphernalia contaminated with HIV- containing blood (i.e., transmission among drug abusers sharing syringes and/or needles), and (3) transmission from an infected mother to her child during pregnancy, at delivery, or via the mother's milk. Although *donated* blood can now be screened for the presence of HIV, before screening became commonplace, a number of hemophiliacs contracted the disease after receiving transfusions of contaminated blood.

Immunology of the HIV Virus

The principal targets of the HIV virus are human cells whose plasma membrane contains a protein known as **CD4** (this protein is involved in the binding of the virus to the prospective host cell's surface). Such cells are said to be "CD4$^+$" (i.e., CD4-positive). CD4 is found in the membranes of helper T-cells, some cytotoxic T-cells, and some macrophages; therefore, these cells are the virus' prime targets.

Once the HIV virus infects its host cell, the virus' core RNA is transcribed into DNA. The viral DNA is then integrated into the DNA of the host cell's nucleus, where it is now referred to as a **provirus**. The HIV provirus may remain in an inactive state for some time (even years) or may become active, causing the host to make new viral proteins, new viral RNA, and from these constituents assemble complete virus particles. Eventually, the host cell lyses (and dies) releasing the virus particles that it produced, which may now infect other cells. Once again, the provirus formed may remain latent for some time before becoming active and repeating the cycle.

As the disease progresses, AIDS patients show a marked depletion of their helper T-cells and losses of macrophage scavenging activity. Strangely, the decline in T-cells is only partially due to virus-induced cell lysis. Even in patients in advanced stages of the disease, the percentage of T-cells that can be shown to be infected by viruses is very small. Clearly, there are some other, as yet poorly understood, ways in which the presence of the virus destroys the immune system's effectiveness.

The immune system of an HIV-infected person does attempt to deal with the infection and both humoral and cell-mediated immune responses are mounted. However, these are of so weak a nature that little protection is afforded against the disease. This is understandable in view of the fact that the helper T-cells required to initiate the immune responses are inactivated or killed by the virus. Antibodies against the HIV virus' coat proteins and some of the core enzymes can be detected in a person's blood plasma between three and 20 weeks after infection. Indeed, it is the presence of these antibodies in a person's blood plasma that is the diagnostic indicator of the virus' presence.

AIDS-Related Diseases

HIV infection may be without serious symptoms for quite some time if the viruses remain in the provirus state. In patients in which the virus is actively replicating, a condition called **AIDS-related complex** (**ARC**) may appear and is characterized by diarrhea, fevers, and progressive loss of weight. These are relatively modest symptoms compared to those that characterize a full-blown case of AIDS.

The progressive suppression of the immune system that characterizes AIDS leads to a number of potentially fatal infections. Among these is *Pneumocystis* **pneumonia**, which is the most common cause of death among AIDS patients. AIDS patients frequently develop malignant growths (known as **Kaposi's sarcoma**) on the skin and in the lymph nodes and internal organs. Cancerous growths that are caused by *other* viruses (i.e., viruses other than HIV) may also occur as resistance to viral infection fails. One of the most common of these is a leukemia known

as **Burkitt's lymphoma**, a cancer of B-cell-producing tissues produced by the Epstein-Barr virus. Finally, in most AIDS patients, there are HIV infections of macrophages in the brain, leading to major disturbances of the central nervous system.

BLOOD TYPING AND BLOOD TRANSFUSION

When blood is transfused from one person to another or when organs are transplanted, the success or failure of the tissue transfer depends upon the behavior of the *recipient's* (i.e., the person receiving the blood or organ) immune system. In this section, we will use blood transfusion to illustrate the relationship between a person's immune system and his (or her) body's capacity to accept or reject the transfer of another person's tissue.

The surfaces of red blood cells contain antigens that serve as the basis for blood typing; the distributions of these antigens also dictate when it is safe (or unsafe) to transfuse blood from one person into another. Although there are many blood type antigen families, we will be concerned here with the two most important families known as the **ABO series** and the **Rhesus factor**.

The ABO Series

Although the surfaces of human red blood cells do not contain MHC proteins, they do contain a variety of other antigenic substances including antigens of the **ABO series**. This series consists of two antigens: **antigen A** and **antigen B**. The presence of one, the other, both, or neither of these antigens dictates each person's blood type, and this is determined genetically (see figure 9.16). That is, each of us has inherited from our parents two genes (one gene from the father and one from the mother) that *indirectly* (see below) determine which (if any) antigens of the ABO series our red cells will possess.

The A and B antigens are not proteins; rather, they are a combination of lipid and carbohydrate (sugar chains) called **glycosphingolipids**. Every (normal) human synthesizes identical core glycosphingolipids to which specific *terminal* sugars are added enzymatically, thereby completing the antigen molecules. The nature of the terminal sugars that are enzymatically added to the core determines the final structure of the antigen and this depends on the enzyme that is available to make the sugar transfer. When the enzyme (called a *glycosyl transferase*) adds the sugar *acetylgalactosamine* to the glycosphingolipid, the result is antigen A. When the enzyme adds the sugar *galactose* to the glycosphingolipid, the result is antigen B. It is the genes for one, the other, both, or neither of these two *glycosyl transferase* enzymes that one inherits from his or her parents.

Three different forms of the *glycosyl transferase* gene exist: (1) the gene for *acetylgalactosamine transferase* (called gene A), (2) the gene for *galactosyl transferase* (called gene B), and (3) a gene encoding a product that is devoid of enzymatic activity (let's call it "silent" gene O). If you inherit only *acetylgalactosamine transferase* genes (one from your father and one from your mother, thereby giving you **genotype** AA), your red blood cells will produce only antigen A; if you inherit only *galactosyl transferase* genes (thereby giving you genotype BB), your red blood cells will produce only antigen B; if you inherit *both* an *acetylgalactosamine*

transferase gene and a *galactosyl transferase* gene (one gene from each parent, giving you genotype AB), your red blood cells will produce both antigen A and antigen B; if you inherit two silent genes (one from each parent, giving you genotype OO), neither terminal sugar is added, and the resulting glycosphingolipid produced by your red blood cells will have *no antigenic activity*.

Two other genotypes are possible: AO (an A gene from one parent and an O gene from the other) and BO (a B gene from one parent and an O gene from the other). A person with the AO genotype produces A antigens (remember, there are no O antigens) and a person with the BO genotype produces B antigens (see figure 9.16).

Note (fig. 9.16), there are two genotypes that produce type A individuals (i.e., genotypes AA and AO) and two genotypes that produce type B individuals (i.e., genotypes BB and BO). In contrast, there is only one genotype that results in type AB (i.e., genotype AB) and only one genotype that results in type O (i.e., genotype OO). The relative frequencies with which the four ABO phenotypes occur in the population are depicted in figure 9.17.

Antigens that are very similar to the human A and B antigens are found in animal and plant tissues and in the walls of certain bacteria living in the human intestine, and this has important physiological consequences. Within a year or two after birth, as these antigens are encountered in food that is eaten and in the air that is breathed (air contains suspended plant pollen), a child's immune system begins to respond to the "nonself" antigen. For example, a child with type A blood does not respond to A antigens in pollen because the A antigens are recognized as "self;" however, B antigens in pollen

BLOOD TYPE	ANTIGEN(S) PRESENT	ANTIBODIES PRESENT	POSSIBLE GENOTYPE(S)	RELATIVE FREQUENCY BLACK	RELATIVE FREQUENCY WHITE
A	A	anti-B	AA AO	27	41
B	B	anti-A	BB BO	20	10
AB	A and B	NONE	AB	7	4
O	neither A nor B	anti-A and anti-B	OO	46	46

Figure 9.16
Antigen and antibody combinations in the blood of individuals belonging to each of the four blood groups of the ABO series.

trigger an immune response in which anti-B antibodies are manufactured and secreted into the bloodstream. The production of these antibodies continues through life as the immune system is repeatedly exposed to the B antigens. Consequently, a person with type A antigens in his red blood cell membranes has anti-B antibodies in his blood plasma. Similarly, a person with type B antigens in his red cells has anti-A antibodies in his blood plasma. A person with type A and type B antigens in his red cells (i.e., blood type AB) will not develop an immune response to either the A or B antigen because both of these are recognized as self. Therefore, a type AB person has neither anti-A nor anti-B antibodies in his blood plasma. Finally, the immune system of a type O person will

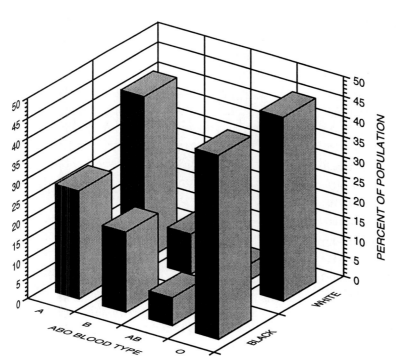

Figure 9.17
Relative frequencies with which each of the four ABO blood types occurs in the population.

respond to both the A and B antigens that are encountered and will have both anti-A and anti-B antibodies in his blood plasma (fig. 9.16).

| 9-2 |

When the combinations of antigens and antibodies present in the red blood cells of two people are *compatible*, it is possible to successfully *transfuse* blood from one person into the other. In a transfusion, the person receiving the transfused blood is called the **recipient** and the person whose blood is used in the transfusion is called the **donor**. When determining whether or not a particular combination of donor and recipient is compatible, one is concerned principally with correctly matching the antigens in the donor's red blood cells and the antibodies present in the recipient's blood plasma. This is known as the **major cross-match**. It is not as crucial to consider the donor's antibodies (or the recipient's antigens) because when modest amounts of donor blood are used, the blood plasma of the donor is so greatly diluted by the recipient's plasma. However, matching the donor's plasma antibodies with the recipient's rbc antigens (called the **minor cross-match**) is important when large volumes of donor plasma are to be used.

For example, the transfusion of blood from one type A person into another type A person is compatible because the recipient's blood plasma does not contain anti-A antibodies. However, the transfusion of type B blood into a type A recipient would not be compatible because the recipient's blood contains anti-B antibodies. These antibodies would agglutinate the red blood cells in the donated blood, creating small masses that can block smaller blood vessels and interfere with the normal flow of blood in the body. Moreover, the removal of products that result from the breakdown of the agglutinated cells (as well

as red cells whose bound antibody induces complement fixation) places an excessive strain on the kidneys and other tissues and can result in organ damage. Using major cross-match criteria only, the compatibility/non-compatibility relationships of other combinations of donor and recipient bloods are shown in figure 9.18.

| 9-3 |

As you examine figure 9.18, note that all of the boxes in the AB recipient column contain plus signs. What this means is that in

RECIPIENT'S
BLOOD TYPE

+ = COMPATIBLE MAJOR AND MINOR CROSS-MATCHES

(+) = COMPATIBLE MAJOR CROSS-MATCH ONLY

− = INCOMPATIBLE

Figure 9.18
Major cross-match compatible (+) and incompatible (−) ABO series blood transfusions. Plus signs in parentheses indicate that it is only the major cross-match that is compatible (minor cross-match is incompatible).

an emergency (when only major cross-matches may be considered), a type AB person can receive blood from type A, B and O donors (as well as type AB donors). This is because a type AB person has no anti-A or anti-B antibodies in his blood plasma. It is for this reason that type AB is known as the **universal recipient** blood type. Also note that all of the boxes in the O donor row contain plus signs. What this means is that in an emergency, type O blood can be given to A, B, and AB (as well as type O individuals) For this reason type O is known as the **universal donor** blood type.

The Rhesus Factor

Another important red blood cell antigen is the **Rhesus factor** (named after the *Rhesus* monkey in which the antigen was first discovered and studied). Every person is either Rhesus positive (i.e., **Rh$^+$**) or Rhesus negative (i.e., **rh$^-$**), and like the ABO series of blood types, one's Rhesus blood type is determined genetically. Two genes are involved–one that is expressed by the production of the Rhesus antigen (i.e., the Rh$^+$ gene) and one whose products, if any, are not antigenic (i.e., the rh$^-$ gene). There are, therefore, three possible genotypes (fig. 9.19): Rh$^+$Rh$^+$ (one Rh$^+$ gene inherited from each parent); Rh$^+$rh$^-$ (an Rh$^+$ gene inherited from one parent and an rh$^-$ gene from the other);

and rh$^-$rh$^-$ (rh$^-$ genes inherited from both parents). So long as at least one Rh$^+$ gene is inherited, the person's blood cells will contain the Rhesus antigen and be classified as Rhesus positive (about 85% of the population). If no Rh$^+$ gene is present, the person is Rhesus negative (about 15% of the population).

A major difference between the ABO series antigens and the Rhesus factor is that a person who is rh$^-$ does not automatically develop antibodies against the Rhesus antigen (recall that the plasma of a person who is type A contains anti-B antibodies; the plasma of a person who is type B contains anti-A antibodies; and so on). In order for an rh$^-$ person to develop antibodies against the Rhesus antigens, Rh$^+$ blood must have entered his body (as might occur during a transfusion). For example, if a person whose blood type is A/rh$^-$ (i.e., his ABO series type is A and he is Rhesus negative) receives a transfusion of A/Rh$^+$ blood, the Rh$^+$ blood cells will not be agglutinated (assuming, of course, that this is the *first* time that he receives a transfusion of Rh$^+$ blood). However, this transfusion **sensitizes** (or "immunizes") him against Rhesus antigens. That is, the Rhesus antigens in the donated red blood cells trigger an immune response in which **anti-Rhesus antibodies** appear in the blood plasma in a week or two. As a result, should he receive another transfusion of Rh$^+$ blood, his anti-Rhesus antibodies will agglutinate the

BLOOD TYPE	ANTIGENS PRESENT	POSSIBLE GENOTYPE	
Rh$^+$	Rhesus	Rh$^+$Rh$^+$	Rh$^+$rh$^-$
rh$^-$	no Rhesus	rh$^-$rh$^-$	

Figure 9.19
The Rhesus blood types.

donated red cells. For this reason, it is advisable that blood transfusions involve blood types that are both ABO compatible and Rhesus compatible.

Erythroblastosis Fetalis

The Rhesus factor takes on special significance during pregnancies in which the mother is rh⁻ and the developing fetus is Rh⁺. As the fetus's vascular system develops, Rhesus antigen-bearing red cells appear in the fetal circulation. These cells pick up oxygen from the mother's bloodstream as the cells circulate through the capillary beds of the **placenta**. (The respective bloodstreams of the mother and fetus do not mix, because they are separated by the placental membranes.) The oxygen picked up by the fetal red cells is then carried back to the fetus, where it is consumed during metabolism.

If some of the fetal red blood cells (or even small fragments of the cells) pass through the placental membranes separating the fetal circulation from the mother's circulation, the Rhesus antigens may trigger an immune response by the mother. This is because the mother's tissues perceive the fetal Rhesus antigens as "nonself," and so her immune system begins to produce antibodies against Rhesus antigens. The first antibodies to be produced against Rhesus antigens are molecules that are too large to cross the placenta. However, these are followed a few weeks later by antibodies that can cross the placenta. Anti-Rhesus antibodies that make their way across the placental membranes from the mother's blood into the fetus' blood will cause agglutination and destruction of the fetus' red blood cells. The fetus becomes anemic, the anemia worsening as more and

more of the fetal red cells are lost. The resulting condition is called **erythroblastosis fetalis** (or *hemolytic disease of the newborn*), and if the anemia is severe, it may be fatal to the fetus.

It should be emphasized that erythroblastosis can occur only when the mother is rh⁻ and the fetus is Rh⁺. If the mother is Rh⁺, then she perceives any fetal Rhesus antigens that enter her bloodstream as "self" and her immune system does not produce anti-Rhesus antibodies. Of course, the mother will not generate an immune response if both she and the fetus are rh⁻ because there are no Rhesus antigens in the fetus' red cells.

Even when the mother is rh⁻ and the fetus is Rh⁺, erythroblastosis fetalis rarely occurs during a first such pregnancy. Among the reasons for this are: (1) fetal Rhesus antigens may not get into the mother's bloodstream at any time during the pregnancy; hence, there would be no opportunity for the mother to become sensitized to the Rhesus antigens; and (2) the Rhesus antigens may enter the mother's bloodstream so late in the pregnancy that the baby is born before a substantial immune response is mounted.

The most likely time at which an rh⁻ mother can be sensitized by the Rh⁺ antigens present in the blood of the fetus that she is carrying is when she gives birth to the child (i.e., at **parturition**). Birth of a child is accompanied by the loss of some blood on the part of both the mother and the child, and this affords the greatest opportunity for fetal blood cells to enter the mother's tissues. If she is sensitized at that time, then a second (or subsequent) pregnancy in which she again carries an Rh⁺ fetus presents the greatest risk of erythroblastosis.

The prospects of erythroblastosis occurring during a second pregnancy in which the rh⁻ mother carries an Rh⁺ fetus can be diminished greatly by injecting anti-Rhesus antibodies into the mother's blood within 3 days of giving birth to the first Rh⁺ baby. The anti-Rhesus antibodies that are injected into the mother destroy any Rhesus antigens and Rh⁺ fetal red blood cells that "leaked" into her bloodstream during parturition. Therefore, she is not sensitized and does not mount an immune response to the antigens.

CANCER

Cancer is a disease in which there is excessive and abnormal growth of certain tissue cells. In the healthy tissues of an adult, growth is regulated in such a way that cell multiplication exactly balances cell loss. Consequently, once a person reaches adult age, the sizes and cellular contents of the various organs of the body remain more or less constant. Various chemical, physical, and viral agents can cause the loss of control of cell growth, and as a result, a normal cell may be *changed* into a cancerous one; such a change is called a **transformation**. The transformed cell undergoes uncontrolled growth and division producing a large clone of transformed cells that is now identified as a **tumor** or **neoplasm**. *Malignant* tumor cells are cancer cells that spread to neighboring tissues–a process that is called **metastasis**. In contrast, *benign* neoplasms consist of cancer cells that do not spread to distant sites and do not pose so great a threat to life. One line of defense against cancer cells is the body's immune system, which may recognize the changes in cell surface antigens that accompany the transformation of a cell. As a result, some cancer cells are eliminated by the immune system before causing serious harm.

Until recently, this theory of "immune surveillance" was widely accepted in the scientific community, but of late, its importance has been seriously questioned.

Agents that cause the transformation of a normal cell into a cancerous one are said to be *carcinogenic* or *oncogenic*. Among these agents are various chemicals, radiation (e.g., X rays), and infection by certain viruses. Carcinogens transform normal cells by directly or indirectly causing changes in the cell's hereditary material (i.e., the cell's DNA). Changes in DNA produced by carcinogens are called **mutations**, and the carcinogen is thus said to be *mutagenic*.

It is now clear that most (perhaps all) cancers stem from (1) changes in the chromosomal *location* of specific genes, (2) changes in the *structure* of specific genes, and/or (3) the *activation* of specific types of genes called **proto-oncogenes**. Proto-oncogenes are present in all normal cells, where they exist either in an inactive state or function in a normal way. Mutations change or activate proto-oncogenes, converting them to **oncogenes**, which then function abnormally.

Cancers of the Immune System

Burkitt's Lymphoma. Some cancers are clearly associated with changes in the locations of specific genes. Normal human tissue cells contain 46 chromosomes (i.e., 23 pairs of chromosomes), and each chromosome has a particular sequence of genes. The 23 different human chromosomes are readily distinguished by microscopy and each has been assigned a particular number (e.g., chromosome 1, chromosome 2, etc.). Mutations that result in the movements of genes to other regions of a chromosome or to a different chromosome altogether may set the stage for the growth of tumors. One of the best illustrations of this is a human leukemia (common among AIDS patients) called **Burkitt's lymphoma**. In this disease, tumors appear in the B-lymphocyte-producing tissues.

In the most common form of Burkitt's lymphoma, a *reciprocal translocation* of a number of genes occurs between chromosome number 8 and chromosome number 14. A segment at the end of chromosome 8 is translocated to chromosome 14 and is replaced by a segment from the end of chromosome 14. As a result, a proto-oncogene from chromosome 8 ends up near genes of chromosome 14 that are responsible for the production of antibodies. This translocation alters the nature of the antibody that is produced and also serves to convert the proto-oncogene to an oncogene. The result is the excessive multiplication of the transformed cell creating the tumors that characterize this disease.

T-Cell Leukemia. It is now clear that a number of human cancers result from infection by a class of viruses called **retroviruses**. The mechanisms by which the retroviruses cause cancer appear to be diverse and include (1) transferring oncogenes to the cells that they infect (the retrovirus oncogenes are believed to have been picked up from previously infected human host cells), (2) activating oncogenes that are already present in the host cell, and (3) transferring genetic material to the host cells that causes the host's cells to undergo excessive multiplication.

Two retroviruses, called **HTLV-I** and **HTLV-II**, are the cause of blood diseases known as **T-cell leukemias**. These leukemias are characterized by the abnormal and excessive production of T-cells. The HTLV viruses do not carry oncogenes (and, therefore, do not transfer oncogenes to the host) and also appear not to activate oncogenes already present in the host. One model proposed to explain the oncogenic actions of the HTLV retroviruses proposes that the virus directs the infected host T-cell to produce a protein that stimulates not only production of new retroviruses but also production of new host cells. This leads to the abnormal and uncontrolled growth of the transformed cells.

The types of cells infected by the HTLV-I and HTLV-II viruses are the same as those infected by the HIV virus (i.e., the cause of AIDS). In contrast to the effects of HTLV-I and HTLV-II, HIV causes T-cell death rather than excessive multiplication.

THE IMMUNE SYSTEM AND CANCER

As noted a number of times in this chapter, in addition to protecting the body against infection, the immune system may serve as a line of defense against the growth of cancerous tumors. Therefore, it is not surprising that suppression of the body's immune system frequently results in cellular transformation and the appearance of tumors. For example, it is not uncommon for individuals with AIDS to develop the cancer called **Kaposi's sarcoma** (see earlier).

Tumor cells are not only functionally different from their normal predecessors (i.e., the normal cells that were somehow transformed) but they may also be antigenically different. Tumor cells may lose their major histocompatibility proteins, may produce new proteins that serve as antigens, or may produce membrane proteins normally found only at much earlier (e.g., embryonic) stages of development. Tumors that are the result of an infection by oncogenic viruses may display antigens that are characteristic of the virus. If the antigens present in the surface of a tumor cell are sufficiently different from those present in the surface of a normal cell, the cell will be regarded as "nonself" and be attacked by the immune system.

SELF TEST *

True/False Questions

1. NK cells are antigen-independent lymphocytes that are able to destroy foreign cells and abnormally functioning human cells.

2. B-cells are derived from embryonic stem cells that migrated to and matured in the thymus gland.

3. Class II MHC antigens are found in the surfaces of nearly all cells of the body.

4. Most of the antibodies produced by the immune system are bivalent.

5. Complement fixation is the major cause of osmotic lysis of bacteria that have infected the body.

6. Plasma cells are the source of active immunity, whereas memory cells are the source of passive immunity.

7. One of the reasons that the AIDS virus is so dangerous is that the human immune system is not able to produce antibodies against this virus.

8. The most common of the ABO series of blood types is type O.

9. Erythroblastosis fetalis can occur only when a pregnant female is Rhesus positive and the fetus that she is carrying is Rhesus negative.

10. It is now believed that some cancers are caused by certain viral infections.

Multiple Choice Questions

1. Antibodies are produced by (A) stem cells, (B) plasma cells derived from B-cells, (C) plasma cells derived from T-cells, (D) the liver, (E) the thymus gland.

2. Antibodies are also called (A) antigens, (B) haptens, (C) immunoglobulins, (D) cytokines.

3. "Cell-mediated" immunity is provided by (A) the B-cells of the immune system, (B) the T-cells of the immune system, (C) reticulocytes, (D) antigens, (E) complement.

4. Which major histocompatibility proteins are found in the surfaces of macrophages? (A) only class I, (B) only class II, (C) both class I and class II.

5. Myasthenia gravis and systemic lupus erythematosis are (A) parts of the body's immune system, (B) diseases caused by viruses, (C) diseases caused by bacteria, (D) autoimmune diseases.

6. Certain T-cell secretions attract leukocytes and macrophages to a site of infection in the body; these secretions are called (A) antibodies, (B) antigens, (C) cytokines, (D) bradykinins, (E) histamines.

7. "Membrane attack complexes" are formed by (A) antibodies, (B) complement proteins, (C) antigens, (D) cytokines.

8. The AIDS virus attacks human T-cells, and as a result (A) the T-cells of the AIDS victim are unable to produce and secrete antibodies, (B) the T-cells of the AIDS victim are unable to produce and secrete antigens, (C) cell mediated immunity is compromised (i.e., severely reduced), (D) all memory cells are lost from the body (i.e., the body loses all of its "immunological memory").

9. Complement proteins are found in (A) B-lymphocytes, (B) T-lymphocytes, (C) red blood cells, (D) blood plasma.

10. Which one of the following blood types belongs to a person known as a "universal recipient?" (A) A, (B) B, (C) AB, (D) O.

THE RESPIRATORY SYSTEM

In this chapter, we will be concerned with the respiratory system. This organ system is responsible for providing the tissues of the body with the oxygen that is needed for metabolic processes that fuel the body's activities. The respiratory system also provides for the excretion of waste carbon dioxide produced during metabolism. As an auxiliary function, movements of the respiratory muscles are also essential for oral communication (i.e., speech).

The term **respiration** refers to the entire process by which the exchange of gases between the atmosphere that surrounds our bodies and the cells that make up our tissues takes place. This involves (1) the transfer of oxygen gas to the blood from the atmosphere; (2) the transport of oxygen in the blood to the tissues; (3) the transfer of the oxygen from the blood to the tissue cells; (4) the transfer of waste carbon dioxide from the tissues to the blood; (5) the transport of carbon dioxide in the blood to the lungs; and (6) the excretion of the carbon dioxide gas into the surrounding atmosphere.

MAJOR ORGANS OF THE RESPIRATORY SYSTEM

The major organs of the respiratory system are depicted in figure 10.1 and include the nose (and/or **mouth**), **pharynx**, **trachea**, **bronchi** and **bronchioles**, **lungs**, **diaphragm**, **ribs**, and **rib muscles** (i.e., the external and internal *intercostal* muscles). During inspiration, air passes through the **nasal cavity** (or through the **oral cavity**, if the nasal cavity is blocked), and down a system of channels within which the air is filtered, warmed, and moistened before reaching the lungs' **air sacs**. At the front of the **nasal cavity**, nasal hairs act to filter the larger airborne particles that enter the cavity through the **nostrils** (or **nares**).

10-1

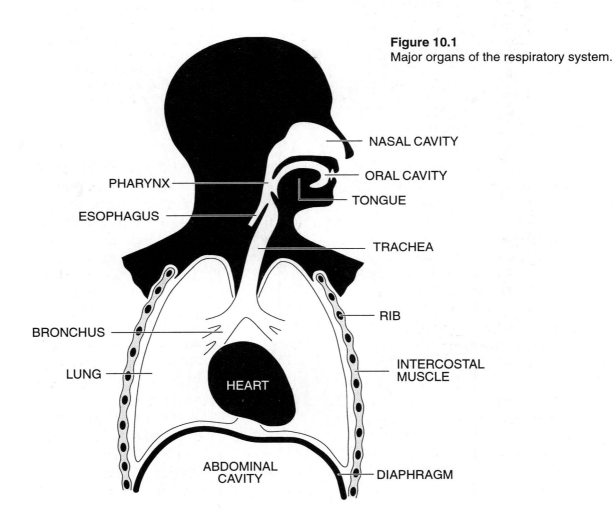

Figure 10.1
Major organs of the respiratory system.

NASAL CAVITY

ORAL CAVITY

PHARYNX

TONGUE

ESOPHAGUS

TRACHEA

RIB

BRONCHUS

INTERCOSTAL MUSCLE

LUNG

HEART

ABDOMINAL CAVITY

DIAPHRAGM

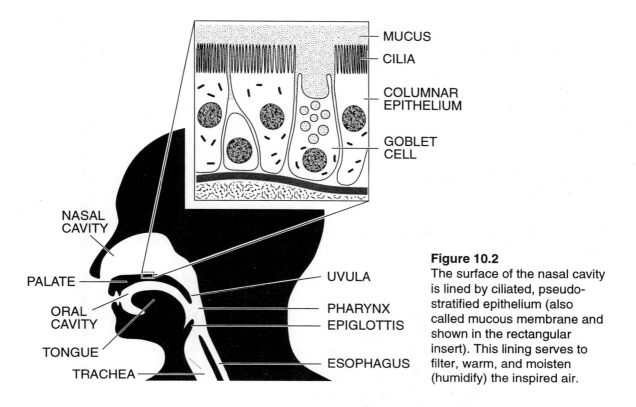

MUCUS
CILIA
COLUMNAR EPITHELIUM
GOBLET CELL

NASAL CAVITY
PALATE
ORAL CAVITY
TONGUE
TRACHEA

UVULA
PHARYNX
EPIGLOTTIS
ESOPHAGUS

Figure 10.2
The surface of the nasal cavity is lined by ciliated, pseudo-stratified epithelium (also called mucous membrane and shown in the rectangular insert). This lining serves to filter, warm, and moisten (humidify) the inspired air.

The nasal cavity and the respiratory passageways into which the nasal cavity leads are lined with a tissue called **ciliated, pseudostratified epithelium** (also known as **mucous membrane**; see figure 10.2). This tissue contains **goblet cells**, which secrete a sticky substance called **mucus** that traps small airborne debris. The hairlike **cilia** that project from the **columnar epithelial cells** of the mucous membrane beat back and forth in synchrony, thereby sweeping the mucus and the entrapped particles toward the throat (where the mucus is swallowed). Some debris inevitably reaches the lungs, but in the lungs **macrophages** ingest and destroy these particles.

The nasal and oral cavities are separated from one another by a shelflike structure called the **palate**. Anteriorly, the palate is supported by bone and is called the **hard palate**; posteriorly, the bone gives way to muscle, thereby forming the **soft palate**. At the rear of the soft palate, the oral and nasal cavities merge to form a single cavity called the **pharynx**. Protruding into the pharynx from the rear of the soft palate is a fingerlike projection called the **uvula**. Air passes downward through the pharynx and into a tube called the **trachea** (also known as the "windpipe"). The trachea is kept open (even when swallowing) by cartilaginous rings (not shown in figures 10.1 and 10.2). The upper end of the trachea forms the **larynx** (or voice box) which contains the **vocal cords**. The opening that leads into the larynx is protected by a muscular flap called the **epiglottis**. The epiglottis folds over the **glottis** at the entrance to

10-2

the larynx whenever you swallow; as a result, food is prevented from entering the trachea and blocking the air passageway.

At its base, the trachea divides into two branches, the left and right **major (or primary) bronchi**. Each major bronchus enters a lung, where it then divides into **minor (or secondary) bronchi** that carry air into the lobes of the lungs. On the right side of the chest, the lung (i.e., the "right lung") consists of three lobes (*superior*, *middle*, and *inferior*); on the left side, the lung (i.e., the "left lung") consists of two lobes (*superior* and *inferior*). The minor bronchi divide into successively smaller branches called **bronchioles**.

The smallest of the bronchioles, called **respiratory bronchioles**, end as grapelike clusters of small **air sacs** or **alveoli** (fig. 10.3). It is estimated that in the average person the lungs contain about 300 million alveoli. The large numbers of air sacs give the lungs a spongy appearance and feel. The wall of each air sac is composed of a thin, delicate layer of epithelial cells; on one side the epithelial surface faces air "drawn" into the lungs from the surrounding atmosphere. On the opposite surface, each air sac faces an extensive network of capillaries formed from branches of the pulmonary arteries. Gases (e.g., oxygen and carbon dioxide) readily pass between the air of the air sacs and the blood of the alveolar capillaries by **diffusion**. Among the cells forming the epithelial lining of the alveoli are **type II alveolar cells**. These cells secrete an oily substance (a *lipoprotein* called **surfactant**) that coats the epithelium and prevents the bubble-like air sacs from collapsing during *expiration*. Also present in the epithelium are wandering macrophages

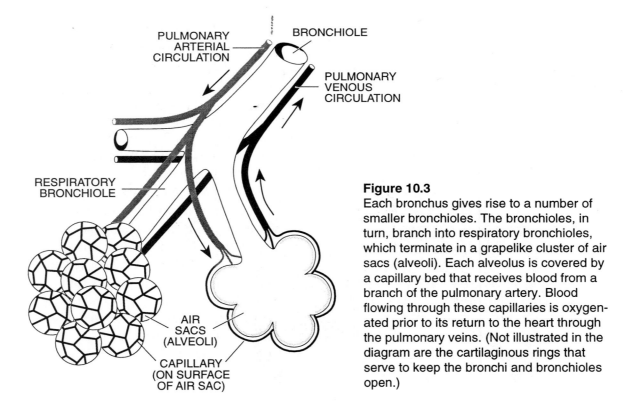

PULMONARY ARTERIAL CIRCULATION

BRONCHIOLE

PULMONARY VENOUS CIRCULATION

RESPIRATORY BRONCHIOLE

AIR SACS (ALVEOLI)

CAPILLARY (ON SURFACE OF AIR SAC)

Figure 10.3
Each bronchus gives rise to a number of smaller bronchioles. The bronchioles, in turn, branch into respiratory bronchioles, which terminate in a grapelike cluster of air sacs (alveoli). Each alveolus is covered by a capillary bed that receives blood from a branch of the pulmonary artery. Blood flowing through these capillaries is oxygenated prior to its return to the heart through the pulmonary veins. (Not illustrated in the diagram are the cartilaginous rings that serve to keep the bronchi and bronchioles open.)

that phagocytose debris and particulate material brought into the air sacs with the inspired air.

The Thoracic and Pleural Cavities

The lungs and heart are housed in the body's chest cavity or **thoracic cavity** (see figure 10.1). The walls of the thoracic cavity are formed by the **rib cage**, and the floor is formed by a thick striated muscle called the **diaphragm**; indeed, the diaphragm separates the thoracic cavity from the cavity below, the **abdominal cavity**. The surface of the lungs and the inside wall of the thoracic cavity do not directly touch one another. Rather, these surfaces are separated by two membranous sheets of a tissue called **pleura** (fig. 10.4). One pleural layer covers

the surface of each lung and is referred to as the **pulmonary pleura**; the other pleural layer, called **parietal pleura**, covers the inside wall of the thorax and also runs over the upper surface of the diaphragm.

A narrow space exists between the pulmonary pleura and the parietal pleura. This space, which is filled with a thin layer of fluid and a small amount of gas, is called the **pleural cavity**. In figure 10.4, the size of the pleural cavity has been deliberately exaggerated so that the cavity can be visualized more readily. Under normal circumstances, even after you have exhaled air, your lungs are in a partially expanded state, thereby limiting the pleural cavity to a narrow, sheet-like gap that follows the contours of the thoracic wall. As you examine figure 10.4, also note that the pleural membranes are arranged in such a manner as to create

Figure 10.4
The pulmonary and parietal pleura cover the surface of the lungs, the wall of the thoracic cavity, and the surface of the diaphragm and give rise to the left and right pleural cavities. For clarity, in this diagram the size of the pleural cavity has been exaggerated (see the text for discussion).

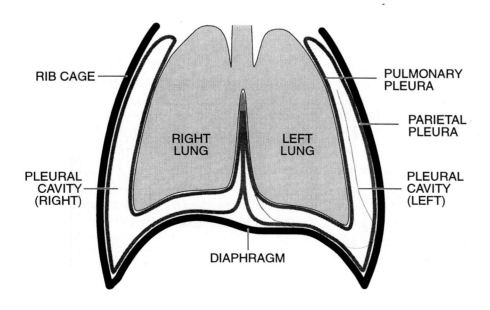

separate pleural cavities on the right and left sides of the thoracic cavity.

Unlike the gaseous contents of the lungs, which are continuous with the surrounding air via the respiratory passageways, the small quantity of gas in the pleural cavity is sealed inside by the parietal and pulmonary pleura. If the right or left wall of the chest is accidentally punctured so that atmospheric air can enter the right or left pleural cavity, it is no longer possible to inflate the lung on that side of the chest. Indeed, the lung within the injured pleural cavity "collapses," a condition known as a **collapsed lung** or **pneumothorax**.

MECHANICS OF BREATHING: INSPIRATION AND EXPIRATION

Air and Atmospheric Pressure

The movements of air *into* and *out of* the lungs are respectively referred to as **inspiration** and **expiration**. These air movements are the consequences of changes in the relative pressures of the air in the atmosphere around us and the air that fills the lungs' air sacs. At sea level, the pressure of the air in the atmosphere is about 760 mm Hg. That is, the air in the atmosphere pushes down on the surface of the earth with a pressure equal to that exerted by a column of mercury that is 760 mm (or about 30 inches) high. The pressure varies somewhat from day to day (or hour to hour) depending upon weather conditions and also diminishes rapidly with increasing altitude.

So long as the nasal passageways are not blocked (or one's mouth is open), an equilibrium is maintained between atmospheric pressure and air pressure within the lungs. If atmospheric pressure exceeds the lung pressure, then air is pushed into the lungs (from the atmosphere) until a new equilibrium is attained. On the other hand, if the lung pressure exceeds atmospheric pressure, then air is pushed from the lungs into the atmosphere.

Air is a mixture of a number of different gases. The most abundant of the gases in air is *nitrogen* gas (i.e., N_2), which accounts for close to 80% of all the gas molecules in air. Another 19-20% of the gas in air is *oxygen* gas (i.e., O_2). Of the remaining gas, about 0.1% is *carbon dioxide* (i.e., CO_2). As noted earlier, the pressure exerted on the surface of the earth by air (i.e., atmospheric pressure) is equal to that exerted by a column of mercury that is 760 mm high. The *concentration* of each of the gases in air is represented by the fraction of this total pressure that is represented by the gas. For example, the concentration of nitrogen gas in air is about 608 mm Hg (i.e., 80% of 760 mm Hg is 608 mm Hg); the concentration of oxygen is about 152 mm Hg (i.e., 20% of 760 = 152); and the concentration of carbon dioxide is about 0.8 mm Hg (0.1% of 760 = 0.8). Each of these values is referred to as the gas' **partial pressure**.

Relationship Between Gas Volume and Gas Pressure

So long as the temperature of a gas remains constant, then the pressure that a given quantity of gas exerts on its surroundings is *inversely* related to its volume (i.e., the smaller the volume, the greater the pressure and the greater the volume, the smaller the pressure). This relationship is illustrated in figure 10.5 and is also known as **Boyle's law**. As you will now learn, Boyle's law plays an important

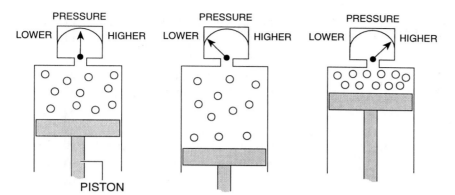

Figure 10.5
An inverse relationship exists between the volume that a given quantity of gas (small spheres) occupies and the pressure that the gas exerts; this relationship is known as Boyle's law.

role in the mechanics of inspiration and expiration.

Inspiration

Inspiration and expiration are the consequences of muscle contractions and relaxations that alter the air pressure within the lungs, causing a new pressure equilibrium to be sought with the surrounding atmosphere. Inspiration results from the contractions of the **diaphragm** and the **external intercostal muscles**. As has already been noted, the diaphragm (fig. 10.1) is a thick sheet of striated muscle that separates the thoracic cavity (which houses the lungs and heart) from the abdominal cavity (which houses the stomach, small intestine, large intestine, and other organs). Prior to contraction, the diaphragm arches upward into the thoracic cavity; that is, when the diaphragm is in a relaxed state, it is dome-shaped. When the diaphragm contracts, the dome is flattened, and this acts to increase the vertical height of the thoracic cavity (it is as though the floor of a sealed chamber were suddenly lowered).

The front, rear, and side walls of the thoracic cavity are formed by the **rib cage** and the two sets of intercostal muscles–the *external* and *internal intercostal muscles*. It is the external intercostal muscles that play a role in inspiration. Each external intercostal muscle originates on the undersurface of a rib where the rib articulates with the vertebral column (i.e., at the rear of the body). The muscle encircles the rib cage and inserts into the upper surface of the rib *below* where that rib joins the **sternum** (or breastbone). When the external intercostal muscles contract, they act to rotate the ribs upward (fig. 10.6). This action pushes the sternum forward (further away from the vertebral column), thereby increasing the anterior-posterior dimensions of the thoracic cavity and increasing the cavity's volume (it is as though the front wall of a sealed chamber suddenly moved forward). Upward rotation of the ribs is accompanied by some lateral movement; that is, some widening of the thoracic cavity also occurs. Thus, the simultaneous contractions of the diaphragm and external intercostal muscles act to enlarge the thoracic cavity. At rest, when breathing quietly, about 75% of the thoracic cavity's volume increase is due to the contraction of the diaphragm. The deeper (and more rapid) breathing that accompanies physical activity involves more extensive contractions of the external intercostal muscles.

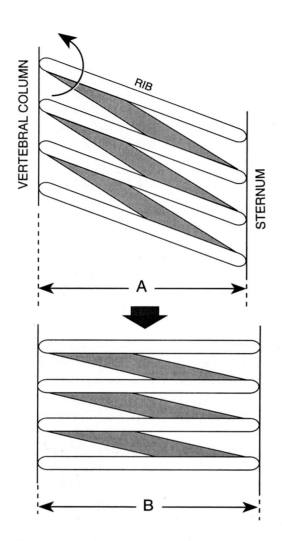

Figure 10.6
Contractions of the external intercostal muscles (represented in grey) rotate the rib cage upward, thereby pushing the sternum forward. This increases the anterior-posterior dimensions of the thorax (i.e., from A to B). Some widening of the rib cage accompanies the upward rotation. These actions increase the volume of the thoracic cavity.

Although the lungs are not attached to the walls of the thorax, the respective downward, forward, and lateral excursions of the thoracic floor and wall are accompanied by expansion of the lungs. Thus, the volume of the lungs increases in sympathy with the increasing volume of the thorax.

As is expected from Boyle's law, the increase in lung volume that results from contractions of the diaphragm and external intercostal muscles drops the air pressure within the lungs to a value lower than atmospheric pressure. Consequently, air moves from the atmosphere (where the pressure is temporarily higher) into the lungs (where the pressure is temporarily lower) in order to bring the two pressures back into equilibrium. It is important to emphasize that air moves from the atmosphere into the lungs *because the lungs were expanded by muscle contractions*.

From the above discussion, it should be clear that the lungs are not expanded by the inward rush of air; rather, the expansion of the lungs causes air to be pushed into the respiratory passageways and into the air sacs from the atmosphere. It is, therefore, not correct to infer that lung volume increases *because of the entry of air*. Recognizing the cause-and-effect relationships between changes in lung volume and movements of air between the atmosphere and the lungs is important to your understanding the mechanics of respiration.

Expiration

Expiration, the movement of air from the lungs back into the surrounding atmosphere, can take either of two forms: **passive expiration** or **active expiration**. Passive expiration is the mechanism that is at work during resting or quiet breathing, whereas active expiration occurs when one is engaged in more strenuous activity (such as walking, running, etc.).

Passive expiration occurs when, at the end of inspiration, the diaphragm and external intercostal muscles are returned to their relaxed state. When the diaphragm relaxes, it resumes its normal shape and it once again arches upward from the abdominal cavity into the thoracic cavity. Relaxation of the external intercostal muscles is followed by the downward rotation of the rib cage. The downward movement of the ribs is assisted by gravity and by the natural elasticity of the thorax wall (which is stretched during inspiration).

When the diaphragm arches upward into the thorax, the volume of the thorax is decreased (i.e., when the floor of the thorax rises, the chamber's vertical height is diminished). When the rib cage rotates downward, the sternum is pulled backward. This too acts to decrease the volume of the thorax by diminishing the chamber's anterior-posterior dimension (i.e., when the front wall of the thoracic cavity moves rearward, the chamber's depth is reduced). These movements squeeze the lungs, causing the volume of the lungs to decrease. The decrease in lung volume raises the air pressure in the lungs to a value that exceeds atmospheric pressure. As a result of the pressure difference that now exists between air in the lungs and air in the atmosphere, air moves out of the lungs.

It is important to note that during expiration the decrease in lung volume *precedes* the movement of air out of the lungs. That is to say, air leaves the lungs *because the lung volume decreases* (i.e., the decreased lung volume raises the pressure, thereby causing air to move in the direction of the lower, atmospheric pressure).

The lungs are not compressed by the outward rush of air; rather, the compression of the lungs causes air to be pushed out of the air sacs and respiratory passageways into the atmosphere. It is, therefore, not correct to infer that lung volume decreases *because of the loss of air*.

In the average person at rest, the cycle of inspiration and expiration that characterizes quiet breathing occurs about 16-18 times each minute. This rate is adequate to supply the body's need to acquire oxygen and eliminate carbon dioxide under resting conditions. During periods of increased muscular activity, both the rate and the depth of inspiration and expiration must be increased. Increasing the depth and rate of inspiration is achieved by more forcible and more rapid contractions of the diaphragm and external intercostal muscles. However, the downward rotation of the rib cage that occurs when the external intercostal muscles relax is not rapid (or forceful) enough to provide the necessary speed and depth of expiration. The *active expiration* that is required under these conditions is achieved as follows.

Attached to the upper surfaces of the ribs where the ribs articulate with the vertebral column are the origins of the **internal intercostal muscles**. Each of these muscles encircles the rib cage and inserts into the undersurface of the rib *above*, where the rib joins the **sternum**. When the internal intercostal muscles contract, they rotate the ribs downward (fig. 10.7). This action pulls the sternum backward (closer to the vertebral column), thereby quickly decreasing the anterior-posterior dimensions of the thoracic cavity and decreasing the cavity's volume. The actions of the internal intercostal muscles may be accompanied by contractions of the walls of the abdominal cavity (which lies below the thoracic cavity). Such contractions force the abdominal organs upward against the undersurface of the diaphragm,

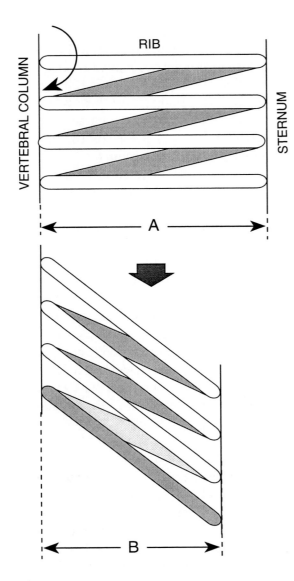

Figure 10.7
Contractions of the internal intercostal muscles (grey) pull and rotate the rib cage downward, thereby drawing the sternum backward. This quickly decreases the anterior-posterior dimensions of the thoracic cavity (from A to B), raising air pressure inside the lungs.

thereby causing the diaphragm to arch further into the thoracic cavity.

GAS EXCHANGES BETWEEN THE AIR SACS, THE BLOOD, AND THE TISSUES

Oxygenation of the Blood

As noted earlier, the concentration of oxygen in inspired air is about 152 mm Hg (i.e., the concentration of a gas is expressed as its partial pressure). An analysis of the oxygen concentration of the air in the lung's air sacs reveals that it is about 105 mm Hg. The lowered partial pressure of oxygen in alveolar air (compared with atmospheric air) is due to the added water vapor content (radiated from the mucous membrane lining the respiratory passageways) and the increased carbon dioxide. By comparison, the concentration of oxygen gas dissolved in the blood plasma and red blood cells entering the pulmonary circulation is about 40 mm Hg. (An equilibrium exists between plasma and red blood cells such that each of the two phases contains the same concentration of dissolved oxygen.) The 65 mm Hg partial pressure difference between alveolar oxygen and pulmonary blood oxygen (i.e., 105 - 40 = 65) causes oxygen gas to diffuse from the air sacs across the alveolar and capillary epithelium and into the bloodstream (see figures 10.8 and 10.9).

Oxygen gas that diffuses into the blood from the alveoli dissolves in the blood plasma and also enters the red blood cells. If there were no hemoglobin in the red blood cells, then the blood leaving the lungs and returning to the left side of the heart would contain about the same amount of oxygen as alveolar air. However, the presence of hemoglobin dramatically alters the blood's oxygen-carrying capacity. Oxygen entering the red blood cells from the plasma reacts with

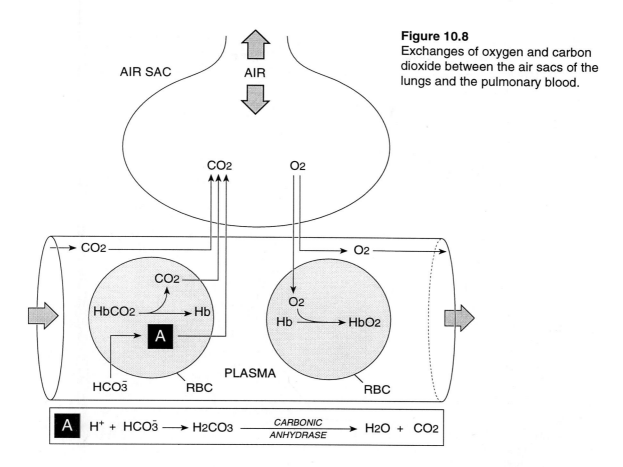

Figure 10.8
Exchanges of oxygen and carbon dioxide between the air sacs of the lungs and the pulmonary blood.

hemoglobin and becomes chemically bound to it. (Each hemoglobin molecule contains 4 atoms of iron and each one of these iron atoms can bind two atoms of oxygen; see chapter 8.) The oxygen that is bound to hemoglobin is no longer in the free, gaseous state and therefore *does not contribute to the oxygen partial pressure*. Consequently, the partial pressure gradient that causes oxygen to diffuse into the blood from the alveoli remains intact as hemoglobin accepts a full oxygen load. It is only when hemoglobin's oxygen binding capacity approaches the saturation point that the partial pressure of oxygen gas in the blood starts to approach the 105 mm Hg level present in the air sacs. The upshot of this is that hemoglobin raises the

oxygen-carrying capacity of the blood about 70-fold.

Were there no hemoglobin in red blood cells, then 100 c.c. of blood could accept about 0.3 c.c. of gaseous oxygen. However, the presence of hemoglobin increases the oxygen-carrying capacity of 100 c.c. of blood to 20 c.c. The additional 19.7 c.c. of oxygen (i.e., 20 - 0.3 = 19.7) is converted from its gaseous state to its hemoglobin-bound state when it reacts with hemoglobin. Thus, each 100 c.c. of blood that returns to the left side of the heart through the pulmonary veins contains two forms of oxygen: (1) oxygen gas dissolved in the plasma (and dissolved in the red cells)–amounting to about 0.3 c.c. and which exerts a partial pressure of about

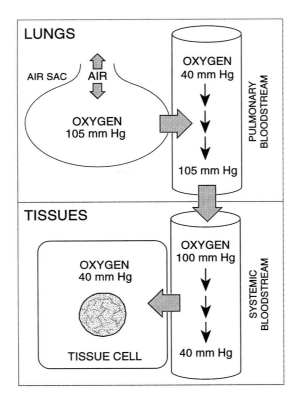

LUNGS

AIR SAC / AIR

OXYGEN
105 mm Hg

OXYGEN
40 mm Hg

105 mm Hg

PULMONARY BLOODSTREAM

TISSUES

OXYGEN
40 mm Hg

TISSUE CELL

OXYGEN
100 mm Hg

40 mm Hg

SYSTEMIC BLOODSTREAM

Figure 10.9
Partial pressure gradients that cause oxygen to diffuse into the blood from the air sacs and diffuse into the tissues from the blood.

105 mm Hg and (2) hemoglobin-bound oxygen, which, if it were converted to gaseous oxygen, would occupy an additional 19.7 c.c.

Delivery of Oxygen to the Tissues

Because they are continuously consuming oxygen during their metabolism, tissue cells (and the tissue fluid that bathes the cells) have a low oxygen partial pressure (about 40 mm Hg in a person who is maintaining a modest level of activity). Although some oxygen is lost to the walls of blood vessels carrying systemic blood away from the left

side of the heart, the partial pressure of oxygen in the blood entering a tissue's capillary beds is not far below 100 mm Hg. The 60 mm Hg partial pressure difference between the blood and the tissue (i.e., 100 - 40 = 60) results in the diffusion of oxygen out of the blood (figures 10.9 and 10.10). The falling partial pressure of oxygen in the blood acts as a trigger for the release of oxygen that has been bound to hemoglobin. Figure 10.11 graphically depicts the relationship between the partial pressure of oxygen in the blood plasma and the *percent saturation* of hemoglobin (i.e., the percent of hemoglobin's oxygen-binding sites that are occupied by oxygen molecules). Between 100 mm Hg and 40 mm Hg partial pressure, hemoglobin unloads about 25% of its oxygen complement. Notice that if the partial pressure drops still further (as it does when a person's level of activity increases), the dramatic fall in hemoglobin saturation continues. For example, between 40 mm Hg and 30 mm Hg partial pressure, hemoglobin releases another 20% of its initial oxygen complement.

Upon its release from hemoglobin, the oxygen returns to its gaseous form, diffuses from the red cells into the plasma, and then passes along the sustained concentration gradient from the blood into the tissues. By the time blood leaves the capillary networks of a tissue and returns to the right side of the heart, the partial pressure of oxygen in the blood has fallen to 40 mm Hg (or lower).

Transport of CO_2 in the Blood

Carbon dioxide is one of the major waste products of tissue metabolism. In a typical, actively metabolizing tissue, the partial pressure of CO_2 is about 50 mm Hg. In contrast, CO_2 dissolved in the systemic blood enter-

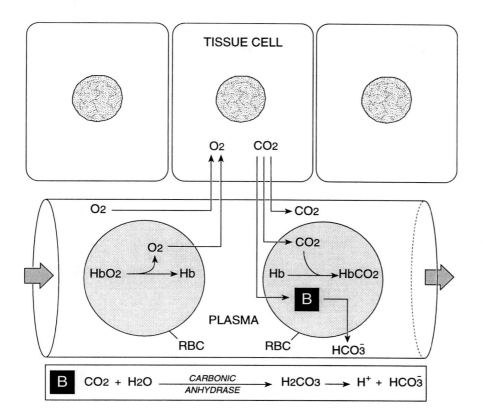

Figure 10.10
Exchanges of oxygen and carbon dioxide between the systemic blood
and the tissues of the body.

ing a capillary bed is about 40 mm Hg. The 10 mm Hg partial pressure difference (i.e., 50 - 40 = 10) causes CO_2 to diffuse from the tissue cells and the tissue fluid that bathes them into the bloodstream (fig. 10.12). CO_2 entering the bloodstream from a tissue is carried back to the right side of the heart and then into the pulmonary circulation in three different chemical forms: (1) as a *dissolved gas*, (2) *chemically combined with hemoglobin*, and (3) as *bicarbonate ions* (figures 10.8 and 10.10).

About 7% of the CO_2 that enters the blood stays in its gaseous form and dissolves in the plasma. Another 20% of the CO_2 diffuses into

the red blood cells where it reacts with hemoglobin to form **carbaminohemoglobin**. Unlike oxygen, which combines with the iron atoms of hemoglobin's heme groups, CO_2 combines with amino acids of this protein. (Do not confuse CO_2 with CO [i.e., *carbon monoxide*]; CO is an extremely poisonous gas that displaces oxygen from hemoglobin and combines almost irreversibly with hemoglobin's iron atoms, thereby destroying hemoglobin's ability to transport oxygen.)

The remaining 73% of the carbon dioxide transported in the blood takes the form of **bicarbonate ions** (HCO_3^-), nearly all of

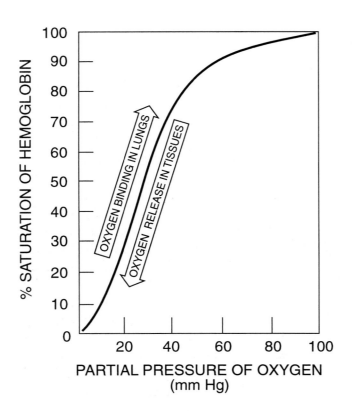

Figure 10.11
Oxygen dissociation curve of hemoglobin.

which are carried in the plasma. The conversion of CO_2 to bicarbonate occurs in the following way. When carbon dioxide and water are mixed together, some of the CO_2 reacts with water to form **carbonic acid** (H_2CO_3); i.e.,

$$CO_2 + H_2O \xrightarrow{\textit{Carbonic anhydrase}} H_2CO_3$$

Nearly all of the carbonic acid that is formed breaks down (i.e., "dissociates") immediately into bicarbonate ions and hydrogen ions (H^+), leaving only a small percentage of the carbonic acid molecules intact; i.e.,

$$H_2CO_3 \longrightarrow HCO_3^- + H^+$$

In water (or in blood plasma), this reaction sequence occurs very slowly and little carbonic acid and bicarbonate are produced. However, red blood cells contain an enzyme called *carbonic anhydrase* that has a dramatic effect on the formation of carbonic acid. When the level of CO_2 in red blood cells is *rising* (as occurs when CO_2 diffuses into the blood from the tissues), *carbonic anhydrase* catalyzes the formation of carbonic acid. In contrast, when the level of CO_2 in red blood cells is *falling* (as occurs in the lungs when CO_2 diffuses out of the blood into the alveoli; see below), *carbonic anhydrase* catalyzes the breakdown of carbonic acid into water and CO_2.

CO_2 diffusing into the blood from the tissues enters red blood cells where *carbonic anhydrase* converts the CO_2 to carbonic acid

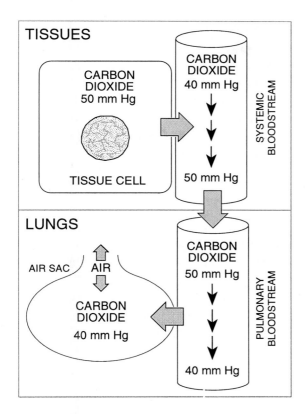

TISSUES

CARBON DIOXIDE 50 mm Hg

TISSUE CELL

CARBON DIOXIDE 40 mm Hg

50 mm Hg

SYSTEMIC BLOODSTREAM

LUNGS

AIR SAC / AIR

CARBON DIOXIDE 40 mm Hg

CARBON DIOXIDE 50 mm Hg

40 mm Hg

PULMONARY BLOODSTREAM

Figure 10.12
Partial pressure gradients that cause carbon dioxide to diffuse from the tissues into the blood and from the blood into the air sacs of the lungs.

(the reaction is driven in the direction of carbonic acid because the CO_2 level is rising). The carbonic acid that is formed dissociates into hydrogen ions and bicarbonate, and the bicarbonate is then secreted from the red cells into the plasma. Therefore, although the bicarbonate originates in the red blood cells, it is transported to the lungs in the blood plasma. Bicarbonate enters the plasma in exchange for chloride ions (Cl^-), which enter the red blood cells. This exchange is referred to as the **chloride shift**.

Elimination of CO_2 in the Lungs

The concentration of CO_2 gas in alveolar air (about 40 mm Hg partial pressure) is less than that in the blood entering the pulmonary capillaries (about 50 mm Hg partial pressure; fig. 10.12). As a result, CO_2 diffuses from the blood into the air sacs. As the CO_2 level of the blood falls, hemoglobin-bound CO_2 is released, diffuses out of the red cells, and passes into the air sacs (see figure 10.8). Moreover, the falling level of CO_2 drives the *carbonic anhydrase* reaction in the direction of CO_2 formation. That is, plasma bicarbonate is taken up by the red cells and reacts with hydrogen ions to form carbonic acid, which *carbonic anhydrase* then quickly converts to CO_2 and water. The CO_2 produced in this way leaves the red cells and diffuses into the alveolar air. Once again there is a "chloride shift;" that is, the uptake of bicarbonate by the red blood cells is accompanied by the expulsion of chloride ions into the plasma.

Earlier it was noted that the oxygen-carrying capacity of the blood is considerably increased through the conversion of most of the dissolved gas to oxygenated hemoglobin. Because most of the carbon dioxide that enters the blood is either converted to bicarbonate ions or combines with hemoglobin to form carbamino hemoglobin, the carbon dioxide-carrying capacity of the blood is similarly increased.

Nitrogen Gas

Although the respiratory system functions to provide an adequate exchange of oxygen and carbon dioxide between the air and the tissues of the body, the major molecular component of air is nitrogen gas (i.e., N_2). Indeed, N_2

accounts for about 80% of all of the gas in air. Therefore, the air sacs of the lungs are rich in N_2. Nitrogen gas enters the bloodstream and establishes a partial pressure equilibrium with the nitrogen present in the alveolar air. However, nitrogen gas is not metabolized by the body and so none is consumed or produced by the tissues. Blood contains no nitrogen carriers and nitrogen gas is not converted to any other molecular form. Consequently, the only N_2 in the blood is the small amount that dissolves in the blood plasma.

THE LUNG VOLUMES

The amount of air that enters and leaves the lungs during each respiratory cycle can be varied according to the body's need to ac-quire additional (or less) oxygen or to eliminate more (or less) carbon dioxide. Changes in the volume of air entering and leaving the lungs are brought about by varying the extent and frequency of contractions of the diaphragm and external intercostal muscles during inspiration and by varying the actions of the internal intercostal muscles during expiration.

The volumes of air entering and leaving the lungs during the respiratory cycle can be measured using an instrument called a **spirometer**. During quiet (i.e., resting) breathing, the air that enters (or leaves) the lungs in one respiratory cycle is called the **tidal volume** (fig. 10.13). In the average, adult male this amounts to about 500 c.c. During maximal inspiration and expiration, the volume of air

10-3

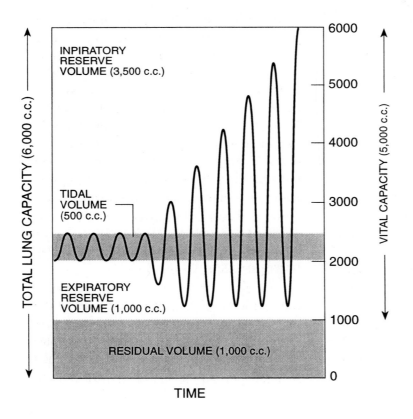

Figure 10.13
The various lung volumes in an average, adult male (see text for explanation). Rising portions of the curve represent air entering the lungs. Descending portions of the curve represent air leaving the lungs.

exchanged with the atmosphere can be increased about tenfold (to about 5,000 c.c. or 5 liters). This maximum volume is called the **vital capacity**. The additional air that can be inspired after a normal inspiration is completed is known as the **inspiratory reserve volume**. In a person whose vital capacity is 5,000 c.c., the inspiratory reserve volume amounts to about 3,500 c.c. The additional volume of air that can be expired after a normal expiration is completed is known as the **expiratory reserve volume**. In a person whose vital capacity is 5,000 c.c., the expiratory reserve volume amounts to about 1,000 c.c. As seen in figure 10.13, the larger the tidal volume, the smaller are the inspiratory and expiratory reserve volumes.

Even after all of the expiratory reserve volume has been expelled from the lungs, the lungs are still not empty. About 1,000 c.c. of air remains in the lungs; this volume is known as the **residual volume**. The sum of the residual volume and the vital capacity is known as the **total lung capacity**. In a person whose vital capacity is 5,000 c.c., the total lung capacity amounts to 6,000 c.c.

REGULATION OF BREATHING

Although we only occasionally exercise conscious control over the respiratory movements, the intercostal muscles and diaphragm do not possess the spontaneous rhythmicity that characterizes heart muscle. Rather, the respiratory musculature must be stimulated by nerve impulses carried by certain spinal nerves. The actions of these nerves are regulated by *respiratory centers* in the **brain stem**.

The overall control of respiration is based on five interactions: (1) motor nerve pathways that take their origin in respiratory centers located in the **medulla** of the brain stem; (2) respiratory centers in the **pons** that can exercise control over the medullary centers, (3) **stretch receptors** (interoceptors) in the lungs that monitor increasing lung volume and which can apprise the brain stem of the lung volume changes, (4) input to the brain stem's respiratory centers from **chemoreceptors** in the brain stem itself and in the **aortic arch** and **carotid bodies**, and (5) conscious (cerebral cortex) intervention in the breathing cycle in order to accommodate such activities as speech and the voluntary holding of one's breath.

Inspiratory and Expiratory Centers

In the medulla of the brain stem are two centers that play important parts in the regulation of breathing. These are known as the **inspiratory center** and the **expiratory center** (fig. 10.14). Impulses that are initiated by the inspiratory center in a periodic (cyclical) manner pass downward into the spinal cord, where they are transmitted to motor fibers of spinal nerves in the cervical (neck) and thoracic (chest) regions. The cervical fibers form the **phrenic nerve**, which innervates the diaphragm, whereas the thoracic fibers become part of the **intercostal nerves** and innervate the external intercostal muscles (fig. 10.14). Contractions of the diaphragm and external intercostal muscles that result from this innervation expand the thoracic cavity and lungs, leading to inspiration. Remaining uncertain is whether the cyclical activities of the inspiratory center are *intrinsic* to the cells forming the center (i.e., although the presence of a built-in rhythmicity is popularly supported, at the

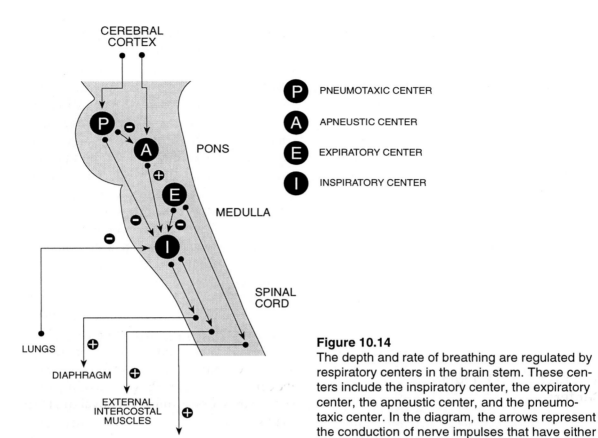

CEREBRAL CORTEX

PONS

MEDULLA

SPINAL CORD

LUNGS

DIAPHRAGM

EXTERNAL INTERCOSTAL MUSCLES

INTERNAL INTERCOSTAL MUSCLES

P PNEUMOTAXIC CENTER

A APNEUSTIC CENTER

E EXPIRATORY CENTER

I INSPIRATORY CENTER

Figure 10.14
The depth and rate of breathing are regulated by respiratory centers in the brain stem. These centers include the inspiratory center, the expiratory center, the apneustic center, and the pneumotaxic center. In the diagram, the arrows represent the conduction of nerve impulses that have either stimulatory (+) or inhibitory (−) effects on their targets.

present time agreement among physiologists is not universal).

The cyclical activity of the inspiratory center is all that is needed to sustain resting breathing. The cessation of the flow of motor impulses to the diaphragm and lungs as the inspiratory center temporarily shuts down is followed by passive expiration as the diaphragm relaxes and once again arches into the thoracic cavity and the ribcage rotates downward. It is worth repeating that during resting breathing, most of the lung volume changes can be attributed to the movements of the diaphragm.

If the level of the body's activity increases (as during walking, running, or exercising),

passive expiration is not rapid (or deep) enough to support the body's respiratory needs. In this case, the expiratory center becomes active. Impulses arising in the expiratory center are conducted into the thoracic region of the spinal cord, where they are transmitted to motor fibers of the intercostal nerves that innervate the internal intercostal muscles (fig. 10.14). Contractions of the internal intercostal muscles lead to more forceful and rapid expiration of air from the lungs. Evidence also exists that the expiratory center has some inhibitory control over the inspiratory center.

The actions of the inspiratory center are also influenced by a feedback mechanism involving the lungs known as the **Hering-Breuer reflex**. In this reflex, expansion of the lungs during inspiration serves to stimulate stretch receptors (interoceptors) located in the walls of the bronchi and bronchioles. These receptors then initiate the flow of sensory impulses along fibers of the vagus nerve to the inspiratory center, where they have an inhibitory effect. Inhibition of the inspiratory center acts to bring inspiration to a halt (thereby halting further stretch within the lungs). The Hering-Breuer reflex is believed to have significant influence only during the more rapid and deeper breathing cycles that characterize high levels of body activity and acts to prevent over inflation of the lungs.

Apneustic and Pneumotaxic Centers

Breathing is also influenced by the activities of two other respiratory centers in the brain stem; these are the **apneustic center** (which is located in the lower region of the **pons**) and the **pneumotaxic center** (which is located in the upper region of the pons; figure 10.14). It has been shown that the apneustic center sends stimulatory impulses to the inspiratory center, thereby prolonging inspiration, but it is uncertain whether this action has any effect during normal, quiet breathing. The pneumotaxic center has an inhibitory effect on inspiration, reducing both the depth and the rate of breathing. The inhibitory effects of the pneumotaxic center are believed to be mediated through the apneustic center; that is, impulses are sent from the pneumotaxic center to the apneustic center, which responds by reducing its stimulatory effects on the inspiratory center.

Influences of Blood Chemistry on Breathing

During vigorous activity, the body's muscles consume more oxygen and produce more carbon dioxide. Oxygen deficiency and carbon dioxide excess act to increase both the depth and rate of respiration through two mechanisms: (1) direct and indirect effects of these two gases on the brain stem, and (2) indirect effects on the brain stem that are facilitated by peripheral chemical receptors.

Medullary Chemoreceptors. Near the ventral surface of the medulla (close to where the glossopharyngeal and vagus nerves exit the medulla) is a cluster of chemoreceptors that are believed to have an important influence on the rate and depth of breathing. These receptors are stimulated by changes in the concentration of hydrogen ions (H^+) in the surrounding tissue fluid. The level of H^+ in the tissue fluid is determined in turn by the H^+ concentration in the medulla's **cerebrospinal fluid** (CSF).

During increased muscle activity, the amount of dissolved CO_2 gas entering the blood from the muscles increases and raises the blood's CO_2 partial pressure. The generalized elevation of circulating CO_2 leads to the diffusion of this gas from the blood into the CSF. In the CSF, CO_2 reacts with water to produce carbonic acid, which then dissociates into bicarbonate ions and H^+ (the chemical conversion of CO_2 to bicarbonate and H^+ was described earlier in the chapter). Because there is little protein dissolved in the CSF, its buffering capacity is limited. As a result, the pH of the CSF falls and so does the pH of the medulla's tissue fluid. This drop in pH is the stimulus for the medullary chemoreceptors. Stimulated in this manner, the chemoreceptors of the medulla initiate the flow of nerve impulses to the inspiratory center, thereby bringing about a more rapid (and deeper) respiratory cycle. The increased respiratory rate that accompanies elevated muscle activity acts to normalize the pH of the blood and CSF by eliminating more CO_2.

Increased muscle activity is also accompanied by a fall in the concentration of oxygen gas dissolved in the blood. The fall in the O_2 level, however, is not as dramatic as the rise in CO_2 because of the oxygen reservoir that is bound to hemoglobin and which is released as O_2 diffuses from the blood into the tissues. Moreover, the increased respiratory rate makes more oxygen available to the pulmonary blood.

Peripheral Chemoreceptors. The aortic arch and carotid arteries contain interoceptors that respond to changes in the levels of oxygen and carbon dioxide dissolved in the blood (recall [chapter 7] that these structures also contain receptors responsive to changing blood pressure; see figure 7.14). The aortic arch receptors are known as the **aortic bodies**, while the carotid artery receptors (located at the division of the carotid arteries into their internal and external branches; figure 7.14) are called the **carotid bodies**.

The aortic and carotid bodies are sensitive to the partial pressure of oxygen gas dissolved in the blood and when this level falls, the chemoreceptors stimulate the dendritic endings of sensory nerve fibers that carry impulses to the respiratory centers of the medulla; this is followed by an increase in motor impulses from the inspiratory center, bringing about an increase in the rate and depth of breathing. Sensory fibers arising from the aortic bodies communicate with the medulla via the vagus nerve, whereas the fibers arising from the carotid bodies become part of the glossopharyngeal nerve. The carotid bodies (but not the aortic bodies) also respond to a fall in the pH of the arterial blood, as occurs when the blood's CO_2 level rises.

SPEECH

The organs of the respiratory tract are not only responsible for gas exchanges between the atmosphere and the blood but also make speech possible. Speaking requires major modifications and interruptions of the respiratory cycle (i.e., you cannot breathe in a normal, cyclical pattern and speak at the same time). The nerve impulses that provide for speech and that simultaneously modify the respiratory pattern arise in the cerebral cortex. These impulses go to the respiratory centers in the medulla and pons and to the musculature of the mouth, tongue, and larynx. Speech occurs during expiration as air is forced past the vocal cords, causing them to vibrate. The frequency of vibration (hence the pitch of the sound; see chapter 6) varies according to the position and tension in the vocal cords, properties that are regulated by the muscles of the larynx. Movements of the tongue and lips further modify the resulting sounds.

SELF TEST[*]

True/False Questions

1. Oxygen and carbon dioxide account for less than one-half of all the gas that enters the lungs during inspiration. Indeed, most of the gas that enters the lung's air sacs from the surrounding atmosphere is nitrogen gas.

2. Vigorous contraction of the internal intercostal muscles can expel all but about 5 percent of the air that is present in the lungs.

3. Carbon monoxide is dangerous because it combines irreversibly with the iron atoms of hemoglobin, thereby rendering the hemoglobin incapable of transporting oxygen.

4. According to Boyle's law, the pressure of a gas increases as its volume increases.

5. The pneumotaxic and apneustic respiratory centers are in the medulla of the brain stem.

6. The aortic bodies respond to falling levels of oxygen in the arterial blood.

Multiple Choice Questions

1. The contractions of which one of the following combinations of muscles brings about inspiration? (A) diaphragm & abdominal muscles, (B) internal intercostals & diaphragm, (C) internal intercostals & abdominal muscles, (D) external intercostals & diaphragm, (E) external intercostals & abdominal muscles.

2. Most of the carbon dioxide released into the bloodstream from the body's tissues is carried to the lungs as (A) carbamino hemoglobin, (B) dissolved CO_2 gas, (C) bicarbonate ions inside red blood cells, (D) bicarbonate ions in the blood plasma, (E) none of these choices is correct.

3. A person's vital capacity is equal to (A) 3% of the total body weight, (B) the sum of the inspiratory reserve volume and the expiratory reserve volume, (C) the sum of the inspiratory reserve volume, the expiratory reserve volume, and the residual volume, (D) the total lung capacity minus the lung's residual volume.

4. The epiglottis (A) warms and filters the inspired air, (B) prevents swallowed food from getting into the esophagus, (C) prevents swallowed food from getting into the pharynx, (D) prevents swallowed food from getting into the trachea, (E) humidifies the inspired air.

5. Blood returning to the heart from the lungs' pulmonary circulation would contain both gaseous oxygen and oxygenated hemoglobin. Approximately what percentage of the oxygen in this blood is in the gaseous state? (A) 0% (i.e., none), (B) 2%, (C) 20%, (D) 70%, (E) 100%.

---

[*] *The answers to these test questions are found in Appendix III at the back of the book.*

THE DIGESTIVE SYSTEM

The organs of the digestive tract break down the food that we eat and convert it into a form that can then be absorbed into the bloodstream and into the lymph. For the most part, the breakdown of the food is a chemical process in which digestive enzymes progressively convert large molecules (principally proteins, carbohydrates, and fats) into simpler, smaller molecules. The breakdown of the food is also assisted by a variety of mechanical processes, such as chewing and the vigorous muscular contractions of the walls of the digestive organs. Once digestion and absorption are completed, body water added to the food during its passage from one organ to another is removed from the residual matter and returned to the bloodstream, with the result that a semisolid, undigested waste (called **feces**) is eliminated from the body.

FOOD AND ITS FATE

The food that we eat contains a great variety of different substances. The major food constituents include the (1) **nutrients**, which are **proteins**, **carbohydrates** (much of it **polysaccharide**), and **fat** (or **lipid**), (2) **minerals**, (3) **vitamins**, and (4) **trace elements**. Among these, only the nutrients undergo major modifications as they pass through the various segments of the digestive tract. The goal of the digestive system is to sequentially convert the protein into **amino acids**, the carbohydrate into **monosaccharides**, and the fats into **sterols** (primarily *cholesterol*), **glycerol**, **monoglycerides**, and **fatty acids**.

The various stages through which the proteins pass during their digestion are summarized in figure 11.1. Proteins consist of one or more unbranched chains of amino acid called **polypeptides** (see chapter 2). The polypeptide chains are first split into smaller chains called **peptides**. The peptides are then split into still smaller chains consisting of just a few amino acids and called **oligopeptides**. Finally, the oligopeptides are broken down into **dipeptides** and **tripeptides** ("chains" of just two and three amino acids) and individual **amino acids**. Although some dipeptides and tripeptides may be absorbed from the digestive tract, most of the digested protein that is absorbed is in the free (individual) amino acid form. About 20 different amino acids are produced by the digestion of the proteins in our food; these amino acids are identified in figure 2.17.

Carbohydrates pass through a similar series of stages during their digestion (fig. 11.1). Most of the carbohydrate in our diet takes the form of long, branched chains of sugars called **polysaccharides**. Much of the dietary polysaccharide is the **starch** that is present in the fruit and vegetables that we eat. However, smaller quantities of polysaccharides are acquired from the glycogen that is present in meats. (Interestingly, the

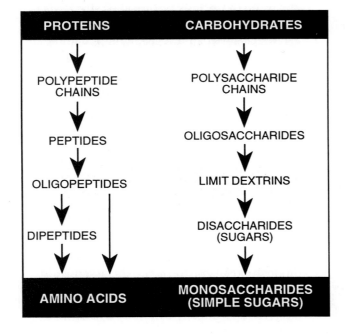

Figure 11.1
Intermediate stages in the digestion of proteins and carbohydrates present in food.

most abundant polysaccharide on the earth is **cellulose**–the main structural constituent of plant tissues. However, the human digestive tract can't digest cellulose and, therefore, can't access its rich sugar content.) The polysaccharides are first broken down into smaller chains of simple sugars called **oligosaccharides**. These are then degraded to **limit dextrins** (very short chains containing "branch points;" see figure 2.20). The limit dextrins are then broken into **disaccharides** ("chains" of just two simple sugars). Finally, the disaccharides are converted to simple sugars or **monosaccharides**. One particular monosaccharide is the most abundant product of polysaccharide digestion; this is **glucose** (also known as **dextrose**). Some of the carbohydrate in our diets consists of disaccharides. For example, "table sugar" (which we add to coffee or tea and which is present in most soft drinks) is the disaccharide called **sucrose**. When sucrose is digested, the products are the monosaccharides glucose and **fructose** (fructose is also present in fruit). Another common disaccharide is **lactose**–the sugar present in milk. When lactose is digested, the products are glucose and **galactose**. It is only at the monosaccharide stage that the products of carbohydrate digestion may be absorbed into

the bloodstream from the digestive tract.

Fats are a heterogeneous collection of nutrients which, unlike proteins and carbohydrates, do not consist of chains of repeating units. Much of the fat present in food is **neutral fat** or **triglycerides** (fig. 11.2). Other fats in food include **sterols** (mostly a particular sterol called **cholesterol**) and **sterol esters** (mainly cholesterol esters). The digestion of triglycerides produces **glycerol**, **monoglycerides**, and **fatty acids**. The digestion of cholesterol esters produces fatty acids and cholesterol (fig. 11.2). **11-1**

MAJOR ORGANS OF THE DIGESTIVE SYSTEM

The digestive system includes the organs of the digestive tract itself and the accessory organs and glands that provide enzyme-rich digestive fluids (figures 11.3 and 11.4). The major organs of the digestive tract include (1) the **mouth** (or **oral cavity**), (2) the **esophagus**, (3) the **stomach**, (4) the **small intestine**, and (5) the **large intestine** (or **colon**). Among the accessory digestive organs are (1) the **salivary glands**, (2) the **pancreas**, (3) the **11-2**

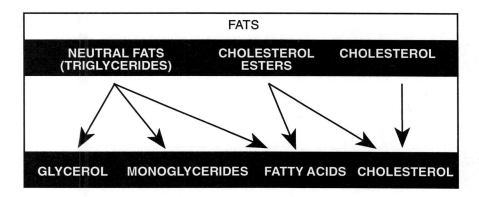

Figure 11.2
Types of fat present in food and the products of fat digestion.

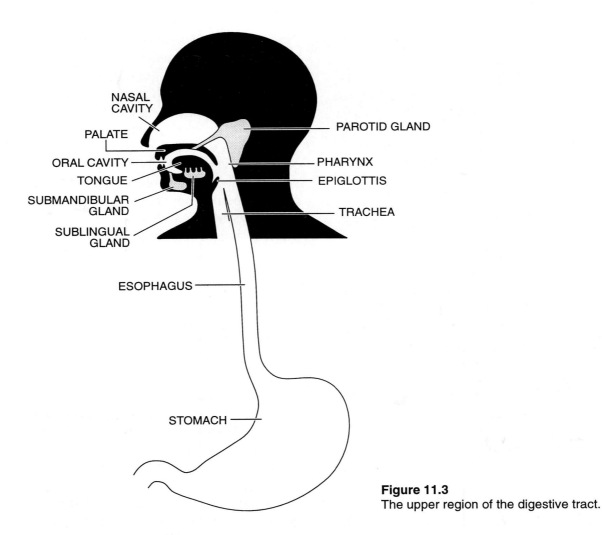

NASAL
CAVITY

PALATE

ORAL CAVITY

TONGUE

SUBMANDIBULAR
GLAND

SUBLINGUAL
GLAND

PAROTID GLAND

PHARYNX

EPIGLOTTIS

TRACHEA

ESOPHAGUS

STOMACH

Figure 11.3
The upper region of the digestive tract.

liver, and (4) the **gallbladder**. With the exceptions of the mouth, salivary glands, pharynx, and esophagus, the organs of the digestive system are confined to the body's abdominal cavity.

| 11-3 |

THE MOUTH

The digestion of food begins almost immediately upon its entry into the mouth. The presence of food in the mouth and its partial dissolution in the saliva serves as a chemical stimulus for receptors in the tongue (the "taste buds" or gustatory receptors; see chapter 6 and figure 6.5) which initiate the flow of sensory impulses to the central nervous system via sensory nerve fibers. The odor of the food also serves as a stimulus for olfactory receptors in the nose (see chapter 6, figure 6.8). Consequently, sensory signals ·sent to the brain from the tongue are usually accompanied by additional signals from the nasal cavity. These actions are reflexively followed by the passage of motor impulses from the **salivation center** in the medulla of

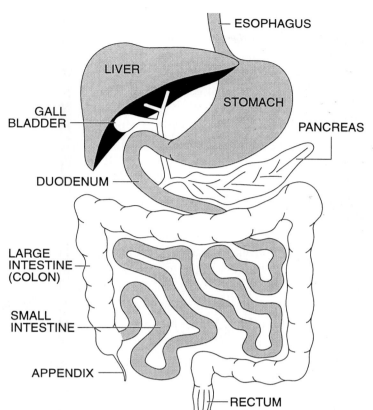

Figure 11.4
Major organs of the digestive system in the abdominal cavity.

the brain stem to the three pairs of **salivary glands**, resulting in an increased salivary flow.

The Salivary Glands

The secretion of saliva occurs continuously and is not confined to those times in which food is present in the mouth. However, the presence of food in the mouth and the action of chewing (called **mastication**) serve to increase the production and secretion of saliva. Control of the salivary glands is effected by the autonomic nervous system, which communicates with the salivary glands via sympathetic and parasympathetic motor fibers.

The action of the parasympathetic pathway is to promote increased output by the glands; the sympathetic pathway acts to reduce the blood supply to the glands, resulting in a less watery, less voluminous, and thicker secretion.

There are three *pairs* (i.e., left and right members) of salivary glands; these are called (1) the **parotid**, (2) the **sublingual**, and (3) the **submandibular** glands (fig. 11.3). The parotid glands are the largest of the three pairs and are located below and in front of the ears. The parotids are responsible for about 25% of the daily secretion of saliva. The submandibular glands lie along the inside of the jaw and are the source of most of the saliva (about 70%). The remaining 5% of the saliva is produced by the sublingual glands, which lie along the floor of the mouth, at the sides of and under the tongue. Altogether, the salivary glands secrete 1,000 to 2,000 c.c. of saliva each day.

Composition of Saliva

The major constituent of saliva is water (about 99%), which serves to extract small molecules from the solid food and dissolve them. Because the gustatory receptors of the tongue respond only to substances that are dissolved in water, dissolution of food materials is fundamental to the sense of taste. In the saliva's water are a number of inorganic substances such as sodium, potassium, chloride, phosphate, and bicarbonate. The resulting pH of the saliva varies from being mildly acidic to mildly alkaline (the pH varies between 6.7 and 7.5).

Also present in saliva is **mucus**, a sticky material that serves several roles. First of all, the mucus keeps the shredded food together in one mass (called a **bolus**), so that it can be manipulated by the tongue and sliced and crushed more effectively during chewing. Salivary mucus also serves as a lubricant, making it easier to swallow the food bolus and drive it toward the stomach. Saliva also contains a very important digestive enzyme called *salivary amylase* (also known as *ptyalin* or *alpha-amylase*).

Salivary amylase is an enzyme that digests carbohydrates like starch and glycogen (i.e., polysaccharides) by breaking specific chemical bonds that link neighboring sugars of the polysaccharides together. Although the degradative action of salivary amylase is very rapid, complete digestion of the carbohydrate is rarely achieved before the food is swallowed. As a result, the partially degraded carbohydrate that passes from the mouth into the stomach consists of short, sometimes branched, sugar chains of reduced length and small linear chains called **oligosaccharides**. Also produced in small numbers are disaccharides (principally **maltose**, which consists of two glucose molecules).

After the food is sufficiently shredded and mixed with saliva it is swallowed. Swallowing (also known as **deglutition**) is achieved when the bolus of food is pushed into the pharynx by the tongue, while muscle contractions raise the soft palate upward to close off the rear of the nasal cavity. Contractions of muscles in the walls of the pharynx then drive the bolus into the esophagus. Entry of food into the esophagus is controlled by a circular band of muscle called an *esophageal sphincter*. Relaxation of this sphincter accompanies swallowing so that entry into the esophagus is permitted. Passage of the swallowed food into the trachea is prevented by the action of the epiglottis, which seals off the larynx (fig. 11.3).

Mechanical Digestion vs. Chemical Digestion. It is possible to identify two types of digestion: **mechanical digestion** and **chemical digestion**. Mechanical digestion refers to the breakdown of food resulting from mechanical actions of the digestive tract, such as chewing in the mouth and the movements of the walls of other digestive organs. Chemical digestion refers to the breakdown of food resulting from chemical reactions initiated by secretions of the digestive organs. For example, the digestion of carbohydrate by *salivary amylase* is an example of chemical digestion. Most, but not all, chemical digestion is achieved through the actions of enzymes.

pH Sensitivity of the Digestive Enzymes

Like virtually all other enzymes of the body, the enzymes produced by the organs of the digestive system are proteins and are especially sensitive to the pH of the fluid into which they are secreted (see chapter 2 for a discussion of pH). Some enzymes function at neutral pH (i.e., around pH 7), some at alkaline pH (above pH 7), and others at an acidic pH (below pH 7). If the pH of the fluid in which digestion is proceeding is altered, enzymatic digestion may be abruptly halted.

In the case of *salivary amylase*, this enzyme functions in a pH range that extends from just below neutrality to just above neutrality (i.e., the pH range of saliva). This pH persists as food is swallowed and passed into the stomach. However, the secretion of acid into the stomach (see below) dramatically lowers the pH, and this serves to halt the action of any *salivary amylase* that makes contact with the acidic stomach secretions. Whereas carbohydrate digestion by *salivary amylase* may be halted, the low pH established in the stomach provides the acidic environment in which the stomach's enzymes can function. Thus, pH changes that turn off some enzymes while turning on others are the norm as food passes from one segment of the digestive tract to another.

THE ESOPHAGUS

The esophagus is a muscular tube about ten inches long that descends through the thoracic cavity, then through a small opening in the diaphragm (the **esophageal hiatus**), and leads into the stomach at the **gastroesophageal junction**. Food is swept from the top of the esophagus to the stomach by wavelike contractions of the esophageal wall called **peristaltic waves** (contractions involving narrowing then shortening of successive segments of the esophagus). Peristalsis is controlled by the autonomic nervous system and is reflexively initiated when mechanoreceptors in the esophageal wall sense the distention created by the swallowed food bolus. The distribution of muscle tissue in the wall of the esophagus is unusual and quite interesting. The uppermost few inches of the esophageal wall are rich in striated muscle tissue, the center section is a mixture of striated and smooth tissue; the lower section contains smooth muscle only.

Peristaltic waves of the esophagus travel at about one inch per second, so that swallowed food remains in the esophagus for no longer than about ten seconds. Passage of the swallowed food from the esophagus into the stomach at the gastroesophageal junction occurs when the **gastroesophageal sphincter** (also called the **cardiac sphincter**) temporarily relaxes.

Since the esophagus secretes no digestive enzymes and does not alter the pH of the swallowed material, the digestion of carbohydrate that began in the mouth is continued while the food travels through the esophagus.

THE STOMACH

The stomach, which is positioned just below the diaphragm in the upper part of the abdominal cavity, is divided into five regions; these are the **cardiac region**, the **fundic region**, the **body**, the **antrum**, and the **pyloric region** (fig. 11.5). The esophagus merges with the cardiac

11-4

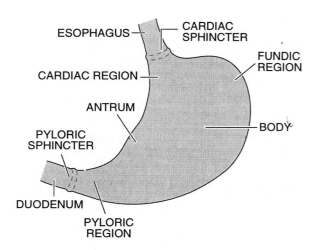

Figure 11.5
Regions of the stomach; note the sphincters that guard access into (cardiac sphincter) and exit from (pyloric sphincter) the stomach.

region of the stomach, the junction protected by the cardiac (or gastroesophageal) sphincter. Contraction of this sphincter prevent the regurgitation of food into the esophagus during contractions of the stomach's walls.

Internally, the walls of the stomach are deeply folded; these folds are called **rugae**. At the microscopic level, there is yet another order of folds and depressions called the **gastric pits** (fig. 11.6). At the base of each gastric pit lie the cells that secrete **gastric fluid** into the food. Four major types of cells form the gastric pits; these are called (1) **chief** cells, which secrete the stomach's digestive enzyme *pepsin* (see below), (2) **oxyntic** (or **parietal**) cells, which secrete **hydrochloric acid** (HCl), (3) **goblet** cells, which secrete mucus, and (4) **G** cells. Unlike the chief cells, oxyntic cells, and goblet cells whose secretions form the gastric fluid, the G cells empty their secretions (a hormone called **gastrin**) into the bloodstream.

The odor and taste of food stimulate the olfactory and gustatory receptors of the nose and tongue, and this results in the flow of sensory impulses to the brain. This promotes reflexive actions on the part of the parasympathetic division of the autonomic nervous system that are characterized by the conduction of motor impulses to the walls of the stomach, where the release of acetylcholine causes an initial secretion of gastric fluid into the stomach and the secretion of gastrin into the bloodstream. Gastrin is then carried back to the right side of the heart, through the pulmonary circulation, and into the systemic circulation. When blood containing gastrin flows through the capillary networks of the stomach's walls it promotes the more quantitative release of gastric fluid by the chief cells, oxyntic cells, and goblet cells.

Mucus secreted by the goblet cells forms a coating on the *epithelial* lining of the stomach wall, protecting the lining from the destructive effects of the hydrochloric acid and digestive enzymes that are also present in the gastric fluid. When the amount of mucus that is secreted is insufficient, parts of the stomach wall may be left unprotected. HCl and *pepsin* may degrade the exposed surfaces, leading to the progressive erosion of the stomach wall and the formation of **ulcers**. Ulcers of the stomach (and the duodenum, see below) may also be formed as a result of the regurgitation of "bile salts" from the small intestine into the stomach. The bile salts act like a detergent and clear away the stomach's mucus layer, exposing the stomach wall to the action of HCl and *pepsin*. Despite the protective action of the mucus layer, it is estimated that each minute some 500,000 epithelial cells are exfoliated (shed from the surface) and must be replaced; the

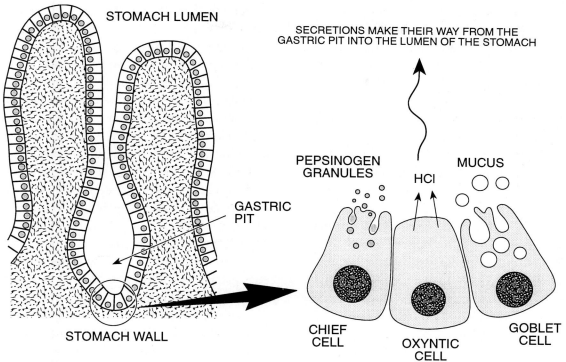

STOMACH LUMEN

SECRETIONS MAKE THEIR WAY FROM THE
GASTRIC PIT INTO THE LUMEN OF THE STOMACH

GASTRIC PIT

STOMACH WALL

PEPSINOGEN GRANULES

HCl

MUCUS

CHIEF CELL

OXYNTIC CELL

GOBLET CELL

Figure 11.6
The walls of the stomach are studded with gastric pits. Each pit's chief cells produce and secrete the protein-digesting enzyme **pepsin**; the oxyntic cells secrete hydrochloric acid, and the goblet cells secrete mucus.

entire epithelial lining of the stomach is replaced every three days!

The hydrochloric acid secreted by the oxyntic cells plays several roles: (1) being a very strong acid, HCl dramatically lowers the pH of the stomach's contents, bringing carbohydrate digestion to a halt and providing the acidic environment in which the stomach's enzymes can function (the pH of the stomach's contents hovers around 2.0); (2) since uncooked or inadequately cooked food frequently contains large numbers of microorganisms, the strong acid of the gastric secretion acts to kill most (but not all) of the microorganisms that are present; (3) HCl denatures protein present in the food, converting the protein to a form that is more easily digested by protein-digesting enzymes; and (4) HCl plays a role in the activation of the stomach's digestive enzyme **pepsin**.

In adults, the stomach secretes a single digestive enzyme called **pepsin**. In acidic surroundings, **pepsin** is a powerful protein-digesting enzyme. However, when it is initially released from the stomach's chief cells, **pepsin** exists in an inactive form called **pepsinogen** (the inactive form of an enzyme is called a **proenzyme** or **zymogen**). *Pepsinogen* is activated (i.e., converted to **pepsin**) by the removal of small peptide fragments from the proenzyme. This is achieved by the action of HCl. Once a small amount of the active enzyme is formed, the enzyme itself serves to convert additional *pepsinogen* to *pepsin*. In a cascadelike series of reactions, the amount of activated enzyme present in the stomach increases rapidly and geometrically (fig. 11.7). (Actually, there are three different forms of *pepsinogen* called *pepsinogen I*, *II*, and *III*; all are converted to *pepsin* by the removal of small sequences of amino acids.)

Peristaltic waves spreading across the walls of the stomach churn the food and act to mechanically break down food particles; the churning action also serves to effectively mix the food with the gastric fluid. In this manner, the food is progressively liquefied, forming what is known as **chyme**. The pyloric region of the stomach leads through the **pyloric valve** into the first segment of the small intestine (i.e., the duodenum). The stomach's peristaltic waves push the chyme against the pyloric valve, and with each wave a small amount of chyme is driven into the small intestine. The remainder of the chyme is refluxed back into the body of the stomach where it is subjected to additional churning and additional chemical degradation. Eventually all of the chyme is swept into the small intestine.

At this point, it might be well to ask how much digestion has been completed by the time the food reaches the small intestine? In fact, only a small amount of digestion is completed at this stage and is limited to the actions of *salivary amylase* on carbohydrate and the actions of HCl and *pepsin* on protein. There has been no fat digestion whatsoever. *Salivary amylase* reduces the carbohydrate in food to smaller, branched chains of sugars (polysaccharides), and *pepsin* reduces the protein in the food to short chains of amino acids (small polypeptides). Neither of these products is in a form that can be absorbed into the blood or lymph. It is in the small intestine that digestion is effectively completed and from which nearly all absorption takes place. (What absorption does take place in the stomach is limited to certain small ions, alcohol [which explains why intoxication can be so rapid], and aspirin.)

Because the role of the stomach in digestion is somewhat limited, and little digested food is absorbed by the stomach, the stomach is not considered essential to survival. When necessary (e.g., as a result of the growth of tumors or the development of large numbers of stomach ulcers), parts (or all) of the stomach may be surgically removed (an operation called a *gastrectomy*) without disastrous consequences. There is, however, one vital function of the stomach that must be taken into account following a gastrectomy. The oxyntic cells of the stomach secrete into the food a glycoprotein called **intrinsic factor**, which is essential for the absorption of **vitamin B**$_{12}$ (also called **cyanocobalamine** or **extrinsic factor**) from the ileum section of the small intestine.

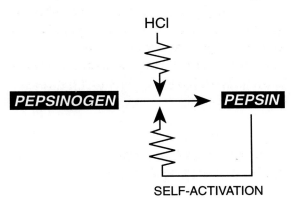

HCl

PEPSINOGEN ⟶ PEPSIN

SELF-ACTIVATION

Figure 11.7
Activation of pepsin in the stomach. HCl promotes the conversion of some *pepsinogen* into *pepsin*, after which *pepsin* serves to activate additional *pepsinogen* molecules (i.e., "self-activation").

Intrinsic factor secreted by the stomach and vitamin B_{12} present in the food form a complex that is later absorbed from the intestine. Vitamin B_{12} is essential for the proper production of red blood cells in the bone marrow; inadequate uptake of vitamin B_{12} causes **pernicious anemia**. Therefore, to survive, individuals who have had a gastrectomy must receive oral doses of intrinsic factor or intravenous injections of vitamin B_{12}.

THE SMALL INTESTINE

There are two organs in the abdominal cavity referred to as *intestines*; these are the **small intestine** and the **large intestine**. The words "small" and "large" refer to the diameter of the intestine, not to the intestine's length. Thus, the small intestine is a narrow tube, whereas the large intestine is a wide tube. Insofar as the lengths of these organs are concerned, the small intestine is very much longer than the large intestine. (The large intestine is commonly referred to as the **colon**.) It is the small intestine into which food passes from the stomach.

The walls of the small intestine are rich in smooth muscle. In a living person, this musculature is always contracted to some degree, so that the length of the organ rarely exceeds 12 feet. In a cadaver, the musculature is no longer contracted so that the small intestine extends to its maximum length. The relaxed length of the small intestine is more than 20 feet.

The first twelve inches of the small intestine is called the **duodenum**. The remaining length is approximately equally divided into the **jejunum** and the **ileum**. The ileum leads into the large intestine (colon). It is in the small intestine that the digestive process is effectively completed, and it is from the small intestine that nearly all of the digested food is removed by absorption into the bloodstream and into the lymphatic circulation.

The Duodenum

Acidic chyme pushed into the duodenum from the stomach requires immediate neutralization. This is achieved in part by the secretion of bicarbonate by the duodenum's epithelial lining (bicarbonate is also provided by the secretions of the pancreas, liver and gallbladder; see below). Failure to neutralize the acidic chyme can lead to the formation of duodenal ulcers. The presence of chyme in the duodenum causes the duodenum to release two substances (i.e., *hormones*) into the bloodstream; these are **secretin** and **cholecystokinin** (abbreviated CCK). Secretin and CCK circulate through the blood, and upon reaching the pancreas, liver, and gallbladder promote the release of **pancreatic juice** and **bile**, which are conveyed by a series of ducts into the duodenum (figures 11.4 and 11.8). Consequently, the partially digested food driven into the duodenum by contractions of the stomach's walls is quickly mixed with bile and pancreatic juice.

Secretions of the Liver, Gallbladder, and Pancreas

The Liver. The liver continuously produces and secretes bile, although secretion is augmented when secretin and CCK reach the liver from the duodenum. A daily bile output exceeding one-half liter is not unusual. The

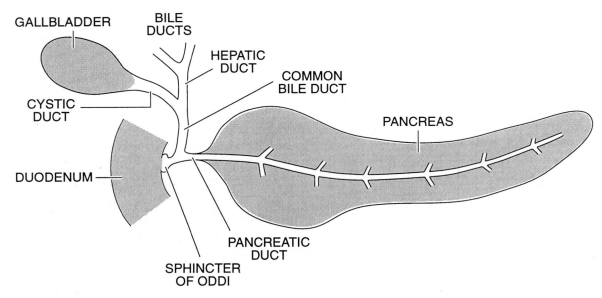

Figure 11.8
Relationships among the ducts that lead into the duodenum from the liver, gallbladder and pancreas.

bile, which leaves the liver through the **bile ducts** and **hepatic duct**, is a watery solution containing a variety of inorganic and organic substances; included in the bile are the bile salts (or **bile acids**), **bile pigments**, **cholesterol**, **bicarbonate**, and sodium and potassium chloride.

The bile salts are especially important in the digestive process. Bile salts are not salts in the conventional sense (such as sodium chloride and potassium chloride); rather, the bile salts are large organic molecules. Upon reaching the small intestine and mixing with the chyme, the bile salts act to **emulsify** fat that is present in the food. Typically, the fat present in chyme takes the form of large droplets called **lipid droplets**. The bile salts emulsify the fat in an action much like that performed by a detergent. The bile salts aggregate at each lipid droplet's surface,

breaking the droplet into successively smaller fragments. The smallest fragments can then be acted on by lipid-digesting enzymes that are also released into the small intestine (see below). Bile salts are reabsorbed from the small intestine along with digested fat and return to the liver through the *hepatic portal circulation* (see later). As a result, the same bile salts are used over and over again in the emulsification, digestion, and absorption of dietary fat. The recycling of the bile salts constitutes what is known as the **enterohepatic circulation** (fig. 11.9).

Also present in bile is **bile pigment** (or **bilirubin**). This colored substance is produced by the reticuloendothelial tissues of the spleen, bone marrow, and liver during the breakdown of hemoglobin (bilirubin consists of a modified heme group that also has lost its central iron atom). Bilirubin produced

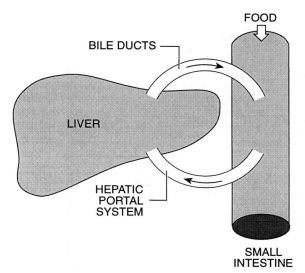

Figure 11.9
The enterohepatic circulation. Bile salts produced and secreted in the bile by the liver participate in the digestion of fat in the small intestine. Eventually the bile salts are absorbed (along with the digested fat) and are returned to the liver where they may once again be secreted in the bile.

in the bone marrow and spleen is shuttled to the liver by the plasma protein albumin. Bilirubin collected in the liver and produced by the liver itself is secreted in the bile and makes its way into the small intestine. Bacteria living in the small intestine convert the bilirubin to another pigment called **urobilinogen**. Thus, it is urobilinogen that colors the feces. Some of the urobilinogen produced in the small intestine is absorbed into the blood, along with digested food. Reabsorbed urobilinogen eventually is removed from the blood by the kidneys and is excreted from the body in the urine. Consequently, the color of urine (as well as feces) is due to its content of the breakdown products of hemoglobin.

Interestingly, bile also contains immuno-

globulins (mainly IgA, see chapter 9) which serve to provide the body with some protection against bacteria and viruses in the small intestine. However, even in a normal, healthy individual, there are large numbers of living microorganisms in the small (and large) intestines (see later).

The Gallbladder. As noted above, the liver produces and secretes bile constantly, not just during digestion. Between periods of digestion, bile leaving the liver through the bile ducts and hepatic duct does not pass directly into the duodenum. Instead, it is directed through the **cystic duct** into the **gallbladder** (see figure 11.8). The diversion of bile into the gallbladder is the result of the closing of the **ampulla of Vater**, which links the **common bile duct** and **pancreatic duct** to the duodenum (fig. 11.10); the ampulla is squeezed shut by contraction of the **sphincter of Oddi**.

In the gallbladder, much of the bile's water is removed and returned to the bloodstream, and mucus is added. When CCK that is secreted into the blood by the duodenum

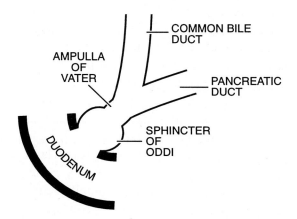

Figure 11.10
Ducts conveying bile and pancreatic secretions into the duodenum.

reaches the gallbladder, it causes the muscular walls of the gallbladder to contract. CCK also causes relaxation of the sphincter of Oddi, thereby opening the ampulla of Vater. Thus, the thickened and concentrated bile is pumped out of the gallbladder, through the cystic duct and common bile duct and into the duodenum. The gallbladder remains in a contracted state (and the sphincter of Oddi remains relaxed) during the course of digestion in the small intestine, so that bile exiting the liver can course directly into the duodenum.

The Pancreas. The pancreas is a large gland located behind (dorsal to) and below (inferior to) the stomach. The pancreas is an unusual gland in that it is *both* an **exocrine gland** and an **endocrine gland**. It's exocrine function involves the production of enzyme-rich **pancreatic fluid** by the organ's **acinar cells**. This fluid is released into fine ducts that ultimately empty into the pancreatic duct; the latter leads through the ampulla of Vater and into the duodenum (fig. 11.10). It is a characteristic of exocrine glands that their secretions are carried from the gland by ducts. (The salivary glands and the gastric pits of the stomach also fall into the category of exocrine glands.) Endocrine glands, on the other hand, are "ductless." Their secretions, which are called **hormones**, are released into the bloodstream and are then carried away from the gland to their eventual "target." (You will learn more about exocrine and endocrine glands in chapter 14.) The pancreas is also an endocrine gland. Scattered through the acinar tissue are clusters of hormone-producing cells called **islets of Langerhans**; the hormones produced by the islets are **insulin** and **glucagon** (chapter 14).

The pancreas produces and secretes **pancreatic fluid** into the duodenum; this fluid contains a variety of substances that are essential to digestive activity. These substances include bicarbonate, which helps to raise the pH of the chyme to about 8.0 and inactivates *pepsin*, and a number of powerful digestive enzymes (table 11.1). The *major* enzymes produced by the pancreas are *pancreatic amylase*, *trypsin*, *chymotrypsin*, *carboxypeptidase*, and *pancreatic lipase*; these enzymes are most effective at the alkaline pH created by the bicarbonate that is mixed with the chyme. The secretion of pancreatic juice is stimulated by CCK and secretin. Pancreatic juice leaves the pancreas through the pancreatic duct and enters the duodenum, along with the bile, through the ampulla of Vater (fig. 11.10). | 11-5 |

Pancreatic amylase digests carbohydrate by degrading polysaccharide chains into successively smaller pieces. Consequently, although the digestion of carbohydrate is temporarily halted in the stomach, it is resumed in the small intestine. *Trypsin*, *chymotrypsin*, and *carboxypeptidase* are protein-digesting enzymes that continue the enzymatic breakdown of protein begun by *pepsin*. These enzymes continue to cleave polypeptide chains into smaller and smaller fragments. *Carboxypeptidase* attacks the *ends* of polypeptide chains, whereas *trypsin* and *chymotrypsin* attack the bonds that link amino acids together *within* a polypeptide chain. *Pancreatic lipase* is a fat-digesting enzyme that attacks the tiny lipid droplets produced by the emulsifying action of the bile salts. | 11-6 |

Like *pepsin*, *trypsin*, *chymotrypsin*, and *carboxypeptidase* enter the duodenum in an inactive (i.e., zymogen) form and must be

---

TABLE 11.1   THE PANCREATIC DIGESTIVE ENZYMES

ENZYME	ACTION
Protein-Digesting Enzymes	
Trypsin	Cleaves bonds *between internal amino acids* of polypeptides
Chymotrypsin	Cleaves bonds *between internal amino acids* of polypeptides
Carboxypeptidase	Cleaves amino acids *from one end* of polypeptide
Elastase	Cleaves bonds *between internal amino acids* of polypeptides
Fat-Digesting Enzymes	
Pancreatic lipase	Splits neutral fats into fatty acids and glycerol
Phospholipase	Splits phospholipids
Other Enzymes	
Pancreatic amylase	Splits long sugar chains into smaller ones
Ribonuclease	Splits RNA into short polynucleotide chains
Deoxyribonuclease	Splits DNA into short polydeoxynucleotide chains

---

activated in order to resume the enzymatic digestion of protein in the chyme. The zymogen forms of these enzymes are called **trypsinogen**, **chymotrypsinogen**, and **procarboxypeptidase**. The activation of these zymogens takes the following course (fig. 11.11). Some of the *trypsinogen* entering the duodenum is converted to *trypsin* by **enterokinase**, an enzyme localized in the plasma membranes of the cells forming the epithelial lining of the duodenum (**enterokinase** activates *trypsin* by removing a small peptide from the *trypsinogen* proenzyme). The *trypsin* formed in this manner then activates additional *trypsinogen* molecules (just as *pepsin* activates *pepsinogen*). Consequently, all of the *trypsinogen* is quickly converted to *trypsin*. **Chymotrypsinogen** and **procarboxypeptidase** are also converted to their active forms by *trypsin* (fig. 11.11).

11-7

Chyme entering the duodenum from the stomach is a very thick (pasty) substance. However, the secretions of the liver and pancreas include very large volumes of water, so that the chyme is quickly diluted as it passes on toward the jejunum and ileum.

Digestion in the Small Intestine

The walls of the small intestine contain large amounts of smooth muscle arranged in two patterns. Some of the smooth muscle runs around the circumference of the intestine and serves to constrict and dilate different regions of the intestine's length. The constrictions occur simultaneously at intervals along the intestine, thereby dividing the intestine's length into a number of consecutive chambers and acting to mix the chyme with the digestive enzymes. The narrow openings that interconnect successive segments are called **valves of Kerckring**.

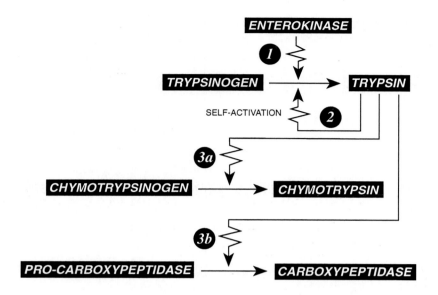

Figure 11.11
Conversion of the zymogens *trypsinogen*, *chymotrypsinogen*, and *pro-carboxypeptidase* to their active enzyme forms: *trypsin*, *chymotrypsin*, and *carboxypeptidase*. The activation cascade is initiated by *enterokinase*, an enzyme in the surfaces of the cells lining the duodenum.

Other smooth muscle runs longitudinally through the intestinal wall (i.e., this smooth muscle is oriented parallel to the long axis of the intestine). Contractions of this muscle are responsible for the peristaltic waves that slowly drive the chyme forward.

The inside surface of the small intestine is not smooth. Instead, millions of tiny finger-like projections extend from the surface into the intestinal lumen. These projections are called **villi** (*singular:* **villus**; figure 11.12), and they greatly increase the surface area of the intestinal wall. The increased surface area maximizes the organ's absorptive properties.

The villi are lined at their surface by epithelium consisting of two types of cells: **goblet** cells and **digestive/absorptive** cells. The goblet cells secrete mucus, which helps to protect the intestinal lining from the ac-

tions of the digestive enzymes. Mucus is also produced and secreted by the **crypts of Lieberkühn** (fig. 11.12), small glandular structures in the wall of the intestine between neighboring villi. The digestive/absorptive cells produce the **intestinal enzymes**, which act to complete the digestion of the food, converting it to forms that can be absorbed into the blood and lymph (table 11.2). The intestinal digestive enzymes are not secreted into the lumen of the intestine; rather, they are anchored in the surfaces (i.e., plasma membranes) of the thousands of **microvilli** that project from each cell (fig. 11.13). (Don't confuse the terms "villi" and "microvilli." Villi are the numerous projections from the wall of the intestine, each villus covered by hundreds of epithelial cells. Microvilli are microscopic projections from the surfaces of individual epithelial cells covering each

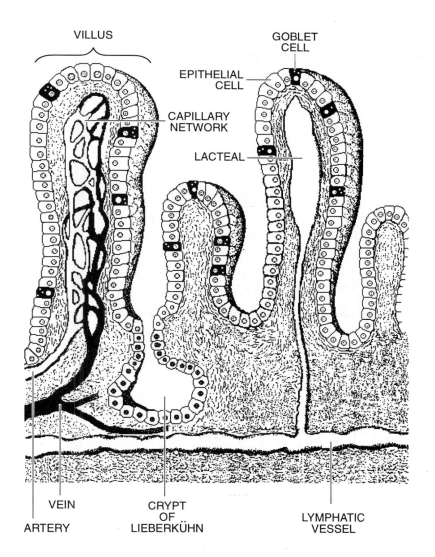

VILLUS

GOBLET CELL

EPITHELIAL CELL

CAPILLARY NETWORK

LACTEAL

VEIN

ARTERY

CRYPT OF LIEBERKÜHN

LYMPHATIC VESSEL

Figure 11.12
Organization of the villi that cover the surface of the small intestine. Each villus contains both a capillary network and a lacteal. Digestive/absorptive epithelial cells line the surface of each villus; scattered among the digestive/absorptive cells are mucus-secreting goblet cells. Mucus is also secreted by the crypts of Lieberkühn.

villus.) Collectively, the microvilli form what is called a **brush border** (fig. 11.13).

The intestinal enzymes (table 11.2) bring the digestive process to completion. The major intestinal enzymes are *aminopeptidase*, *dextrinase*, *maltase*, *sucrase*, and *lactase*. *Aminopeptidase* breaks the short polypeptide fragments (i.e., oligopeptides) into individual amino acids (and small numbers of dipeptides and tripeptides); these are quickly taken up by the digestive/absorptive epithelium. *Dextrinase* digests small, branched sugar chains that were produced by the enzymatic action of *pancreatic amylase*, thereby yielding simple sugars along with a number of disaccharides (mainly maltose). *Maltase* cleaves maltose into individual sugars (i.e., monosaccharides). *Lactase* breaks the disaccharide *lactose* (which is abundant in milk) into individual sugars, and *sucrase* breaks the disaccharide *sucrose* (i.e., "table sugar") into individual sugars.

11-8

The emulsification of lipid that is achieved in the chyme through the actions of the bile salts converts large lipid droplets into much

289

TABLE 11.2 THE INTESTINAL DIGESTIVE ENZYMES

ENZYME	ACTION
Carbohydrate-Digesting Enzymes	
Dextrinase	Splits "limit dextrins" (small, branched sugar chains)
Sucrase	Splits sucrose (table sugar) into glucose and fructose
Lactase	Splits lactose (milk sugar) into glucose and galactose
Maltase	Splits maltose into two glucose units
Protein-Digesting Enzyme	
Aminopeptidase	Splits oligopeptides into amino acids, dipeptides and tripeptides
Fat-Digesting Enzyme	
Enteric lipase	Splits triglycerides into fatty acids and monoglycerides

smaller ones whose contents (mainly triglycerides, cholesterol, and cholesterol esters) can now be digested by *pancreatic lipase* (fig. 11.14). *Enteric lipase*, a lipid-digesting enzyme present in the surfaces of the intestinal epithelium, also plays a minor role in fat digestion. The products of these digestive activities (e.g., **monoglycerides, fatty acids, glycerol** and **cholesterol**) then combine with bile salts to form tiny molecular aggregates called **micelles** (a micelle is many thousands of times smaller than an individual lipid droplet). The micelles migrate to the surfaces of the villi where their contents are taken up by the digestive/absorptive epithelial cells (fig. 11.14). The bile salts of emptied micelles reenter the intestinal lumen where they are used again.

Intestinal Absorption

Digestion is completed in the small intestine, and it is from the small intestine that the products of digestion are passed into the blood and the lymph that circulate through each intestinal villus. As seen in figures 11.12 and 11.13, each villus contains a capillary network formed by branches of the **mesenteric arteries** that carry blood into the walls of the intestine. These capillary networks eventually empty their blood into the **mesenteric veins**. Each villus also contains a blind-ended branch of the lymphatic system that courses through the intestinal wall. These bulb-like structures that rise up through the center of each villus are called **lacteals**. Within each villus, the capillary network is wound around the central lacteal (fig. 11.13).

Absorption of Amino Acids

Amino acids, dipeptides, and tripeptides produced by the digestion of protein in the stomach and small intestine are actively taken up by the digestive/absorptive epithelium lining the intestinal villi. Within the intestinal epithelial cells, the dipeptides and

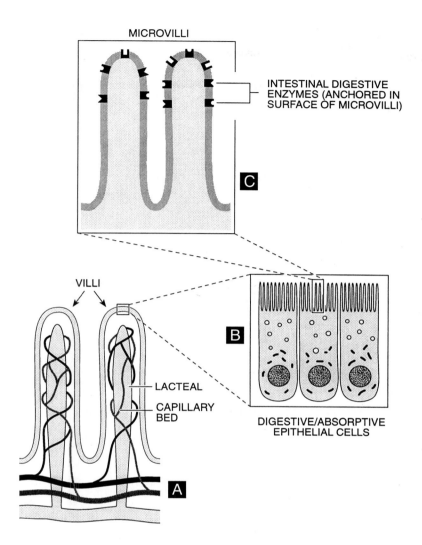

MICROVILLI

INTESTINAL DIGESTIVE ENZYMES (ANCHORED IN SURFACE OF MICROVILLI)

C

VILLI

LACTEAL

CAPILLARY BED

B

A

DIGESTIVE/ABSORPTIVE EPITHELIAL CELLS

Figure 11.13
The villi of the small intestine (A) are covered by a layer of digestive/ absorptive epithelial cells (B). The exposed surfaces of the epithelial cells form numerous, microscopic projections called microvilli (B and C). Anchored in the microvilli and facing the intestine's lumen are the intestinal digestive enzymes (C).

tripeptides are converted into individual amino acids. Amino acids are then actively transported into the blood that is circulating through each villus' capillary network (fig. 11.15).

Although virtually all protein in food is converted into amino acids before it is passed into the bloodstream, there are a few proteins that can be absorbed intact. For example, the intestinal epithelium of a newborn baby can absorb undigested antibody molecules. Since antibodies are present in the milk produced by the mother's mammary glands, a degree of immunity is conferred to a young baby that is breast feeding. The uptake of whole proteins by intestinal epithelium involves an entirely different cellular process than that involved in the uptake of amino acids.

Absorption of Sugars

Sugars produced by the digestion of carbohydrate are also actively taken up by the digestive/absorptive epithelium. Although glucose is the predominant sugar that is

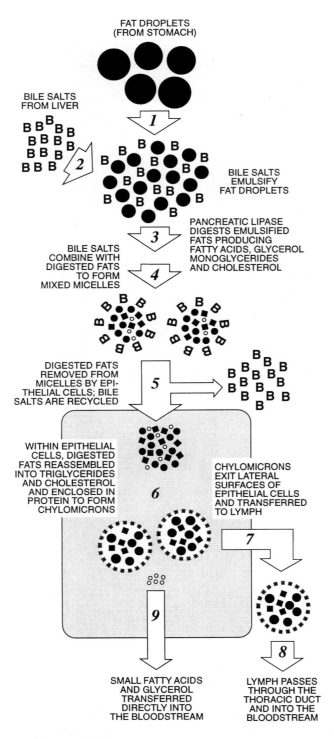

FAT DROPLETS
(FROM STOMACH)

BILE SALTS
FROM LIVER

1

2

BILE SALTS
EMULSIFY
FAT DROPLETS

PANCREATIC LIPASE
DIGESTS EMULSIFIED
FATS PRODUCING
FATTY ACIDS, GLYCEROL
MONOGLYCERIDES
AND CHOLESTEROL

3

BILE SALTS
COMBINE WITH
DIGESTED FATS
TO FORM
MIXED MICELLES

4

DIGESTED FATS
REMOVED FROM
MICELLES BY EPI-
THELIAL CELLS; BILE
SALTS ARE RECYCLED

5

WITHIN EPITHELIAL
CELLS, DIGESTED
FATS REASSEMBLED
INTO TRIGLYCERIDES
AND CHOLESTEROL
AND ENCLOSED IN
PROTEIN TO FORM
CHYLOMICRONS

CHYLOMICRONS
EXIT LATERAL
SURFACES OF
EPITHELIAL CELLS
AND TRANSFERRED
TO LYMPH

6

7

9

8

SMALL FATTY ACIDS
AND GLYCEROL
TRANSFERRED
DIRECTLY INTO
THE BLOODSTREAM

LYMPH PASSES
THROUGH THE
THORACIC DUCT
AND INTO THE
BLOODSTREAM

Figure 11.14
Emulsification, digestion, and absorption of fats
in the small intestine.

actively absorbed by the intestinal epithelium, small amounts of fructose and galactose are also absorbed. Like the amino acids, glucose, fructose, and galactose are then passed into the blood circulating through each villus' capillary network.

Absorption of Fats

As illustrated in figure 11.14, glycerol, fatty acids, and monoglycerides removed from micelles by the intestinal epithelium are used within the epithelial cells in the production of new triglycerides. These are then combined with cholesterol and enclosed in a layer of protein, thereby forming small particles called chylomicrons. The chylomi-

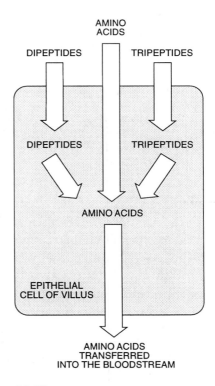

AMINO
ACIDS

DIPEPTIDES   TRIPEPTIDES

DIPEPTIDES   TRIPEPTIDES

AMINO ACIDS

EPITHELIAL
CELL OF VILLUS

AMINO ACIDS
TRANSFERRED
INTO THE BLOODSTREAM

Figure 11.15
Absorption of amino acids, dipeptides, and
tripeptides in the small intestine .

crons are then transferred from the epithelium to the lacteals at the center of the villi (fig. 11.14). Small amounts of fatty acids (principally small ["short-chain"] fatty acids) and glycerol are transferred directly to the bloodstream. *(Note that the bulk of the products produced by the digestion of fats end up in the lymphatic circulation, whereas amino acids and sugars move directly to the bloodstream.)*

The chylomicrons of the lymph eventually enter the bloodstream through the *thoracic duct* (see chapter 7 and the following text); within the bloodstream, chylomicrons are attacked by the enzyme *lipoprotein lipase*, which is present in the endothelial cells that line the inside surfaces of certain blood vessels. *Lipoprotein lipase* frees fatty acids and glycerol from triglycerides making these products available to the tissues through which the blood circulates.

VLDLs, LDLs, and HDLs. Removal of fatty acids and glycerol from the chylomicrons produces so-called "remnant particles," which are rich in cholesterol and which eventually are taken from the circulation by the liver and disassembled. In the liver, the cholesterol is combined with triglycerides synthesized anew from fatty acids and glycerol and combined with liver proteins to form **very-low-density lipoproteins** (**VLDLs**). The hallmark of a VLDL is its very high triglyceride content (up to 80% of the lipoprotein's weight) in comparison with cholesterol. (Triglycerides, fatty acids, sterols, and other lipids have a very low density [lower than the density of water, which is 1.0], whereas proteins have a higher density (around 1.3). As a result, when proteins are combined with lipids (thereby forming *lipoproteins*), the resulting molecular complexes have a density that is lower than uncomplexed protein. Indeed, the more lipid that is present in the lipoprotein, the lower is its density.) The VLDLs are dispatched from the liver and supply the body's tissues with lipids. As triglycerides are removed from the VLDLs (again by *lipoprotein lipase*, the same enzyme that acts on the chylomicrons), the VLDLs are converted to **low-density lipoproteins** (**LDLs**). LDLs are characterized by a high cholesterol percentage. A third group of blood lipoproteins are the **high-density lipoproteins** (**HDLs**). Like the VLDLs and LDLs, the HDLs are synthesized in the liver, but their lipid content is the lowest (they are about 50% lipid and 50% protein). In recent years, there has been an accumulation of evidence that HDLs can pick up cholesterol from the body's tissues and return it to the liver where the cholesterol is then degraded. Since it is recognized that high blood cholesterol levels may be linked to heart disease and atherosclerosis, high blood HDL and low blood LDL levels are considered desirable–hence the popular expressions "good cholesterol" (i.e., HDLs) and "bad cholesterol" (i.e., LDLs).

THE HEPATIC PORTAL SYSTEM

The fact that sugars and amino acids are passed into the blood, whereas lipids are passed into the lymph, has some very important implications. As illustrated in figure 11.16, blood leaving the wall of the small intestine via the **mesenteric veins** does not return directly to the right side of the heart. Instead, this blood is shunted into the **hepatic portal vein**, which carries the blood into the liver. This arrangement in which systemic blood flows through *two successive organs* (and their capillary beds) is an exception to the general rule that systemic blood flows through a single organ during each cycle through the systemic circulation. The implication of this arrangement is that the liver has an opportunity to screen the composition of blood leaving the intestine before that blood enters the general circulation.

As a result of digestion and absorption, the chemical composition of blood leaving the wall of the intestine is dramatically altered; for example, the blood's content of sugars and amino acids is markedly elevated. The sugar- and amino acid-rich blood is directed via the mesenteric and hepatic portal veins into the liver, where sugars and amino acids are removed from the blood and temporarily stored. Between meals, as the various tissues of the body require additional sugars and amino acids for their metabolism, stored sugar and amino acids are released into the blood by the liver in order to meet tissue demands. In this way, the actions of the liver ensure that the chemical composition of the blood in the general circulation does not fluctuate dramatically.

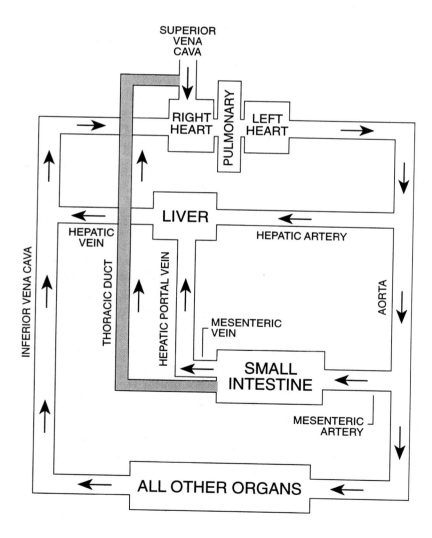

Figure 11.16
The hepatic portal system. The hepatic portal system is an exception to the general rule that systemic blood passes through only one organ before returning to the right side of the heart. Blood leaving the walls of the small intestine is directed through the liver before returning to the heart. ("Pulmonary" = pulmonary circulation between the right and left sides of the heart.)

Although it may be rich in sugars and amino acids, the blood entering the liver through the hepatic portal vein is poor in oxygen (having given up oxygen to the tissues that comprise the walls of the small intestine). The oxygen needs (and other needs) of the liver are met by the **hepatic artery** (a direct branch of the descending aorta, see figure 11.16) which carries fresh systemic blood to the liver. (About 75% of the blood entering the liver enters through the hepatic portal vein, while the remaining 25% enters through the hepatic artery.) Portal and systemic blood circulating through the liver are carried out of this organ by the **hepatic vein,** which directs the blood into the inferior vena cava and then the right atrium.

In sharp contrast to the fate of absorbed sugars and amino acids (which are passed into the blood), the lipid-rich chylomicrons are passed into the lacteals and are carried away from the small intestine with the lymph. This lymph is carried upward through the abdominal cavity into the thoracic cavity by lymphatic vessels. Within the thoracic cavity, the **thoracic duct** empties the chylomicron-rich lymph into the bloodstream at the junction of the **left brachiocephalic**, **left subclavian**, and **left internal jugular veins** (see figure 7.16). From there, the chylomicrons enter the right atrium and become part of the general circulation. Consequently, beginning near the heart, all tissues of the body have immediate access to the lipids of the chylomicrons. This creates a potentially harmful scenario in which lipids are removed from the blood and may accumulate in excess in the walls of blood vessels near the heart and in the muscle tissues of the heart itself (producing heart disease–the major cause of death in the United States).

Duration of Digestion and Efficiency of Absorption

By the time that the chyme reaches the end of the jejunum, the digestion of the food and its absorption into the blood and lymph are largely complete. In a normal, healthy individual who has consumed an average meal, digestion and absorption extend over a period of 3 to 4 hours. Digestion and absorption are remarkably efficient; nearly 100% of all digested carbohydrate is absorbed and more than 95% of all digested protein and fat. For the major nutrients, there is no regulation of the amount of digested food that is absorbed. In other words, the more that is eaten, the more will be digested and absorbed.

THE LARGE INTESTINE (COLON)

Chyme passes into the **large intestine** (also called the **colon**) from the ileum of the small intestine and begins to make its way through the large intestine's three major sections: the *ascending colon* (on the right side of the abdominal cavity), the *transverse colon* (oriented horizontally, just below the stomach and pancreas), and the *descending colon* (on the left side of the abdominal cavity; see fig. 11.4). The surface of the colon is lined by epithelial cells, some of which secrete mucus that serves as a lubricant for the expulsion of the feces. Also present are crypts of Lieberkühn; there are, however, no villi and there are no digestive enzymes secreted by the epithelial cells. The colon's lining secretes small amounts of bicarbonate, which acts to neutralize the acids that are produced during the metabolism of bacteria that live in the colon (see below). The bicarbonate ions are exchanged for chloride ions, which along with sodium ions, are transported into the blood. Consequently, during its passage through the colon, the chyme is altered in several ways, ultimately forming the semi-solid **feces** that are eliminated from the body through the **rectum** and **anus**.

Absorption of Water by the Small and Large Intestine

A great deal of water passes through the digestive tract each day. This water includes (1) water present in the food that one eats, (2) water (or other liquids) consumed along with a meal, and (3) water that is added to the food by the digestive organs. Typically, the water ingested in food and drink amounts to about 2 liters each day. To this amount is added 2 liters that are provided by the salivary glands (i.e., saliva). Another 3 liters is provided by the stomach (i.e., gastric fluid), an additional 2.5 liters by the pancreas (i.e., pancreatic fluid), and a final 0.5 liters by the liver (i.e., bile). Therefore, taken altogether, the average person's digestive tract must deal with about 10 liters of water each day.

About 80% of the water reaching the small intestine is absorbed into the blood, along with amino acids and sugars (and other substances) produced by the digestive process. The remaining 20% (about 2 liters per day) passes into the large intestine. Of this amount, about 90% (i.e., 1800 c.c.) is absorbed from the colon into the blood, while the remainder (i.e., about 200 c.c.) is eliminated from the body with the feces. (Additional water is lost from the body by its excretion in the urine; this amounts to about 1.2 liters each day [see chapter 13].) Most of the water returned to the bloodstream from the colon is absorbed in the first half of the colon's length; the second half of the colon serves principally in storing the feces until they are expelled.

Colonic Bacteria

A number of bacteria live and grow in the large intestine. These colonic bacteria (e.g., *E. coli*) live off the undigested and unabsorbed food that enters the colon from the small intestine (principally undigested cellulose). During their metabolism, the colonic bacteria produce and excrete a variety of substances including several that are useful to humans. Included among the useful excretory products of the colonic bacteria are a number of vitamins such as B_{12}, K, thiamin, and riboflavin (the functions of these vitamins [and others] are considered further in chapter 12). These vitamins are absorbed by the epithelial lining of the colon and transferred to the bloodstream. The amount of vitamin K in a normal diet is usually below the body's needs; as a result, we rely on the colonic bacteria to provide the requisite additional amount of this vitamin (which is important for proper blood coagulation).

The colonic bacteria also produce gaseous waste products including methane, hydrogen, and hydrogen sulfide. While some of the gas is absorbed through the walls of the colon, most of it exits the body through the anal opening along with the feces.

It should be noted that not all *E. coli* serve a useful purpose; indeed, certain strains of *E. coli* ingested in "fast food" are believed to have been the cause of serious gastric disturbances, including several deaths.

SELF TEST[*]

True/False Questions

1. No enzymatic digestion of food occurs in the mouth, because chewing is considered to be "mechanical" digestion.

2. The digestion of the disaccharide sucrose produces the monosaccharides glucose and fructose.

3. Hydrochloric acid activates *pepsin*, the protein-digesting enzyme of the stomach.

4. The cardiac sphincter separates the esophagus from the heart.

5. Oxyntic cells of the gastric pits are the source of the stomach's hydrochloric acid.

6. A substance called "intrinsic factor" is required for the absorption of vitamin B_{12} from the small intestine. Intrinsic factor is secreted into the partially digested food by the stomach.

7. The small intestine is several feet shorter than the large intestine (or colon).

8. The liver produces and secretes bile only when there is partially digested food present in the small intestine.

9. The actions of the bile salts and *pancreatic lipase* are to emulsify the fat, digest it, and enclose the digestive products in small particles called micelles.

10. Most of the bile salts secreted into the small intestine end up in the feces that are eliminated from the body.

11. The crypts of Lieberkühn are the principal source of the small intestine's digestive enzymes.

12. Even though they may both be absorbed from the small intestine at exactly the same time, absorbed lipid will reach the right side of the heart faster than absorbed sugar.

Multiple Choice Questions

1. The epiglottis serves to (A) prevent air from accidentally entering the esophagus, (B) prevent air from accidentally entering the trachea, (C) prevent swallowed food from accidentally entering the esophagus, (D) prevent swallowed food from accidentally entering the trachea.

2. The enzymatic digestion of polysaccharides occurs within the (A) mouth, stomach, and small intestine, (B) mouth, small intestine, and pancreas, (C) stomach and large intestine, (D) mouth and small intestine, (E) mouth, liver, and small intestine.

3. Pepsinogen is secreted by the (A) chief cells, (B) oxyntic cells, (C) goblet cells, (D) G cells of the gastric pits.

4. The partially digested food that is pumped into the duodenum from the stomach is called (A) bile, (B) chyme, (C) bolus, (D) pdf, (E) regurgin.

5. Which one of the following is not secreted by the pancreas? (A) *chymotrypsinogen*, (B) insulin, (C) bicarbonate, (D) *pancreatic amylase*, (E) all of these substances are secreted by the pancreas.

6. Bile enters the gallbladder through (A) the hepatic duct, (B) the cystic duct, (C) the common bile duct, (D) the thoracic duct,

---

[*] *The answers to these test questions are found in Appendix III at the back of the book.*

(E) bile does not enter the gallbladder–it only leaves the gallbladder.

7. The mesenteric veins (A) carry oxygen-rich blood into the small intestine, (B) carry oxygen-poor blood from the small intestine into the inferior vena cava, (C) carry sugar-rich blood into the hepatic portal vein, (D) carry oxygen-rich blood into the liver.

8. If the body produces and secretes too little "intrinsic factor," the result is likely to be an abnormality called (A) pernicious anemia, (B) iron-deficiency anemia, (C) hepatitis, (D) sickle cell disease, (E) hypoglycaemia.

9. Which one of the following chemical substances is secreted into the bloodstream by cells in the walls of the duodenum? (A) enterokinase, (B) gastrin, (C) bilirubin, (D) cholecystokinin (CCK), (E) insulin.

10. Which of the following lipid-containing particles is released into the lymph by the intestinal epithelium? (A) micelles, (B) chylomicrons, (C) LDLs, (D) HDLs.

CHAPTER **12**

METABOLISM AND NUTRITION

Special Contributions by Annette Sheeler

Chapter Outline

In this chapter, we will be concerned with the essential features of the body's metabolism and nutrition. We will consider the ultimate fates of the foods that are ingested and digested and the ways in which the digestive products are used by the tissues of the body as raw materials for repair and growth and as sources of metabolic energy. We will consider the elements of a healthy diet and the roles that are played by minerals and by small molecules such as vitamins in supporting the body's metabolism.

AN OVERVIEW OF METABOLISM

By "metabolism" is meant all of the chemical reactions that take place in the body. Included are those reactions that take place *intracellularly* (i.e., the metabolism that goes on within the cells and tissues) and those that take place *extracellularly* (e.g., during digestion within the lumens of the digestive organs). Digested food is the source of raw materials that the body needs and uses. The utilization of these raw materials takes two principal forms (fig. 12.1): (1) the raw materials are degraded (usually to CO_2 and water), yielding heat and energy, and (2) energy from the degradation of raw materials is used, along with additional raw materials, to support growth, physiological activity, and tissue repair.

Catabolism and Anabolism

The chemical reactions of metabolism fall into two major categories: (1) reactions in which molecules are made *simpler* by the removal of specific atoms or groups of atoms and (2) reactions in which *more complex* molecules are formed by combining smaller molecules (or parts of smaller molecules). Generally speaking, the term **catabolism** refers to either individual reactions in the body or *sequences* of reactions (referred to as **metabolic pathways**) that make molecules smaller and simpler; an example of a catabolic pathway is the series of cellular reactions that converts the sugar glucose to CO_2 and water. Usually, catabolic processes are **exergonic**; that is, they are associated with the *release of energy* that may be lost as heat or may be "trapped" by making ATP (see below).

In contrast, **anabolism** refers to chemical reactions and reaction sequences that make molecules larger and more complex; an example of an anabolic pathway would be the series of cellular reactions that incorporates amino acids into new tissue protein. Usually, anabolic processes are **endergonic**; that is, they require an input of energy, which usually implies that they consume ATP (again, see below).

Nearly all of the materials needed to sustain metabolism are acquired through the

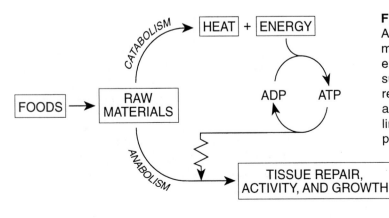

Figure 12.1
An overview of metabolism. Raw materials are the source of heat and energy. The energy is used to support physiological activity, tissue repair, and growth. ATP and ADP act as the cellular currency that links these catabolic and anabolic processes.

digestive system and its associated structures. (Oxygen, also an essential participant in metabolism, is acquired through the lungs.) Carbohydrates, proteins, and fats in ingested foods are digested to produce sugars, amino acids, fatty acids, sterols, and so on (chapter 11), and these materials are employed by the tissues to create tissue proteins, carbohydrates, and lipids and to fuel the body's energy needs. The scheme shown in figure 12.2 summarizes the major metabolic processes of the body. In this figure, the catabolic pathways are shown with black arrows, whereas the grey arrows represent anabolic pathways. As you continue through this chapter, you will learn more and more of the details of this overall scheme.

Role of ATP (Adenosine Triphosphate)

It is the energy that is made available during exergonic processes that "drives" or fuels the energy needs of the reactions of endergonic processes. However, in the tissues of the body endergonic, anabolic processes (such as protein synthesis) are rarely coupled *directly* to exergonic, catabolic processes (such as the breakdown of glucose). Rather, in one place at one moment, the energy of an exergonic process is temporarily trapped by using it to make ATP; whereas, in another place at another moment, ATP is broken down and the energy that is made available is used to drive an endergonic process. Because of this "middle-man"-like role that is played by ATP during metabolism, ATP is often described as the "energy currency" of cells.*

The usual way in which energy is trapped to make ATP is by the addition of a phosphate group to an existing molecule of ADP (adenosine diphosphate). The chemical structures of ATP and ADP and the nature of this chemical conversion are shown in figure 4.20. Usually, when the energy of ATP is used to drive an endergonic reaction, the ATP is converted to ADP. The cells of the body do have other ways of trapping and using energy. For example, other compounds such as *guanosine triphosphate* [GTP] and *uridine triphosphate* [UTP] can play metabolic roles similar to those of ATP; however, the interconversions of ATP and ADP are by far the most common methods used by the body's cells to trap and recruit energy.

Energy and Calories

In the same sense that an automobile will not operate without an input of energy (i.e., the energy provided by the motor's combustion of gasoline), the functions of the body (indeed, life itself) requires a continuous energy input. Since energy fuels the body's metabolism, it is important at the outset to understand what is meant by energy and the units in which energy is measured.

Energy may be defined as *the capacity to do work*, and all molecules, indeed all matter, is a potential source of energy. One of the ways in which the potential energy of

---

* The analogy between ATP and currency is a fitting one. Paper money itself has no intrinsic value; however, for performing some job function [say, assembling an automobile] a person is paid currency; the currency is then spent to acquire food, clothing, and shelter. The currency is the convenient and efficient link between the labor and the resulting acquisitions. The alternative–namely, being "paid" directly in food, clothing, or shelter–would be unwieldly.

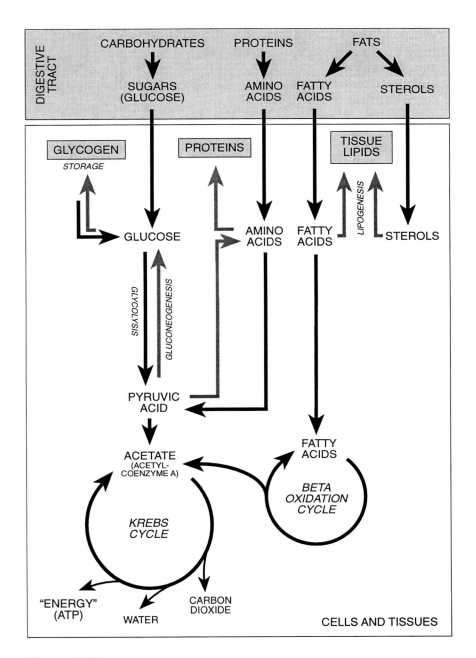

Figure 12.2
A general scheme of the body's metabolism. Digested carbohydrates, proteins, and fats produce the raw materials required for tissue metabolism (e.g., sugars, amino acids, fatty acids, and sterols), which are eventually taken up by the body's tissues. In these tissues, the raw materials will be used to assemble new carbohydrates, proteins, and fats. Additionally, some interconversion among raw materials takes place. The energy requirements for tissue metabolism are met by fully degrading some of the raw materials, especially sugars and fats. In this diagram, black arrows represent catabolic processes, whereas grey arrows represent anabolic processes.

molecules can be made available is by breaking the bonds that link the atoms of these molecules together. In the case of human metabolism, only certain molecules can serve as sources of energy; this is because the body is capable of breaking down certain substances and not others. For example, molecules of cellulose are rich in the sugar glucose. However, the human digestive system cannot degrade cellulose and therefore cannot extract the energy from its sugar content. On the other hand, starch (which also is rich in glucose) can be processed by the human digestive tract and is therefore an important source of sugars and the inherent energy of the sugar molecules.

One of the ways in which the potential energy of a substance can be measured is by determining the amount of heat that is released when the substance is burned. The most popular unit of expressing this energy is the **calorie**, which is defined as the amount of heat energy required to raise the temperature of one gram (or one cubic centimeter [i.e., 1 c.c.]) of water from 15°C to 16°C. For most substances that are important body energy sources, one gram of the substance yields thousands of calories, and this makes

TABLE 12.1	CALORIFIC VALUES OF SELECTED FOODS		
SUBSTANCE	UNITS	CALORIES	CALORIES/GRAM
General			
Water	-	-	0.0
Vegetable fibers	-	-	0.0
Carbohydrates	-	-	4.1
Glucose	-	-	3.8
Proteins	-	-	4.1
Fats (Average)	-	-	9.3
Specific Foods			
Orange	1 oz.	10	0.35
Milk (Nonfat)	1 cup	90	0.37
Apple	1 oz.	15	0.53
Milk (Regular)	1 cup	150	0.61
Potato	1 oz.	18	0.63
Wine	1 oz.	24	0.85
Green Peas	1 oz.	30	1.06
Fish (Salmon)	1 oz.	40	1.41
Eggs	one (50 *g*)	80	1.60
Meat	1 oz.	72	2.54
White Bread	1 oz.	80	2.82
Cheese	1 oz.	105	3.70
Chocolate	1 oz.	150	5.28
Butter	1 oz.	203	7.16
Margarine	1 oz.	205	7.23

the calorific values of many substances awkward to express in calories. Therefore, another unit is used–also called a *Calorie* (but notice that the letter "C" is capitalized)–which is equal to 1,000 calories.* Typically, a calorie is known as a "small calorie," whereas a Calorie is known as a "big calorie" (thus a Calorie can raise the temperature of 1 *liter* of water from 15°C to 16°C). The calorific values of a sampling of human nutrients are listed in table 12.1.

CARBOHYDRATE METABOLISM

Digestion of Carbohydrate

The body's major sources of dietary carbohydrate are *glycogen* in meats, *starch* in vegetables, and certain sugars such as *sucrose* (table sugar) and *lactose* (milk sugar). Digestion of these carbohydrates within the digestive tract (see chapter 11) produces

simple sugars (monosaccharides), principally glucose, and it is glucose that is absorbed into the bloodstream. Glucose removed from the bloodstream by the body's tissues is used in several ways: (1) it may be converted to glycogen and stored; (2) it may be used for the synthesis of cellular glycoproteins, including certain enzymes and membrane receptors, (3) it may be completely degraded, producing CO_2 and water, an exergonic process that is directly coupled to the production of ATP; or (4) it may be partially degraded and the products used for the synthesis of other substances including amino acids and fatty acids. These alternative fates of blood glucose are summarized in figure 12.3.

Glycogenesis and Glycogenolysis

By **glycogenesis** is meant the conversion of sugar (principally glucose) into **glycogen** for temporary storage; as needed, the glucose is

---

* One Calorie = 1 kilocalorie = 1000 calories.

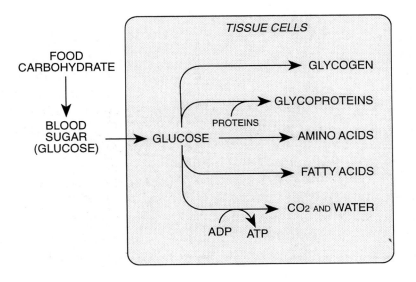

Figure 12.3
Ways in which glucose produced during the digestion of carbohydrates is used inside the cells of the body's tissues.

later withdrawn from the glycogen and used for other purposes (see above). Glycogen consists of large numbers of branched chains of glucose molecules (see figure 2.20). During periods of active glycogenesis (e.g., after digesting a meal that is rich in carbohydrate), it is not unusual for individual molecules of glycogen to reach sizes that consist of hundreds of thousands of glucose units. Glycogen molecules are assembled by sequentially adding glucose units to the growing ends of chains, accompanied by the introduction of branch points. As a result, a complex and compact treelike structure is progressively established.

The addition of a glucose unit to a growing glycogen molecule involves the following series of reactions. Glucose entering a cell is first converted into *glucose-1-phosphate*, an endergonic process that requires the energy input from one molecule of ATP for each glucose converted (much more energy will be regained later when the glucose unit is removed from the glycogen molecule and is degraded to CO_2 and water). Using UTP (uridine triphosphate, a molecule similar to ATP), the glucose-1-phosphate is then converted to *uridine diphosphoglucose* (UDPG). UDPG then reacts with the growing glycogen molecule, transferring its glucose to the

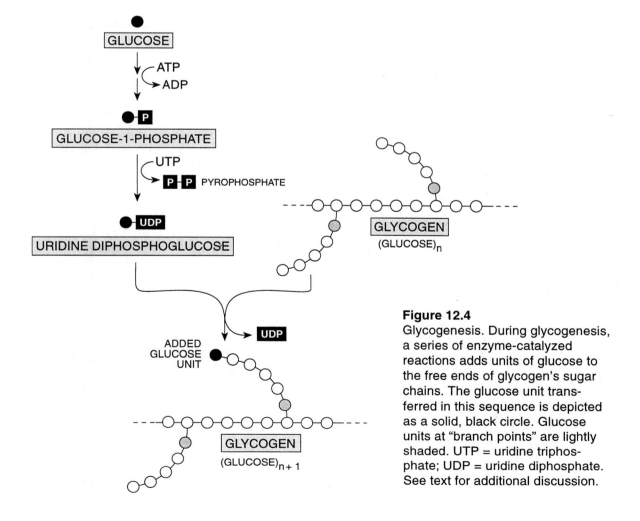

Figure 12.4
Glycogenesis. During glycogenesis, a series of enzyme-catalyzed reactions adds units of glucose to the free ends of glycogen's sugar chains. The glucose unit transferred in this sequence is depicted as a solid, black circle. Glucose units at "branch points" are lightly shaded. UTP = uridine triphosphate; UDP = uridine diphosphate. See text for additional discussion.

end of an elongating (or branching) chain. The sequence of reactions is illustrated in figure 12.4. (For simplicity, the names of the enzymes that catalyze each of the steps of glycogenesis have been omitted. However, it is important to recognize that virtually every step of a metabolic pathway is catalyzed by a particular enzyme and that reverse reactions (if, indeed, the reverse reactions can occur) are usually catalyzed by *different* enzymes.

While nearly all cells of the body are capable of producing and storing glycogen, little glycogenesis occurs in organs other than the muscles of the body and the liver. Of the total amount of glycogen stored in these tissues, about 80% is stored in muscles and the remaining 20% stored in the liver .

Glycogenolysis refers to the breakdown of glycogen, freeing its glucose content (which may then be further degraded or used in the synthesis of other molecules). Glycogenolysis is not simply the reverse of glycogenesis. During glycogenolysis, molecules of glucose are cleaved away from the ends of glycogen's sugar chains; the reactions consume phosphate and free each glucose unit from the glycogen molecule as *glucose-1-phosphate*. Glucose-1-phosphate is then converted to *glucose-6-phosphate*, which can now enter any of a variety of other metabolic pathways. The stages of glycogenolysis are depicted in figure 12.5. When glycogenolysis occurring in the liver is aimed at releasing sugar back into the bloodstream, glucose-6-phosphate must be converted to glucose before it leaves the liver cells.

12-1

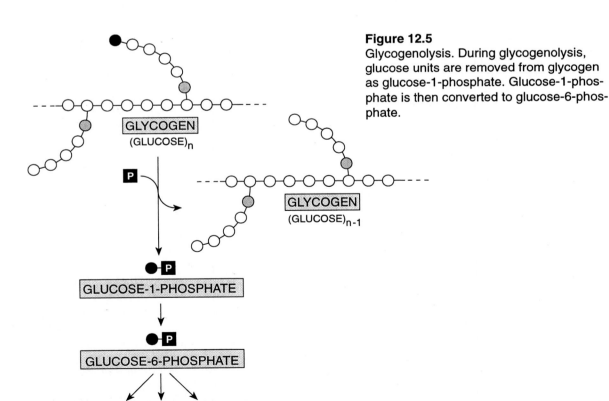

Figure 12.5
Glycogenolysis. During glycogenolysis, glucose units are removed from glycogen as glucose-1-phosphate. Glucose-1-phosphate is then converted to glucose-6-phosphate.

Glycolysis

Essentially all cells of the body are capable of using glucose in an extremely important metabolic pathway called **glycolysis**. Indeed, the glycolytic pathway is common to most living things on this planet. During glycolysis, a series of nine enzyme-catalyzed reactions progressively convert one molecule of glucose into two molecules of pyruvic acid (fig. 12.6). The original glucose molecule consists of six carbon atoms, 12 hydrogen atoms, and six oxygen atoms (i.e., the formula for glucose may be represented as $C_6H_{12}O_6$). The two molecules of pyruvic acid that are produced during glycolysis account for all but four of the hydrogen atoms in the original glucose molecule. The four hydrogens end up attached to a special hydrogen carrier and **coenzyme** (see chapter 2) known as **nicotinamide adenine dinucleotide** (abbreviated NAD *before* it picks up hydrogen [i.e., when it is in the "oxidized" state] and $NADH_2$ *after* it has picked up hydrogen [and is in the "reduced" state]). As you will learn later, the vitamin **niacin** is the source of the body's NAD. You will also learn later that the hydrogen atoms of $NADH_2$ are eventually combined with oxygen to form water in an exergonic series of reactions called *oxidative phosphorylation,* which generates large amounts of ATP.

The steps of glycolysis are summarized in figure 12.6. First, glucose is converted to *glucose-6-phosphate*, the reaction consuming a molecule of ATP. Glucose-6-phosphate is then converted to *fructose-6-phosphate*. In a third step, another molecule of ATP is consumed converting fructose-6-phosphate to *fructose-1,6-diphosphate*. In the fourth step of glycolysis, the 6-carbon fructose-1,6-diphosphate molecule is split into two 3-carbon molecules: *dihydroxyac-*

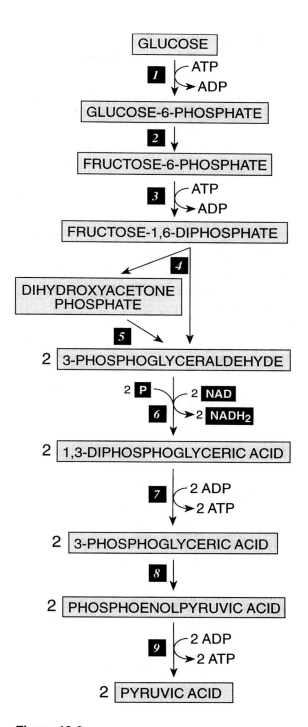

Figure 12.6
The glycolytic pathway. During glycolysis, one molecule of glucose is converted to two molecules of pyruvic acid. Overall, two molecules of ATP are produced from ADP.

etone phosphate and *3-phosphoglyceralde-hyde*. The dihydroxyacetone phosphate is immediately converted to a second molecule of 3-phosphoglyceraldehyde. In the next step, the two molecules of 3-phosphoglyceralde-hyde are converted to two molecules of *1,3-diphosphoglyceric acid*. In this reaction, two molecules of phosphate are incorporated into the product and two molecules of NAD pick up four hydrogen atoms. In the next step of the glycolytic pathway, the two molecules of 1,3-diphosphoglyceric acid are converted to two molecules of *3-phosphoglyceric acid*; this reaction is coupled to the production of two molecules of ATP (one molecule of ADP is converted to ATP for each molecule of 1,3-diphosphoglyceric acid that is *dephosphorylated*). The two molecules of 3-phosphoglyceric acid are now converted to two molecules of *phosphoenolpyruvic acid* and, in what is the last reaction of the glycolytic pathway, the phosphoenolpyruvic acid molecules are converted to two molecules of *pyruvic acid*. Accompanying the latter reaction is the conversion of two more molecules of ADP to ATP (fig. 12.6).

12-2

Glycolysis may be summarized as follows: (1) one molecule of glucose is converted to two molecules of pyruvic acid; (2) two molecules of NAD are converted to $NADH_2$; and (3) there is a direct *net* production of two molecules of ATP (i.e., two molecules of ATP are consumed, but four molecules of ATP are produced). All of the reactions of glycolysis occur in the cytoplasm of cells.

Fate of Pyruvic Acid

Usually, one of two things happens to the pyruvic acid that is formed during glycoly-sis. Either the pyruvic acid is converted to **lactic acid** or it is converted to **acetyl-coenzyme A**, which enters the metabolic pathway known as the **Krebs cycle** (fig. 12.7). The conversion of pyruvic acid to lactic acid usually occurs when either (1) there is a shortage of cellular NAD or (2) the lack or absence of oxygen blocks entry into the Krebs cycle.

The conversion of pyruvic acid to lactic acid, requires two atoms of hydrogen; these are provided by $NADH_2$, which is converted to NAD in the process. As a result, the NAD is now available to participate in glycolysis (i.e., step 6 in figure 12.6).

Because the conversion of pyruvic acid to lactic acid occurs when there is a lack of oxygen, the reaction pathway beginning with glucose and ending with lactic acid is sometimes referred to as **anaerobic respiration**. Anaerobic respiration is to be compared (see below) with **aerobic respiration**, which occurs when oxygen is not limited. During aerobic respiration, glucose is first converted to pyruvic acid and the pyruvic acid is then converted to carbon dioxide and water.

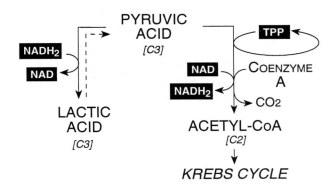

Figure 12.7
Alternate fates of pyruvic acid. Either pyruvic acid is converted to lactic acid (anaerobic) or it is converted to acetyl-coenzyme A and enters the Krebs cycle (aerobic).

We have considered anaerobic respiration previously in connection with the physiology of striated muscle tissue (chapter 4). For example, actively contracting white (fast) muscle tissue utilizes glycolysis as a major source of ATP. The needed ATP is produced, despite the limited oxygen available by employing glycolytic reactions; the reaction pathway is sustained by converting pyruvic acid to lactic acid, thereby recycling the required NAD. Once the level of contractile activity is reduced, lactic acid that has accumulated in the tissue can be converted back into pyruvic acid (fig. 12.7), which may then be converted to CO_2 and water.

The alternative fate of pyruvic acid is its conversion to **acetate** (or **acetic acid**) in the form of **acetyl-Coenzyme A** (abbreviated *acetyl-Co A*) prior to its entry into the Krebs cycle (see chapter 2 for a discussion of coenzymes). This conversion is known as **pyruvic acid oxidation**. As seen in figure 12.7, the reactions that convert pyruvic acid (a 3-carbon compound) into acetate (a 2-carbon compound) involve the loss of one carbon atom (as carbon dioxide) and two hydrogen atoms (as $NADH_2$). Two coenzymes participate in these reactions: **thiamine pyrophosphate** (**TPP**), derived from **thiamine** (vitamin B_1) and **Coenzyme A**, derived in part from **pantothenic acid** (vitamin B_3, see later in chapter). Coenzyme A ends up linked to the acetic acid. Pyruvic acid oxidation occurs within a cell's mitochondria.

Gluconeogenesis

Lactic acid produced in excess of that which can be further metabolized in the tissues passes into the bloodstream. Indeed, this is a common occurrence in actively contracting striated muscle tissue. Lactic acid is carried in the blood to the liver which converts the lactic acid back into pyruvic acid and then converts the pyruvic acid into glucose. (This cycle in which glucose is degraded in muscle and reformed from lactic acid by the liver is known as the **Cori cycle** and is described in chapter 4; see figure 4.24). The liver can also convert amino acids into pyruvic acid and then into glucose. The production of glucose from these metabolites is called **gluconeogenesis**. Figure 12.8 summarizes the pathways of glycogenesis, glycogenolysis, glycolysis, and gluconeogenesis.

Krebs Cycle and Oxidative Phosphorylation

Overview of the Krebs Cycle. The Krebs cycle is a central and extremely important metabolic pathway in nearly all cells of the body. The Krebs cycle is also a complicated pathway, and it is, therefore, a good idea to consider the cycle's overall effects before examining the individual reactions of the pathway itself. The Krebs cycle can be said to meet three principal goals: (1) the cycle degrades the acetic acid in acetyl-Co A to CO_2, thereby completing the breakdown of sugar that was begun during glycolysis, (2) the cycle produces one molecule of ATP (for each turn of the cycle), and (3) the cycle produces additional reduced hydrogen carrier molecules (e.g., $NADH_2$), whose oxidation during *oxidative phosphorylation* (see later) generates additional quantities of ATP. As we consider the Krebs cycle, it should be kept in mind that the acetyl-Co A that starts the cycle can be derived through the breakdown of glucose (or other sugars) *and* from the catabolism of fats, proteins, and certain amino acids.

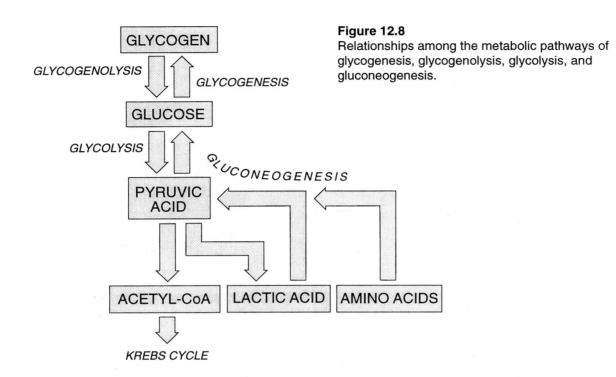

Figure 12.8
Relationships among the metabolic pathways of glycogenesis, glycogenolysis, glycolysis, and gluconeogenesis.

Reactions of the Krebs Cycle. The reactions of the Krebs cycle are summarized in figure 12.9. The cycle begins (figure 12.9, step 1) when acetyl-Co A reacts with *oxaloacetic acid*, forming *citric acid* and simultaneously frees the coenzyme. Acetic acid is a 2-carbon compound and oxaloacetic acid is a 4-carbon compound; therefore, their combination to form citric acid creates a 6-carbon compound. Citric acid is then converted to *cis-aconitic acid* (figure 12.9, step 2), which is converted to *isocitric acid* (figure 12.9, step 3). No changes in the numbers of carbon atoms accompany these reactions (i.e., isocitric acid is a 6-carbon compound). In the fourth step of the cycle (figure 12.9, step 4), one of the carbon atoms of isocitric acid is removed as CO_2, producing the 5-carbon compound *alpha-ketoglutaric acid*. Also removed are two hydrogen atoms, which are picked up by NAD (which is thereby converted to $NADH_2$).

In step 5 of the Krebs cycle (fig. 12.9), *alpha*-ketoglutaric acid is converted to *succinic acid*–a 4-carbon compound. Therefore, in this step, another carbon atom is eliminated as CO_2. Additionally, two more atoms of hydrogen are removed and accepted by NAD. The Krebs cycle reaction that produces succinic acid is indirectly coupled to the synthesis of one molecule of ATP from ADP.

Succinic acid is then converted to *fumaric acid* by the loss of two hydrogen atoms (figure 12.9, step 6). This time the hydrogen acceptor is not NAD; rather, it is **flavin adenine dinucleotide** (abbreviated FAD)–a coenzyme that is derived from vitamin B_2 (**riboflavin**, see later). By accepting the two hydrogen atoms, FAD is converted (i.e., reduced) to $FADH_2$.

Fumaric acid is converted to *malic acid*

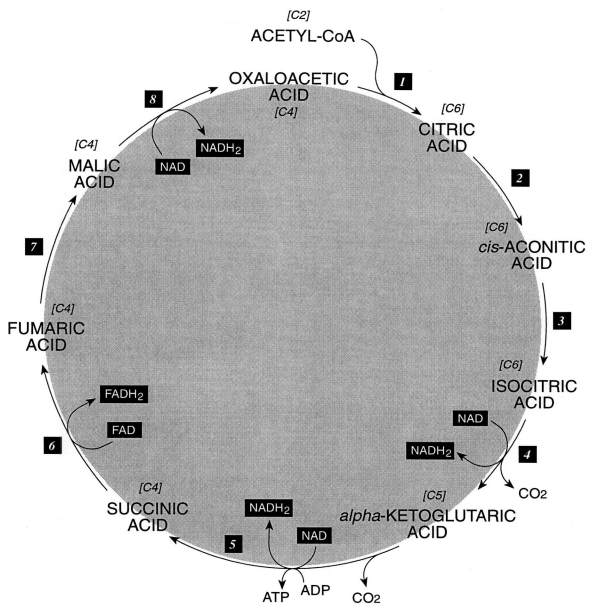

Figure 12.9
The reactions of the Krebs cycle .

(figure 12.9, step 7). No carbon atoms are lost (or gained) during this reaction. Finally, in the last reaction of the Krebs cycle (figure 12.9, step 8), malic acid is converted to *oxaloacetic*

12-3

acid. This requires the removal of two hydrogen atoms, which are accepted by NAD. The reactions of the Krebs cycle occur within the mitochondria of a cell.

The reactions of the Krebs cycle may be

summarized as follows:

(1) the two carbon atoms of the acetic acid molecule that enters the cycle end up in two molecules of carbon dioxide. Since glycolysis produces *two* molecules of pyruvic acid, each of which yields a molecule of CO_2 and a molecule of acetyl-Co A, there will be two turns of the Krebs cycle per glucose molecule; this implies that altogether six molecules of CO_2 result from the complete degradation of one glucose molecule;

(2) oxaloacetic acid consumed in the first reaction of the Krebs cycle is restored in the last reaction (that's why the pathway is a *cycle*);

(3) one molecule of ATP is generated directly by each turn of the cycle, but since each molecule of glucose gives rise to two turns of the cycle, two molecules of ATP are generated;

(4) three molecules of $NADH_2$ and one molecule of $FADH_2$ are produced by each turn of the cycle. Therefore, during the complete degradation of one molecule of glucose, the resulting two Krebs cycles produce

six molecules of $NADH_2$ and two molecules of $FADH_2$.

Oxidative Phosphorylation. Oxidative phosphorylation refers to the production of ATP from ADP that accompanies the oxidation of reduced hydrogen carriers like $NADH_2$ and $FADH_2$. The reason that ATP is generated by oxidation of these carriers is because the oxidation reactions (in which molecular oxygen is combined with hydrogen to form water) are highly exergonic. The cell traps much of the energy of these reactions by adding phosphate to ADP (i.e., by "phosphorylating" ADP).

Oxidation of each molecule of $NADH_2$ is coupled to the phosphorylation of *three* molecules of ADP, and oxidation of $FADH_2$ is coupled to the phosphorylation of *two* molecules of ADP. For the two turns of the Krebs cycle that occur for each glucose molecule that is degraded to acetyl-Co A, 18 molecules of ATP are produced from $NADH_2$ oxidation and four ATP are produced from $FADH_2$ oxidation (see table 12.2). Oxidative

TABLE 12.2	ATP PRODUCTION PER MOLECULE OF GLUCOSE		
PATHWAY	ATP PRODUCED BY PATHWAY	REDUCED CARRIER PRODUCED	ATP PRODUCED BY REOXIDIZING CARRIER
Glycolysis	2	2 $NADH_2$	6
Pyruvic Acid Oxidation	0	2 $NADH_2$	6
Krebs Cycle	2	6 $NADH_2$	18
		2 $FADH_2$	4
TOTAL ATP PRODUCTION PER MOLECULE OF GLUCOSE = 38			

phosphorylation occurs within a cell's mitochondria.

As seen in table 12.2, when one molecule of glucose is processed by glycolysis, pyruvic acid oxidation, the Krebs cycle, and oxidative phosphorylation, 38 molecules of ADP are converted to ATP. The overall reaction sequences may be summarized by a single comprehensive equation; namely,

$$C_6H_{12}O_6 + 6\ O_2 \longrightarrow 6\ CO_2 + 6\ H_2O$$

$$\underset{\text{ADP}}{38} \quad \underset{\text{ATP}}{38}$$

Efficiency of Glucose Utilization. Each mole* of glucose contains 686 Calories of energy. The conversion of one mole of ADP to ATP serves to trap 12 Calories of energy. Because 38 moles of ATP are generated from ADP for each mole of glucose that is consumed, altogether 456 Calories (i.e., 38 x 12) are trapped as ATP during glycolysis, pyruvic acid oxidation, the Krebs cycle and oxidative phosphorylation, What this means is that the body is about 66% (i.e., 456/686) efficient in trapping the energy inherent in glucose. This is a surprisingly high percentage, testifying to the efficiency of these cellular processes. The remaining 34% of the energy of glucose that is not trapped in ATP is lost as heat.

PROTEIN METABOLISM

If we exclude water, then more than 70% of the body's weight is represented by protein. Tissue proteins are manufactured from amino

---

* A mole is a molecule's molecular weight in grams. For glucose, one mole is 180 grams.

acids derived from the blood and these amino acids are acquired from the digestion of protein in the food that we eat (chapter 11). Unlike carbohydrate, which is turned over very rapidly in the body because it is a major energy source, the turnover of protein is much slower. (In a healthy individual, proteins do not serve as an energy source.) Although tissue proteins are relatively stable, the body does degrade and excrete the equivalent of 20 to 30 grams of tissue protein each day. Amino acids derived from the digestion of food protein are used to replace this small amount of daily protein loss. If the daily intake of amino acids is greater than that needed to replace the lost proteins, then the excess amino acids are chemically altered and enter metabolic pathways that lead to their storage as carbohydrate or fat (amino acids are not converted to proteins for "storage" purposes).

Essential and Nonessential Amino Acids

Whereas most of the monosaccharide that is present in tissue carbohydrates is the particular sugar *glucose*, tissue proteins contain about 20 different amino acids. The chemical structures of these amino acids are shown in figure 2.17.

Ten of the 20 amino acids regularly found in tissue proteins are called **essential amino acids** because the body cannot synthesize them from raw materials or by chemically altering other amino acids. The essential amino acids must be acquired by the digestion of dietary protein. The other ten amino acids are called **nonessential amino acids**, because they can be produced using the essential amino acids and intermediates of

TABLE 12.3 ESSENTIAL AND NONESSENTIAL AMINO ACIDS	
ESSENTIAL AMINO ACIDS	NONESSENTIAL AMINO ACIDS
Arginine	Alanine
Histidine	Asparagine
Isoleucine	Aspartic acid
Leucine	Cysteine
Lysine	Glutamic acid
Methionine	Glutamine
Phenylalanine	Glycine
Threonine	Proline
Tryptophan	Serine
Valine	Tyrosine

carbohydrate metabolism. Table 12.3 lists the essential and nonessential amino acids.

Transamination, Deamination, and Amination

The synthesis of a nonessential amino acid is usually achieved by a metabolic process called **transamination**. Transamination reactions typically involve certain intermediates of the Krebs cycle (see earlier). By accepting an amino group from one type of amino acid, these Krebs cycle intermediates (called *keto acids*) are converted into a different type of amino acid. This process is illustrated in figure 12.10 in which the amino group of the amino acid *glutamic acid* is transferred to the Krebs cycle intermediate *oxaloacetic acid*. The reaction produces the amino acid *aspartic acid* and another Krebs cycle intermediate–namely, *alpha-ketoglutaric acid*. Each transamination reaction is catalyzed by a specific enzyme and also involves the coenzyme **pyridoxal phos-phate** (derived from pyridoxine [vitamin B_6]).

Unneeded or excess amino acids can be **deaminated**, thereby producing either pyruvic acid, acetic acid, or certain Krebs cycle intermediates. Once deamination has taken place, further catabolism may lead (via the Krebs cycle and oxidative phosphorylation) to CO_2 and water. It may be noted that the complete oxidation of amino acids does not trap as much energy in the form of ATP as does the complete oxidation of an equivalent amount of glucose.)

Deaminated amino acids can also enter an anabolic pathway leading to gluconeogenesis (or even fatty acid synthesis). The amino groups released from deaminated amino acids take the form of **ammonia** (NH_3), which usually is converted to the excretory product **urea** (the metabolism of ammonia and urea are discussed further in chapter 13, which deals with the body's excretory system).

The liver is capable of combining free amino groups derived from ammonia with *alpha*-ketoglutaric acid, thereby forming

314

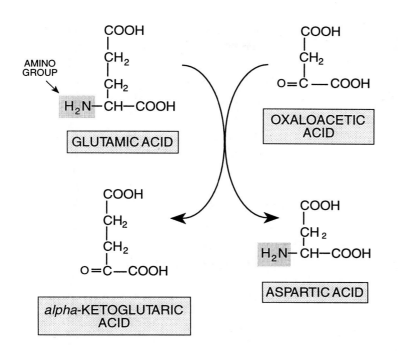

Figure 12.10
Nonessential amino acids may be formed by transamination reactions in which the amino group of one type of amino acid is transferred to a Krebs cycle intermediate, thereby forming a different amino acid and another Krebs cycle intermediate. Illustrated here is the transfer of an amino group from the amino acid glutamic acid to oxaloacetic acid; this produces the amino acid aspartic acid and the Krebs cycle intermediate *alpha*-ketoglutaric acid.

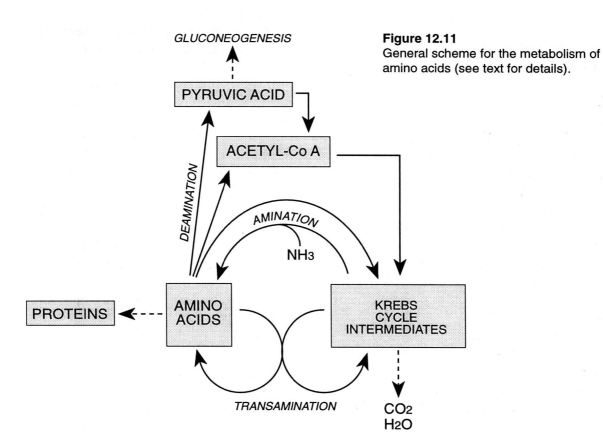

Figure 12.11
General scheme for the metabolism of amino acids (see text for details).

glutamic acid. This form of amino acid synthesis is called **amination**. Figure 12.11 summarizes the alternative metabolic fates of amino acids.

LIPID METABOLISM

Digestion, Absorption, and Circulation of Lipids

As described in chapter 11, by the time that partially digested food reaches the small intestine, fat exists as droplets that are soon emulsified by the action of the liver's bile salts. The emulsified fat droplets are converted to monoglycerides, fatty acids, glycerol, and cholesterol by the lipid-digesting enzymes produced by the pancreas and intestine. These products then combine with bile salts to form micelles and are absorbed by the intestinal epithelium. (See figures 2.24, 2.25, and 2.27 for the chemical structures of fatty acids, triglycerides, and cholesterol; also, reexamine figure 11.14.)

Within the intestinal epithelium, the contents of the micelles are converted to triglycerides, combined with cholesterol, enclosed in protein to form *chylomicrons*, and transferred to the lymphatic circulation. Eventually, the chylomicrons enter the blood through the thoracic duct.

As the chylomicrons circulate through the body's fat (i.e., **adipose**) tissues and liver, *lipoprotein lipase* (an enzyme present in the surfaces of the cells forming the capillary walls) converts the triglycerides into fatty acids and glycerol. The fatty acids are then taken up by the adipose tissue and the liver and combined with cellular glycerol to form new triglycerides. By this process much of the contents of the chylomicrons is removed

after one or two hours in the bloodstream. Triglycerides stored in adipose tissue serve as the major energy source for the body (see later). About half of all of the fat stored in adipose tissue is found in the fat cells of the **subdermis**.

Short-chain fatty acids and glycerol may be absorbed directly into the blood from the small intestine. These substances enter the hepatic portal circulation, are removed from the blood by the liver, and are converted into triglycerides. If large quantities of glucose are absorbed into the blood following digestion of a carbohydrate-rich meal, glucose will not only be removed by the liver for glycogenesis, but will also be used for the

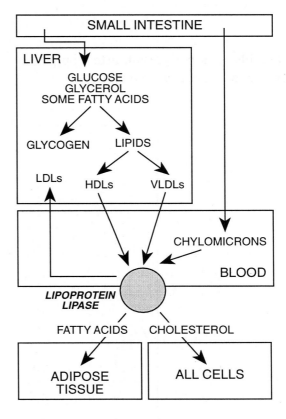

Figure 12.12
Fates of lipids absorbed from the intestine.

synthesis of lipids. Some of the lipid synthesized in the liver is released into the bloodstream as lipoproteins called *VLDLs* (very-low-density lipoproteins, see chapter 11), which provide lipids for all tissues of the body. Figure 12.12 summarizes the fates of absorbed lipids.

Triglycerides stored in fat cells serve as a reserve energy source for other tissues of the body (the metabolic pathway yielding this energy is described later in the chapter). To mobilize this energy source, the triglycerides of fat cells are converted into fatty acids and then released into the blood where they combine with molecules of plasma albumin. Fatty acids bound to albumin are known as **free fatty acids**, to distinguish them from the fatty acids that are chemically bound to glycerol and cholesterol in circulating chylomicrons and VLDLs. Utilization of free fatty acids by the body results in the complete turnover of stored fats every two or three weeks.

Lipogenesis

In addition to acquiring fatty acids from digested fat, the body can synthesize many fatty acids from smaller metabolic intermediates and also convert one type of fatty acid into another (this process is called **lipogenesis**). Among the most common fatty acids found in the body are *myristic acid*, *palmitic acid*, and *stearic acid*; their chemical formulas are given in figure 2.24. Fatty acids consist of a long chain (usually called a **hydrocarbon chain**) of **methylene groups** (i.e., $-CH_2-$) that ends with a single carboxyl (or acid) group. The general chemical formula for a fatty acid is, therefore, $CH_3-(CH_2)_n-COOH$, where n usually is an even number. (For myristic acid, palmitic acid, and stearic acid, n equals 12, 14, and 16, respectively.)

Myristic, palmitic, and stearic acids are classified as **saturated fatty acids** because all of the carbon atoms of the hydrocarbon chain are linked by *single* bonds. When double bonds occur in the hydrocarbon chain, the fatty acid is said to be **unsaturated**. Some unsaturated fatty acids can be synthesized from corresponding saturated fatty acids by the removal of hydrogen atoms from certain neighboring methylene groups, thereby introducing double bonds (see figure 2.24). However, humans cannot produce the double bonds present in the unsaturated fatty acids *linoleic acid* and *linolenic acid*. Since these fatty acids are important constituents of cells and tissues, they are considered **essential fatty acids** and must be acquired directly from digested fats (linoleic and linolenic acid are present in vegetable oils).

In effect, the synthesis of fatty acids by human cells is achieved by successively adding the 2-carbon *acetate* portions of a number of acetyl-Co A molecules to an elongating hydrocarbon chain. For example, the synthesis of the 16-carbon fatty acid palmitic acid would require the combination in sequence of eight molecules of acetyl-Co A. (Note that this mechanism produces chains containing even numbers of carbon atoms). Since acetyl-Co A can be derived from the oxidation of pyruvic acid that has been acquired by glycolysis (see figures 12.6 and 12.7), it is clear that the carbon skeleton of a newly synthesized fatty acid may be derived entirely from glucose. It should also be noted that the condensation of three fatty acids with glycerol derived from 3-phosphoglyceraldehyde (another glycolysis intermediate) generates a complete triglyceride.

Although upon first examination the chemical structure of cholesterol looks quite different from that of a fatty acid (see figure 2.27), cholesterol is, nonetheless, a long-chain hydrocarbon whose carbon atoms, like those of fatty acids, are derived entirely from the acetate of acetyl-Co A. Figure 12.13 summarizes the ways in which fatty acids, triglycerides, and cholesterol are synthesized. Finally, because acetyl-Co A is obtained during the deamination of certain amino acids (fig. 12.11), this implies that excess proteins or amino acids may also be converted to fats.

Oxidation of Fatty Acids

The calorie content of fat is more than twice that of carbohydrate (see table 12.1). Therefore, it is not surprising that most of the body's stored energy takes the form of fat; indeed, the average person stores 200 times as much energy in the form of fat (mainly triglycerides) as in the form of carbohydrate (i.e., glycogen).

Figure 12.14 shows the metabolic pathway, called **beta-oxidation**, in which fatty acids are converted to acetyl-Co A, $FADH_2$ and $NADH_2$. Notice that the pathway involves a cycle, so that the quantities of the products formed depend on the number of cycles required to reduce the hydrocarbon chain length of the original fatty acid to acetate units. For example, an 18-carbon stearic acid molecule would yield nine molecules of acetyl-Co A and require *eight* turns of the cycle. (Whereas the first seven turns of the cycle each yields a single acetyl-Co A molecule, the eighth turn produces two acetyl-Co A molecules.)

Each turn of the beta-oxidation cycle also produces one molecule of $FADH_2$ and one molecule of $NADH_2$. The reoxidation of these hydrogen acceptors yields 5 molecules of ATP (2 for each $FADH_2$ and 3 for each

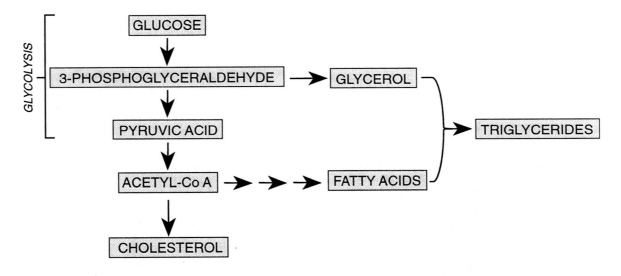

Figure 12.13
Some fatty acids, glycerol, and cholesterol may be synthesized from acetyl-Co A and intermediates of glycolysis.

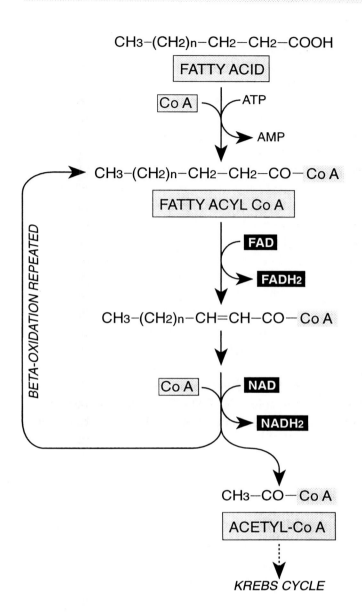

$CH_3-(CH_2)_n-CH_2-CH_2-COOH$

FATTY ACID

CoA ATP

AMP

$CH_3-(CH_2)_n-CH_2-CH_2-CO-CoA$

FATTY ACYL CoA

FAD

FADH2

$CH_3-(CH_2)_n-CH=CH-CO-CoA$

CoA NAD

NADH2

$CH_3-CO-CoA$

ACETYL-CoA

KREBS CYCLE

BETA-OXIDATION REPEATED

Figure 12.14
Beta oxidation of fatty acids, yielding acetyl-Co A, $FADH_2$ and $NADH_2$. Entry of acetyl-Co A into the Krebs cycle generates additional $FADH_2$ and $NADH_2$ as well as ATP. Re-oxidation of $FADH_2$ and $NADH_2$ leads to additional ATP production via oxidative phosphorylation.

$NADH_2$). Therefore, conversion of one molecule of stearic acid to 9 molecules of acetyl-Co A generates 40 molecules of ATP. Because one ATP molecule is consumed in order to initiate *beta*-oxidation (see figure 12.14), there is a *net* production of 39 molecules of ATP.

Even more ATP is generated if the acetyl-Co A enters the Krebs cycle. One molecule of ATP will be produced directly during each cycle (see figure 12.9) and an additional 11 molecules of ATP will be obtained when the one $FADH_2$ and three $NADH_2$ molecules produced during each cycle are reoxidized (see table 12.2). Therefore, the nine turns of the Krebs cycle will produce 108 molecules of ATP (i.e., [9 x 1] + [9 x 11]). Consequently, the complete oxidation of one molecule of stearic acid to CO_2 and water yields 147 (i.e., 39 + 108) molecules of ATP.

Phospholipids and Cholesterol

In phospholipids, one of the fatty acids of a triglyceride is replaced by a phosphate-containing group (see figure 2.26). Nearly all cells of the body are capable of forming phospholipids from triglycerides, and the phospholipids produced play an integral role in the structure of cellular membranes. Phospholipids are also present in lipoproteins circulating in the blood and are derived by secretion from the liver.

Like phospholipids, cholesterol also plays an important intracellular role as a constituent of cell membranes. In addition to the cholesterol derived through the digestion of fat in food, the liver and other tissue of the body manufacture cholesterol from acetyl-Co A.

VITAMINS

Vitamins are organic compounds that are essential to the proper functioning of the body, but because they are not consumed by the metabolic reactions in which they participate, they are required in the diet in only minuscule amounts in comparison with dietary protein, carbohydrate, and fat. Although vitamins are needed in only small amounts, a deficiency in their acquisition can have disastrous consequences for the body. Altogether, there are about a dozen major vitamins divided into two groups–those that are **water-soluble** (table 12.4) and those that are **fat-soluble** (table 12.5). Whereas the water-soluble vitamins are transferred directly into the bloodstream from the small intestine, the fat-soluble vitamins are transferred to the lymph (along with lipids from micelles formed in the small intestine by the action of the bile salts). Many of the water-soluble vitamins become components of *coenzymes* that are involved in critical enzyme-catalyzed

TABLE 12.4		WATER-SOLUBLE VITAMINS	
VITAMIN		ACTIVE FORM(S)	EFFECT OF DEFICIENCY
SYMBOL	NAME		
B1	Thiamine	Thiamine pyrophosphate	Beriberi
B2	Riboflavin	FAD and FMN	Deterioration of epithelium
B3	Niacin	NAD and NADP	Pellagra
B5	Pantothenic acid	Coenzyme A	Retarded growth
B6	Pyridoxine	Pyridoxal phosphate	Retarded growth
B12	Cobalamin	Cobalamin	Pernicious anemia
C	Ascorbic acid	Ascorbic acid	Scurvy
H	Biotin	Biocytin	Impaired fat metabolism
M	Folic acid	Tetrahydrofolic acid	Anemia

reactions of important metabolic pathways. By combining loosely with the appropriate enzyme, a coenzyme completes the structure of the enzyme's active site and, in so doing, confers full catalytic activity to the enzyme. (It might be noted that not all coenzymes are derived from vitamins.) Some of the vitamins require chemical modification before they assume their physiologically active form. What follows is a brief survey of the vitamins and their actions.

Water-Soluble Vitamins

Vitamin B₁ (Thiamine). Once within the body, vitamin B_1 is converted to **thiamine pyrophosphate** (TPP), which acts as a coenzyme in important reactions that remove a carbon unit (in the form of carbon dioxide) from organic substrates. Probably the most important example of the coenzyme action of TPP is in the conversion of pyruvic acid to acetyl-Co A (see figure 12.7 and also note that the latter stage of this conversion involves another coenzyme–coenzyme A). Deficiencies of vitamin B_1 lead to disorders of the nervous, cardiovascular, and gastrointestinal systems–a condition known as **beriberi.**

Vitamin B₂ (Riboflavin). Once absorbed into the tissues of the body, vitamin B_2 is used in the formation of the coenzymes *flavin adenine dinucleotide* (FAD) and *flavin mononucleotide* (FMN), the former being an important coenzyme of the Krebs cycle (see step 6 of figure 12.9). Since the Krebs cycle operates in nearly all cells of the body, adequate intake of this vitamin is very important. Although deficiencies of this vitamin are rare, the most common symptoms are deterioration of the body's epithelial and mucosal tissues.

Vitamin B₃ (Niacin or Nicotinic Acid). Vitamin B_3 is used by the body in the formation of the coenzymes *nicotinamide adenine dinucleotide* (NAD) and *nicotinamide adenine dinucleotide phosphate* (NADP). These coenzymes act as hydrogen acceptors in important oxidation reactions, such as in step 6 of glycolysis (fig. 12.6) and steps 4, 5 and 8 of the Krebs cycle (fig. 12.9). Deficiencies of this vitamin are reflected by abnormalities of the skin (a condition called **pellagra**) and malfunctions of the nervous and gastrointestinal systems.

Vitamin B₅ (Pantothenic Acid). Pantothenic acid contributes to the formation of coenzyme A, which participates in one of the most important metabolic reactions in the body–the conversion of pyruvic acid into acetyl-Co A (see figure 12.7). Deficiencies of this vitamin lead to retarded growth and disturbances of the central nervous system.

Vitamin B₆ (Pyridoxine). Within the body, vitamin B_6 is converted to *pyridoxal phosphate,* which acts as a coenzyme in such important metabolic reactions as transamination of amino acids (fig. 12.10). Although deficiencies of this vitamin are rare, their occurrences are characterized by retarded growth and anemia.

Vitamin B₁₂ (Cobalamin). Vitamin B_{12} is absorbed from the small intestine in combination with **intrinsic factor**, a substance that is secreted by the lining of the stomach (see chapter 11). Once absorbed into the body's tissues, vitamin B_{12} forms an important coenzyme that acts as a hydrogen acceptor in the metabolism of nucleic acids. Vitamin B_{12} is also essential for the proper production and development of red blood cells, a deficiency of this vitamin being responsible for **pernicious anemia**.

Vitamin C (Ascorbic Acid). Ascorbic acid affects the growth of cartilage, bone, teeth, and skin. A deficiency of this vitamin has varied effects, including the failure of wounds to heal properly, impaired bone growth, and fragility of blood vessel walls, leading to unnecessary losses of blood (especially bleeding gums). The condition is referred to as **scurvy**.

Vitamin H (Biotin). Biotin is a coenzyme in certain reactions in which a carboxyl group (i.e., COOH) is transferred from one metabolic intermediate to another. Biotin is important in fatty acid synthesis (lipogenesis).

Vitamin M (Folic acid). Folic acid (or folacin) is important in the synthesis of nucleic acids and the development of red and white blood cells. Whereas some folic acid is acquired from bacteria inhabiting the colon, a person's diet is the primary source. Retarded

12-4

growth and anemia are symptomatic of a deficiency of this vitamin.

Fat-Soluble Vitamins

Vitamin A (Retinol). Vitamin A is derived from the substance *beta*-carotene, a pigment present in certain vegetables (e.g., carrots and squash). The best known function of this vitamin is in vision, where it is a component of the eye's visual pigments (see chapter 6). Deficiencies of retinene lead to a lack of vision under conditions of poor lighting; this is commonly referred to as **night-blindness**. Unlike most other vitamins whose excesses are excreted by the body, excess retinol is stored in the skin where it creates an orange-yellow discoloration.

Vitamin D (Calciferol). Vitamin D can be produced in the skin when the skin is exposed to the ultraviolet rays in sunlight. Deficiencies of this vitamin are not uncommon in those parts of the world where there

TABLE 12.5	FAT-SOLUBLE VITAMINS		
VITAMIN		ACTIVE FORM(S)	DEFICIENCY EFFECT
SYMBOL	NAME		
A	Retinol	Retinol	Night-blindness
D	Calciferol	Dihydroxycholecalciferol	Rickets
E	Tocopherol	Tocopherol	Oxidation of lipids
K	Phylloquinone	Phylloquinone	Poor blood clotting

are long, dark winters; therefore, sources of this vitamin in the diet are an important supplement. Vitamin D facilitates the absorption of calcium from ingested food and its deposition in bone. Vitamin D deficiencies in children lead to improper development and weakness of the bones–a condition known generally as **rickets**.

Vitamin E (Tocopherol). Vitamin E plays an important role in fatty acid metabolism by preventing the oxidation of unsaturated fatty acids. Although deficiencies of this vitamin are rare, they are characterized by abnormal growth (and, in experimental animals, by sterility; hence, vitamin E is also known as the "anti-sterility" vitamin).

Vitamin K (Phylloquinone). Vitamin K is essential for the synthesis of several blood coagulation factors (prothrombin and factors VII, IX, and X; see chapter 8). Therefore, vitamin K deficiency is characterized by failure of the blood to clot properly. An important source of vitamin K are the bacteria living in the colon. Taking oral antibiotics in response to an infection can so reduce the colonic bacteria as to create a temporary vitamin K deficiency.

| 12-5 |

Today, most people acquire sufficient vitamins in their diets and deficiencies are uncommon. Consequently, for most people, taking vitamin supplements is a waste of time and expense. Excess vitamin intake can produce a condition known as **hypervitaminosis**. For most of the water-soluble vitamins, hypervitaminosis does not occur because excess vitamins are excreted via the kidneys. However, for the fat-soluble vitamins, excesses may be stored in the tissues

and can attain toxic levels. It is, therefore, important that supplements of fat-soluble vitamins be taken only under the supervision of a physician.

MINERALS

The elements carbon, hydrogen, oxygen, and nitrogen account for more than 95% of the total weight of the body and form nearly all of the body's organic substance. There is, however, a variety of other elements that are vital to the body but which occur in far smaller amounts; these are the body's **minerals**. The minerals, which are listed in table 12.6 and account for less than 5% of the body's weight, are divided into two main groups. The more common minerals are called "macrominerals," while those that are much rarer are called the "microminerals" (also referred to as "trace elements"). Essentially all of the minerals are derived through the diet and nearly all of them are metals (the exceptions are sulfur, chlorine, fluorine, and iodine).

About 80% of the body's mineral content is in the bones and teeth, which contain large quantities of **calcium** (the most abundant mineral), **phosphorus**, and **manganese**. Although 99% of the body's calcium is in the bones and teeth, the small amounts of calcium found in the tissues bound to proteins and as calcium ions are nonetheless important. In addition to its passive, structural role in bone, calcium plays active roles in muscle contraction and blood coagulation (see chapters 4 and 8).

While most of the body's phosphorus is also present in bones and teeth, about 20% plays vital roles in such important tissue constituents as the nucleic acids, phospho-

TABLE 12.6	MINERALS
MINERAL (SYMBOL, AMOUNT[1])	**ROLES**
Macrominerals	
Calcium (Ca, 1500 *g)*	Mostly structural, in bone and teeth; muscle contraction
Phosphorus (P, 860 *g)*	Mainly in bone, but important in nucleic acids, ATP, and ADP
Potassium (K, 180 *g)*	Important in electrolyte balance; membrane potentials
Sulfur (S, 175 *g)*	Important in tissue proteins, coenzymes, and cartilage
Chlorine (Cl, 74 *g)*	Important in electrolyte balance
Sodium (Na, 64 *g)*	Electrolyte balance; electrical potential of membranes
Magnesium (Mg, 25 *g)*	Abundant in bone and as intracellular ion; enzyme cofactor
Microminerals	
Iron (Fe, 4.5 *g)*	Active component in hemoglobin, myoglobin, and enzymes
Fluorine (F, 2.6 *g)*	Important constituent of bones and teeth
Zinc (Zn, 200 *mg)*	Enzyme cofactor
Silicon (Si, 24 *mg)*	Required for proper bone calcification
Selenium (Se, 13 *mg)*	Participates in enzymic antioxidant activity
Manganese (Mn, 12 *mg)*	Enzyme cofactor
Iodine (I, 11 *mg)*	Thyroid hormone component
Copper (Cu, 10 *mg)*	Enzyme cofactor
Nickel (Ni, 10 *mg)*	Affects absorption of iron; may be enzyme cofactor
Molybdenum (Mo, 9 *mg)*	Enzyme cofactor
Chromium (Cr, 6 *mg)*	Red blood cell component; enzyme cofactor
Cobalt (Co, 1.5 *mg)*	Component of vitamin B12
Vanadium (V, 0.1 *mg)*	Enzyme cofactor

[1]. Amount present in tissues of an average 170 *lb* adult male.

lipids, and ATP (see chapters 2 and 4 and earlier in this chapter). Most of the body's magnesium, like calcium and phosphorus, is present in bone, but magnesium also plays an important intracellular role as a cofactor for certain enzyme-catalyzed reactions.

The macrominerals **potassium**, **sodium**, and **chlorine** play important roles as free *ions* both intracellularly and extracellularly (i.e., in tissue fluid, lymph, and blood plasma). Sodium and potassium ions are especially important in creating the electrical gradients

across membranes that are fundamental to muscle contraction and nerve conduction (see chapters 4 and 5). **Sulfur**, the remaining macromineral, is an important constituent of tissue proteins, cartilage, and certain coenzymes (e.g., coenzyme A).

Among the microminerals, the most abundant is **iron**. Iron's principal role is in hemoglobin where it forms a loose combination with oxygen during oxygen transport in the bloodstream (see chapters 8 and 10). Iron also plays an important oxygen-binding role in muscle myoglobin and is a constituent of certain cellular enzymes. Iron is stored passively in the body as ferritin in the liver, spleen, and bone marrow.

Nearly all of the body's **fluorine** is found in bones and teeth where it is found in combination with calcium and phosphorus. Small amounts of fluorine are found in ionic form intracellularly and in body fluids.

Zinc, **manganese**, **copper**, **nickel, molybdenum, chromium**, and **vanadium** are trace metals that exist in the tissues in ionic form and which play roles as cofactors for certain enzymes. Zinc is a cofactor for the erythrocyte enzyme *carbonic anhydrase*, which is important in the interconversion of carbon dioxide and bicarbonate in the bloodstream (see chapter 10). Manganese is a cofactor in the all-important reactions that convert pyruvic acid to acetyl-Co A (see earlier in this chapter). Copper is an important coenzyme in a number of so-called "oxidation-reduction reactions" in which electrons are transferred among metabolic intermediates.

Although **silicon** is the most abundant element in the earth's surface, it is present in the body in only trace amounts, where it is believed to play a role in the process of calcification in growing bones. **Selenium**, another trace mineral, is believed to be a constituent of certain enzymes that act to prevent the accumulation of peroxides in tissue.

Nearly all of the body's **iodine** is found in the thyroid gland where it is temporarily stored in combination with the protein *thyroglobulin*. Most of this iodine ends up in the thyroid hormone *thyroxine*, which is released into and circulates in the bloodstream and affects the metabolic rate of the body's tissues (see chapter 14). Finally, **cobalt** is an important constituent of cobalamin (vitamin B_{12}), which is essential for the proper production of red blood cells in the bone marrow.

NUTRITION

The goal of good nutrition is to provide the body with the types and quantities of food that contribute to one's health and well-being. The food consumed daily may be referred to as the **diet**, and, ideally, this diet should contain the proper amounts and kinds of proteins, fats, carbohydrates, vitamins, minerals, and water. If the choices of what one eats are prudent, then the variety of proteins and fats that are eaten will provide all of the *essential* amino acids and fatty acids. Of course, the diet must also provide a particular caloric value, namely the calories that are needed to meet the energy needs arising from activity (work, play, etc.). If the caloric value of the food that is consumed exceeds that needed to sustain the body and its activity, the excess calories *may* be stored as fat, and this can lead to **obesity** (having too much body fat and usually [but not always] being "overweight"). Indeed, for every 9.3 Calories of excess energy that enters the body, the body can make and store one gram of fat. On the other hand, if the caloric value of the consumed food does not meet the body's needs, this will lead to losses of body weight as tissue carbohydrate, fat, and protein are consumed to provide the needed calories.

It should be noted that increases in body fat can occur even when the caloric intake is normal if one's level of activity (e.g., exercise) is reduced. Under these conditions, a person may for some time (months to years)

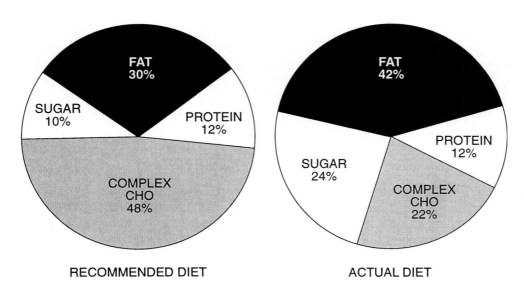

Figure 12.15
Caloric contributions of protein, fat, complex carbohydrate (CHO), and sugar in a healthy (recommended) diet and the actual (average) diet in the United States.

show no weight gain as lost muscle protein is replaced by an equal weight of fat. However, eventually, increased fat production and storage will be reflected by an increase in body weight.

With regard to the relative amounts of proteins, carbohydrates, and fats in the diet, the recommended balance is depicted in figure 12.15. In the recommended diet, 12% of the caloric intake is represented by proteins, 30% by fats, 48% by *complex* carbohydrates (e.g., polysaccharides), and 12% by sugars. In the United States, the actual average diet differs considerably from the desired diet (again, see figure 12.15) in that it contains too much fat and too much sugar.

Dietary Protein, Fat, and Carbohydrate

The recommended daily allowance of pro-tein is about 0.8 grams per kilogram of body weight (e.g., a male weighing 70 *Kg* [154 *lb*] should consume about 56 *g* of protein each day; see table 12.7). Each of the proteins synthesized by the body is comprised of a particular number and sequence of the 20 different amino acids. The proper dietary intake of protein provides all of these amino acids, especially the **essential amino acids** (see table 12.3)–those amino acids that the body cannot synthesize from other amino acids (and which must be acquired, there-fore, through the diet).

Since the proteins of other animals are biochemically very similar to the proteins of our own tissues, eating meat, fish, and eggs provides us with amino acids in approxi-mately the proportions needed for the syn-thesis of new tissue protein. For this reason, the proteins of meat, fish, and eggs are con-sidered *grade I proteins*. Whereas some plants (i.e., vegetables and fruits) do contain grade

TABLE 12.7	RECOMMENDED DAILY ALLOWANCES OF NUTRIENTS	
NUTRIENT	MALES [1]	FEMALES [2]
PROTEIN	56 *g*	44 *g*
VITAMINS		
WATER-SOLUBLE		
B1	1.5 *mg*	1.1 *mg*
B2	1.7 *mg*	1.3 *mg*
B3	19 *mg*	14 *mg*
B6	2.2 *mg*	2.0 *mg*
B12	3.0 μ*g*	3.0 μ*g*
M	0.4 *mg*	0.4 *mg*
FAT-SOLUBLE		
A[3]	6.0 *mg*	4.8 *mg*
D	7.5 μ*g*	7.5 μ*g*
E	10 *mg*	8.0 *mg*
MINERALS		
Ca	0.8 *g*	0.8 *g*
P	0.8 *g*	0.8 *g*
Mg	0.35 *g*	0.3 *g*
Fe	10 *mg*	18 *mg*
Zn	15 *mg*	15 *mg*
I	0.15 *mg*	0.15 *mg*

[1]. Values are for males 19-22 years old and weighing about 70 *Kg* (i.e., 154 *lb*).
[2]. Values are for females 19-22 years old and weighing about 55 *Kg* (i.e., 120 *lb*).
[3]. Vitamin A in the form of *beta*-carotene.

I proteins, most contain *grade II proteins*, that is proteins whose amino acid compositions are quite different from our own and which are lacking in a number of the essential amino acids. Because of this, it is not surprising that a proper vegetarian diet must include legumes, nuts, and even dairy products in order to ensure that all of the essential amino acids are acquired.

Although the recommended diet includes 30% of calories from fat (fig. 12.15), the body can function normally on a much smaller intake of fat; this, of course, presumes that what fat is present in the diet includes the essential unsaturated fatty acids *linoleic acid* and *linolenic acid*. Other fatty acids can be produced from glucose derived from carbohydrates and from amino acids derived from proteins (see figures 12.11 and 12.13). Rich sources of the essential fatty acids include nuts (e.g., walnuts and pecans). Fats are abundant in most meats and are present in smaller amounts in fish, vegetables, and fruit.

Fats play a variety of important roles in the body, serving as constituents of cellular membranes, constituents of lipoproteins, steroid hormones, and as components of bile salts. However, much of the fat in the body represents stored energy and in a healthy, fit male fat should account for no more than 15% percent of the body's weight; in a fit female, the percent body fat should be no more than 22%. Unfortunately, the average American male has 23% body fat and the average female 32%.

Most of the stored fat is found in the body's **adipose tissue** (tissue in which the cells are specialized for the storage of lipid molecules, especially triglycerides) located in the skin's *hypodermis* (where it insulates the body, preventing excessive heat loss), around the kidneys, around the small and large intestines, on the surface of the heart, within muscles (where it is called "marbling"), and in the mammary glands of mature females. A smaller but appreciable amount of fat is also stored in the liver. It is estimated that the body's fat stores can sustain life, even when food (but not water) is deprived for 30 days or more.

Obesity is *excess* storage of fat and is usually the outcome of excessive fat or carbohydrate intake in the diet, typically combined with a failure to do enough exercise. Obesity represents a true health problem by placing additional burdens on the body's physiology. Excess body fat has been shown to be associated with high levels of cholesterol in the bloodstream and this has been linked to *atherosclerosis* and *coronary artery disease* (as cholesterol forms *plaques* in the walls of arteries, slowly occluding them).

Dietary carbohydrate is the source of the body's sugars and takes two principal forms: *complex carbohydrates* (polysaccharides), such as starch and glycogen, and *sugars* themselves, such as sucrose (table sugar produced from sugar cane or beets), glucose, and fructose. The recommended diet prescribes that about 48% of all calories should be derived from complex carbohydrates and only 10% from sugars, whereas the average diet falls far short on complex carbohydrates (only 22%) and is far in excess in sugars (24%; see figure 12.15).

| 12-6 |

Carbohydrates are present in most foods. The principal carbohydrate in meat and fish is the glycogen stored in muscle tissue; this represents a modest source of complex carbohydrate. Vegetables are a rich source of complex carbohydrates (mainly starch), and fruit is a source of simple sugars (e.g., fructose).

Although sugars are constituents of glycoproteins, including a number of important enzymes and hormones, much of the sugar conserved by the body takes the form of glycogen stored in muscles and the liver. In the average, fit male, total glycogen stores amount to about 500-1000 grams and represent a small but readily and rapidly mobilizable energy source. If dietary carbohydrate yields glucose in excess of that which can be stored as glycogen, the excess sugar will be converted to (and stored as) fat.

Dietary Vitamins and Minerals

The diet may provide an adequate caloric intake and be comprised of acceptable amounts of protein, fat, and carbohydrate, yet it may fall short of the body's needs

TABLE 12.8	SOURCES OF VITAMINS AND MINERALS
NUTRIENT	**SOURCES**

VITAMINS

WATER-SOLUBLE

B1	pork, green peas, lima beans
B2	liver, whole milk, whole grains
B3	meats, enriched breads and cereals
B5	liver, eggs
B6	beef, liver, vegetables, whole grain cereals
B12	meats, poultry, fish, milk, cheese, eggs
C	citrus fruit, strawberries, cantaloupe
H	eggs, liver
M	liver, spinach, broccoli

FAT-SOLUBLE

A	liver, carrots, dark-green leafy vegetables
D	milk (fortified), eggs, sunlight on skin
E	milk, eggs
K	green vegetables, liver

MINERALS

Ca	milk, yogurt, cheese
P	meats
Na	table salt, meats, eggs, whole milk
Cl	table salt, meats, chlorinated water
K	bananas, orange juice, potatoes
S	meats, eggs
Mg	cereal grains, dark-green vegetables
I	iodized table salt, seafood
Fe	liver, beef, whole grain cereals
Zn	seafood, milk, eggs
Cu	grains, beans, fruit
F	fluorinated water

if it lacks the necessary vitamins and minerals. As noted earlier, the vitamins and minerals are nutrients needed in minuscule amounts that provide no direct calorific value but are essential to life, growth, and well-being. The recommended daily allowances of some of the vitamins and minerals are given in table 12.7, and the sources of some of the vitamins and minerals are listed in table 12.8.

WATER

Water is the most abundant substance in the body, accounting for about 60% of the body's weight. Despite its abundance, we rarely pay any conscious attention to the important roles that water plays in the body. Whereas the body can survive for weeks or even months without nutrients like carbohydrates, fats, and proteins, survival without a supply of fresh water is limited to just a few days. Water is the predominant component of the body's fluids, including the blood and the lymph, and serves as the vehicle for transporting nutrients to the body's tissues and carrying away the tissues' wastes.

Within cells, water acts as the cytoplasmic solvent, serving in the dissolution of many of the cell's molecular constituents. Cellular water is also the medium within which most of the cell's metabolic reactions take place. In many instances, water does not simply play the passive role of medium; instead, water molecules may participate directly in cellular metabolic reactions, serving as a source of hydrogen ions (H^+) and hydroxyl ions (OH^-) or being formed as a reaction product from other chemical intermediates that release these ions (for example, see figures 2.18 and 2.19).

In addition to its chemical roles as a transport medium, intracellular solvent, and metabolic intermediate, water's special properties endow it with a number of physiological roles. Because water is not compressible, it acts as a good lubricant around joints; water also serves to absorb mechanical shock to certain organs (e.g., the protective roles played by pericardial fluid around the heart and the aqueous and vitreous humors of the eyes). In the case of the eyes, the aqueous and vitreous humors serve to sustain the eye's spherical shape (which is essential to clear vision).

Although there is some variation from one individual to another (depending on the amount of water consumed with and between meals), the average person loses about 5 pints (about 2.3 liters) of water each day. Most of the water that is lost from the body is in the urine produced by the kidneys and the feces produced in the colon. Water in urine acts as the solvent for ridding the blood of waste chemicals; fecal water represents water that is not removed from the feces during their transit through the colon. About half of the daily water loss takes the form of urine (see chapter 13), whereas the feces account for about 10% of the total loss. The remaining water losses are through the skin (perspiration and diffusion accounting for about 15%) and through the lungs during expiration (about 25%).

To replace the body's daily water losses, the average person must consume about 5 pints of water each day. This is equivalent to drinking ten 8-ounce glasses of water each day. However, it should be noted that the solid food consumed each day also contains quantities of water (typically the equivalent of about three 8-ounce glasses of water). There is no danger in consuming larger amounts of water; indeed, it is healthier to consume a quantity in excess of that which is needed to replace lost body water, since the excess will readily be voided through the kidneys and is healthy to kidney function. It is generally recommended that an adult drink at least six to eight 8-ounce glasses of water each day.

During hot weather and during vigorous exercise, water losses via perspiration may exceed 3 pints per hour and can quickly dehydrate the body. Needless to say, much larger quantities of water must be consumed during exercise to sustain the body's normal water level.

The nature of one's water consumption has an interesting effect on body weight. If a person does not drink enough water, the body will automatically make an effort to conserve (i.e., retain) water, and this can manifest itself as increased body weight. On the other hand, when large amounts of water are consumed, the body does not sense the need to conserve water and will excrete excesses, thereby lowering the body weight.

FOOD GROUPS AND A BALANCED DIET

In order to make the selection of a balanced (and nutritious) diet possible, foods may be divided into different groups. In the United States, four food groups are generally recognized; these are (1) the **meat** group, (2) the **milk** group, (3) the **fruits and vegetables**

group, and (4) the **breads and cereal grains** group. Different foods are assigned to a particular group *on the basis of their composition and nutritional content*. For example, meats are foods that are rich in protein, iron, and vitamin B_3 (niacin); members of the milk group are rich in protein, calcium and vitamin B_2 (riboflavin); fruits and vegetables are rich in vitamins A (retinol) and C (ascorbic acid) and pectin fiber; and members of the breads and cereals group are foods that are rich in various B vitamins, cellulose fiber, and trace minerals. A sampling of the variety of foods assigned to each of the four food groups is given in table 12.9.

Sometimes the foods included in a particular food group are surprising. For example, even though they are vegetables, split-peas are placed in the meat group because they are so rich in protein. Beans are placed in the meat group for the same reason. Corn is a vegetable, but because it is a grain, it is placed in the bread group along with other grains like rye, wheat, and rice. Eggs

TABLE 12.9	THE FOUR FOOD GROUPS
FOOD GROUP	**EXAMPLES**
MEATS	Beef (hamburger, steak, etc.), pork (bacon, ham), poultry (chicken), fish , (salmon, tuna), eggs, split-peas, beans, nuts, seeds
MILKS	Milk (nonfat, low-fat, whole milk), cottage cheese, yogurt, cheese, ice cream, butter
BREADS AND CEREALS	Bread, bagels, muffins, rolls, rye, taco shells, tortillas, wheat, rice, corn, oats, barley
FRUITS AND VEGETABLES	Apples, pears, cantaloupe, oranges, peaches, plums, raisins, carrots, cucumbers, lettuce, cabbage, broccoli, eggplant

are included in the meat group because they are high in protein and contain the vitamins and minerals of meats.

A balanced diet is one that includes a variety of foods from each of the four food groups. However, the fact that a diet is balanced does not guarantee that it promotes good health. This is because in addition to being balanced a diet must meet the following additional requirements: (1) it must meet (but should not exceed) the calorific needs of the body (remember that excess calories will be converted to fat), and (2) it should be low enough in fat and sugars and rich enough in protein and complex carbohydrate to approximate the "recommended" diet depicted in the left half of figure 12.15. Accordingly, within each of the four food groups, certain foods are healthier choices than others. Among the meats and milks, the healthiest choices of foods are those that are lowest in fat (remember that high dietary fat has been linked to the principal causes of premature death in the United States–heart disease and atherosclerosis). Among the fruits, vegetables, breads and cereals, the healthiest choices of foods are those that are lowest in sugar and highest in fiber. For example, within the meat food group, beans and fish are much healthier choices than are beef (steak, hamburger, etc.) or nuts, because the former are lower in fat. Within the milk food group, nonfat milk and nonfat yogurt are much healthier choices than whole milk or ice cream. Within the fruits and vegetables food group, raw fruits and vegetables are better choices than are the juices of fruit or vegetables (which are richer in sugar and in which the fiber has been partially degraded). Finally, within the bread and cereal grains food group, whole grains and breads are healthier choices than cake and cookies,

which not only contain added sugar but also contain some fat.

Eating Abnormalities

Kwashiorkor. In order for dietary protein to be adequate it must contain the 20 amino acids in the relative amounts with which these amino acids occur in the tissue proteins that the body continuously makes. This requirement is most easily met by consuming animal proteins (e.g., by eating meat). However, if the protein in the diet comes principally from vegetable sources, it may not provide sufficient amounts of all of the amino acids. **Kwashiorkor** is a disease that results from the lack of certain amino acids in the diet and is particularly common in those parts of Africa where the principal source of dietary protein is corn meal. Corn proteins lack the amino acid *tryptophan*. As a result, the production of human tissue proteins that contain one or more tryptophan positions comes to a halt when this amino acid is not available. In kwashiorkor, low plasma protein levels reduce the blood's colloid osmotic pressure (see chapter 7) and this ultimately leads to the bloating of the tissue spaces (especially in the stomach) with water. The bloatedness or flacidness of a tissue is called **edema** and results when blood pressure forces more water into the tissue spaces than can be withdrawn by the plasma's protein-based osmotic pressure.

In some African children, the characteristic extended belly gives the mistaken impression that the child is well fed, whereas the bloatedness is the edematous condition of the stomach. Kwashiorkor is common among African children who have just been weaned off breast milk when a new sibling is born. Whereas the newborn baby now acquires the full complement of amino acids present in the mother's milk, the older sibling has to contend with the amino acid deficient corn meal.

Anorexia Nervosa. Anorexia nervosa is an extreme form of underweight seen most often in young women (usually teenage girls) and is more common in the developed countries of the world (such as the United States) where there is an excessive emphasis on thinness (i.e., where feminine "beauty" is associated with being thin). It has been estimated that as many as 1 in 100 adolescent females may have some form of anorexia (the percentage is even higher among well-educated, college women). In anorexia, the extreme underweight is the result of a preoccupation with weight loss in an effort to maintain a slim figure; the problem is intimately related to self-image). In order to stay slim and not gain weight, the anorexic deliberately fails to consume a diet that is sufficient in calories and content to sustain normal weight, normal activity, and normal health. The key indicator to this extreme form of underweight is that the condition is *self-inflicted* in that the anorexic deliberately denies herself the needed amounts of nutrients in a purported effort to keep her weight under control. Indeed, despite already being underweight and continuing to lose additional body weight, the anorexic usually perceives herself as "fat."

Anorexia nervosa is a serious condition that usually requires medical treatment. Anorexia can be treated temporarily by forced feeding, but a lasting cure usually requires medical and psychological intervention. The continuing weight loss of a person suffering from anorexia nervosa can quickly lead to a variety of harmful physiological symptoms and effects; these include wasting of muscle tissue, arrested sexual development (and a halt in the menstrual cycle, see chapter 14), anemia, brain damage, lowered blood pressure, and irreversible damage to the heart. In many cases, anorexia nervosa has led to premature death.

Bulimia. Bulimia is a condition in which a person gorges himself (more frequently herself) and soon afterwards forces regurgitation of the food, so that there will be no weight gain. This cycle of abnormal activity is usually referred to as "binge and purge." Like anorexia, bulimia is most common among adolescents and is especially marked among college women, where the percentage of bulimics may be as high as 20%. The eating binges of the bulimic person usually involve large quantities of carbohydrate-rich food, typically "junk food." The frequent purging of the ingested food is accompanied by a number of harmful symptoms; these include disturbances of the body's salt balance (which in turn adversely affects the heart) as purging produces extraordinary losses of potassium and extraordinary decay of the teeth as regurgitated stomach acid erodes the tooth enamel. Finally, it is not uncommon for someone suffering with anorexia nervosa also to be bulimic.

SELF TEST[*]

True/False Questions

1. Anabolic reactions (or pathways) are frequently *endergonic*, whereas catabolic reactions (or pathways) are frequently *exergonic*.

2. Unlike cellulose (which cannot be degraded by the human digestive system), the digestion of starch is a rich source of sugar molecules.

3. The conversion of one molecule of glucose to 2 molecules of pyruvic acid is associated with the production of 38 molecules of ATP from ADP.

4. Thiamine pyrophosphate (TPP) is a coenzyme derived in part from vitamin B1.

5. Less than one-half of the energy inherent in glucose is trapped in the form of ATP when glucose is converted to CO_2 and water by the reactions of glycolysis, pyruvic acid oxidation, the Krebs cycle, and oxidative phosphorylation.

6. Intermediates of the Krebs cycle play important roles in reactions that convert one type of amino acid to another.

7. Beta oxidation is the name of a metabolic cycle in which fatty acids are converted to acetyl-Co A.

8. The so-called "grade II" proteins present in certain vegetables and fruits lack a number of the amino acids that are considered essential in the human diet.

9. It is generally recommended that an adult drink at least six to eight 8-ounce glasses of water each day.

10. Table sugar, glucose, and fructose are important complex carbohydrates.

Multiple Choice Questions

1. The conversion of glucose to pyruvic acid is called (A) glycogenesis, (B) glycogenolysis, (C) glycolysis, (D) the Krebs cycle.

2. Which one of the following coenzymes serves as an acceptor of hydrogen during glycolysis? (A) oxidized nicotinamide adenine dinucleotide, (B) reduced nicotinamide adenine dinucleotide, (C) coenzyme A, (D) oxidized flavin adenine dinucleotide, (E) reduced flavin adenine dinucleotide.

3. If water is excluded from the calculation, then approximately what percent of the body's weight is represented by protein? (A) about 15%, (B) about 25%, (C) about 50%, (D) about 70%, (E) nearly 100%.

4. Linolenic and linoleic acid are (A) Krebs cycle intermediates, (B) acidic amino acids, (C) essential amino acids, (D) saturated fatty acids, (E) essential fatty acids.

5. A deficiency of which one of the following vitamins causes the disease *beriberi*? (A) B1, (B) B2, (C) B6, (D) C, (E) M.

6. Which one of the following is not a fat-soluble vitamin? (A) A, (B) D, (C) E, (D) H, (E) K.

7. Which one of the following vitamins is essential for the proper development and production of red blood cells? (A) thiamine, (B) pantothenic acid, (C) cobalamin, (D) ascorbic acid, (E) calciferol.

8. The most abundant of the body's minerals is (A) carbon, (B) oxygen, (C) nitrogen, (D) calcium, (E) iron.

---

[*] *The answers to these test questions are found in Appendix III at the back of the book.*

CHAPTER 13

THE EXCRETORY SYSTEM

Excretion is defined as the process by which the body rids itself of chemical substances (usually wastes) that were part of a tissue, the bloodstream, or the lymph. Accordingly, several different organs could be regarded as playing a role in excretion. For example, the **lungs** could be considered excretory organs because they act to rid the body of **carbon dioxide** that is formed in the tissues during metabolism. The **skin** might be regarded as an excretory organ because the water and other chemical constituents secreted onto the body surface by the skin's sweat and oil glands were formerly in the tissues or in the bloodstream. Similarly, the **liver** might be regarded as an excretory organ because of its role in ridding the body of the breakdown products of hemoglobin (i.e., the **bile pigments**, see chapter 11). Usually, however, the lungs are considered part of the respiratory system, the skin is considered part of the "integumentary" system, and the liver part of the digestive system. It is the two **kidneys** that represent the major excretory organs of the body. These vital organs serve to remove most of the chemical wastes emptied into the blood during metabolism, and working in concert with the urinary bladder, discharge these wastes from the body.

The two kidneys are bean-shaped organs about four inches long that lie on either side of the vertebral column, just beneath the diaphragm and liver, in the upper part of the abdominal cavity (fig. 13.1). The kidneys receive systemic blood from branches of the aorta called **renal arteries**, and blood leaving the kidneys enters the inferior vena cava through the **renal veins**. (Whenever the word "renal" is used in connection with an anatomical structure, it implies that the structure is related to the kidneys; for example, the

adrenal glands ["ad" = "toward" and "renal" = "kidney"] are the two endocrine glands seated on the upper surface of the kidneys; see figure 13.1.) About 1,000 c.c. (i.e., one liter) of blood enter the two kidneys each minute. Wastes are removed from this blood by the *renal tubules* (see later) and form the **urine**. Although the volume of urine produced by the kidneys varies according to the state of hydration of the body, typically 1-2 c.c. of urine are formed each minute.

Extending from the kidneys downward to the **urinary bladder** are the two tubular **ureters**. These ureters convey the urine from the kidneys into the urinary bladder, which acts as a temporary reservoir. Urine does *not* descend through the ureters by gravity.

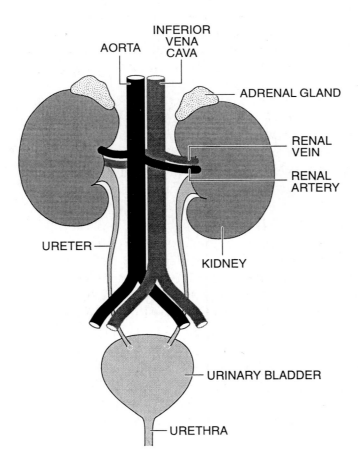

Figure 13.1
The major organs of the excretory system (kidneys, ureters, urinary bladder, and urethra). This view is from the rear (dorsal) side of the body.

Rather, *peristaltic waves* (about four waves each minute) beginning at the upper, flared ends of the ureters progressively sweep the urine downward. Periodically, the muscular walls of the urinary bladder contract and expel the urine from the body through a duct called the **urethra**. In males, the urethra must travel the length of the penis and is, therefore, longer than in females.

13-1

ORGANIZATION OF A KIDNEY

The organization of a kidney is depicted in figure 13.2. Each kidney is covered at its surface by a thick layer of connective tissue called the renal **capsule**. Internally, the tissues of the organ give rise to two distinct regions: the outer renal **cortex** and the inner renal **medulla** (fig. 13.2). The medulla is further subdivided into 7 to 15 renal **pyramids**, each pyramid branching from a small

chamber called a **minor calyx**. Neighboring renal pyramids are separated from one another by renal **columns**. The minor calyxes merge to form **major calyxes**, which collectively open into a large chamber called the renal **pelvis**. The urine produced by the millions of nephrons that make up the kidneys (see below) is emptied into the pelvis. The urine is then withdrawn from the pelvis by the peristaltic actions of the ureters and is pushed down into the urinary bladder.

13-2

Systemic blood enters each kidney through a renal artery. As already noted, the kidneys receive about one liter of blood per minute. Because the average person has only five to six liters of blood altogether, this implies that a volume of blood equal to that of the entire body passes through the kidneys every five to six minutes. This is an especially rich supply of blood and attests to the importance of the kidneys to the body's survival and well-being.

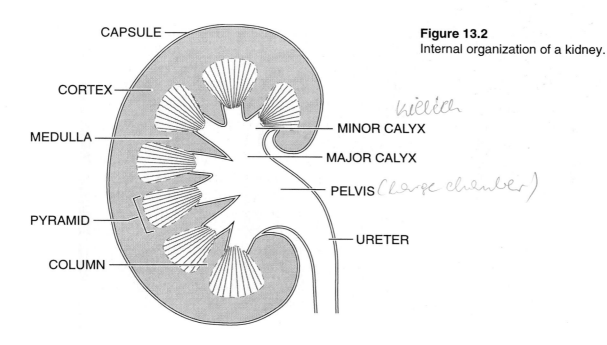

Figure 13.2
Internal organization of a kidney.

CAPSULE

CORTEX

MEDULLA

PYRAMID

COLUMN

MINOR CALYX

MAJOR CALYX

PELVIS

URETER

Within the kidneys, each renal artery subdivides into a number of *interlobar arteries* that carry the blood radially through the renal columns. *Arcuate arteries* branch from each interlobar artery at the boundary between the renal cortex and medulla. The arcuate arteries give rise first to *interlobular arteries* and then *afferent arterioles,* which deliver the blood to individual kidney tubules for filtration (see below). Having circulated through the *glomerular* and *peritubular capillaries* (again, see below), blood is collected by *interlobular veins, arcuate veins,* and *interlobar veins* (i.e., the venous circulation parallels the arterial circulation). Within each kidney, the interlobar veins descend through the renal pyramids and form the renal vein, which carries blood out of the kidney and into the inferior vena cava.

13-3

NEPHRONS

The 1-2 c.c. of urine that are formed from each liter of blood circulating through the kidneys are a product of the kidneys' *functional units,* which are called **nephrons** or **kidney tubules**. Nephrons are microscopic structures, each kidney containing about one million nephrons. An understanding of how the kidneys function is based on an appreciation of how the nephrons work. Since all nephrons function in a more-or-less identical manner, we will focus on the actions of a single nephron.

Organization of a Nephron

The structure of a nephron is depicted in figure 13.3. Each tubule begins as a blind-ended, cup-shaped structure known as a **Bowman's capsule**. The capsule houses and interfaces with a capillary bed known as a **glomerulus** (fig. 13.4). Blood enters the glomerular capillary bed through a small **afferent arteriole** and leaves the capillary bed through an **efferent arteriole** (i.e., "afferent" = "toward" and "efferent" = "away from"). (Note that the glomerular capillaries are drained by an *arteriole* and not a *venule*; the significance of this will become apparent later.) Each Bowman's capsule leads into a twisted segment of the nephron called the **proximal convoluted tubule** ("proximal" means "nearby," so that the proximal convoluted tubule is nearby the Bowman's capsule). The proximal convoluted tubule leads into the **loop of Henle**, a U-shaped segment of the kidney tubule. The first half of the loop of Henle is referred to as the descending limb, whereas the second half is called the ascending limb; each limb has "thick" and "thin" segments. The thick segment of the ascending limb of the loop of Henle leads into the **distal convoluted tubule** (i.e., "distal" implies "distant," the distal convoluted tubule being the twisted segment of the nephron that is *further from* the Bowman's capsule). The distal convoluted tubule merges with a **collecting duct**; each collecting duct receives the fluid contents of a number of neighboring nephrons.

13-4

The efferent arteriole emerging from the Bowman's capsule gives rise to a second capillary network that weaves its way around the proximal convoluted tubule, loop of Henle, and distal convoluted tubule. This array of capillaries is known as the **peritubular capillary network**. The portion of the peritubular capillary network surrounding the loop of Henle is known as the

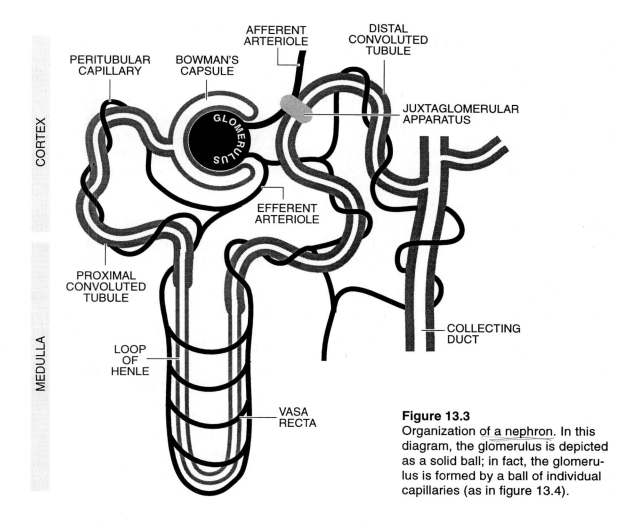

AFFERENT
ARTERIOLE

DISTAL
CONVOLUTED
TUBULE

PERITUBULAR
CAPILLARY

BOWMAN'S
CAPSULE

CORTEX

GLOMERULUS

JUXTAGLOMERULAR
APPARATUS

EFFERENT
ARTERIOLE

PROXIMAL
CONVOLUTED
TUBULE

MEDULLA

LOOP
OF
HENLE

COLLECTING
DUCT

VASA
RECTA

Figure 13.3
Organization of a nephron. In this diagram, the glomerulus is depicted as a solid ball; in fact, the glomerulus is formed by a ball of individual capillaries (as in figure 13.4).

vasa recta. Blood is collected from the peritubular capillary network by venules that conduct the blood into the interlobular veins.

Distribution of Nephrons in the Kidney's Cortex and Medulla

Figure 13.5 illustrates the manner in which the nephrons are arranged in the renal cortex and medulla. The Bowman's capsules and proximal and distal convoluted tubules of the nephrons are located in the renal cortex,

whereas the loops of Henle (and collecting ducts) extend into the medulla. The arrangement resembles the manner in which the spokes of a wheel are arranged around the wheel's hub (i.e., the hub being analogous to the renal pelvis; see figure 13.5).

It should be noted that there are two types of nephrons: *cortical nephrons* (which are the more numerous) and *juxtamedullary nephrons*. The Bowman's capsules, proximal convoluted tubules, and distal convoluted tubules of the cortical nephrons are positioned in the marginal region of the

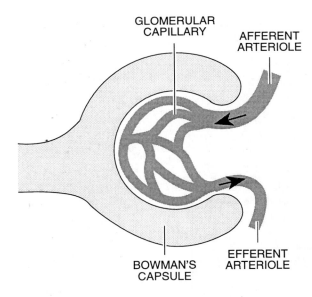

kidney's cortex, and their loops of Henle descend only a short distance into the medulla. In contrast, the Bowman's capsules, proximal convoluted tubules, and distal convoluted tubules of the juxtamedullary nephrons are located closer to the boundary between the cortex and the medulla. The loops of Henle of the juxtamedullary nephrons are longer than those of the cortical nephrons and descend further into the medulla (again, see figure 13.5).

Mechanism of Nephron Action

The forces at play in the Bowman's capsule and glomerular capillary network result in the filtration of the blood. Two pressures characterize the blood in the glomerular capillaries: *hydrostatic pressure* (i.e., blood pressure) and *colloid osmotic pressure* (see

Figure 13.4
Within the Bowman's capsules the afferent arterioles gives rise to the glomerular capillaries; the glomerular capillaries are then drained by the efferent arterioles. (The arrows show the direction of flow of the blood.)

Figure 13.5
Within the kidneys, the nephrons are distributed in a radial pattern (like the spokes of a bicycle wheel). Note that there are two types of nephrons: cortical nephrons and juxtamedullary nephrons.

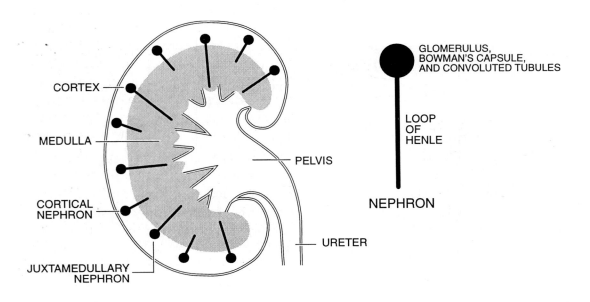

chapter 8 for a discussion of blood pressure and colloid osmotic pressure). As depicted in figure 13.6, the blood pressure (which is directed outward, against the capillary wall) is about 50 mm Hg and the colloid osmotic pressure (which is directed inward from the capillary wall) is about 30 mm Hg. There is, therefore, a pressure gradient of 20 mm Hg (i.e., 50 mm Hg minus 30 mm Hg) acting to express fluid through the walls of the glomerular capillaries; this pressure is called the **glomerular filtration pressure**. The glomerular filtration pressure is uniform over the entire length of the glomerular capillary network. This is in contrast with capillary beds in other tissues of the body in which the

blood pressure falls precipitously. The maintenance of a constant blood pressure over the length of the glomerular capillaries is due in part to the differential sizes of the afferent and efferent arterioles. The afferent arterioles, which conduct blood into the glomeruli, are wide vessels; in contrast, the efferent arterioles are narrow. The narrowness of the single opening through which the blood can exit the glomerular capillary bed contributes to the high pressure that is maintained within the glomeruli.

The Bowman's capsule also contains fluid (i.e., the filtrate of the blood) and this fluid exerts a hydrostatic pressure (see chapter 8) of about 10 mm Hg. Because this pressure is directed against the walls of the Bowman's capsule, it opposes the glomerular filtration pressure. The difference between the glomerular filtration pressure (20 mm Hg) and the hydrostatic pressure of the Bowman's capsule is 10 mm Hg (i.e., 20 minus 10); it is this pressure difference (called the **net filtration pressure**) that results in the movement of fluid from the glomerular capillaries into the Bowman's capsules.

The glomerular capillaries and Bowman's capsules of the nephrons are especially suited for filtration. The capillary walls are highly *fenestrated* (i.e., they are littered with pores) and the Bowman's capsule tissue against which the capillaries are appressed (called the visceral epithelium) consists of a single layer of delicate, thin cells called **podocytes**. Sandwiched between the surfaces of these two cell layers is a **basement membrane** consisting of molecules of *collagen* (a protein) and *proteoglycans*. The basement membrane and cell layers through which filtration occurs are freely permeable to molecules with molecular weights up to about 5,000 daltons. Even molecules with

Figure 13.6
The blood pressure (BP) in the glomerular capillaries (50 mm Hg) exceeds the colloid osmotic pressure (COP, 30 mm Hg) by 20 mm Hg. This difference (i.e., 50–30) is known as the glomerular filtration pressure (GFP). The GFP also dominates the hydrostatic pressure (HP) exerted by fluid in the Bowman's capsules (10 mm Hg). There is, therefore, a net filtration pressure of 10 mm Hg (i.e., 20–10) that drives the blood filtrate into the Bowman's capsules.

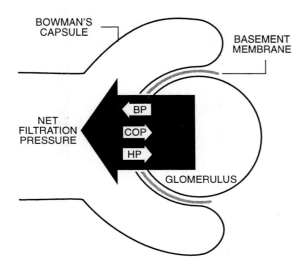

molecular weights as high as 30,000 daltons may be filtered. Interestingly, the range of molecular sizes over which filtration occurs is lower than that expected on the basis of the pore sizes of the glomerular endothelium. Apparently, the molecular size threshold is based not only on pore size but also on certain anatomical specializations of the interface between the glomerular endothelium and the visceral epithelium of the Bowman's capsules.

Most of the small molecules of the blood are readily filterable (e.g., water, salts, sugars, amino acids, vitamins, etc.). Nearly all of the plasma proteins, however, exceed this size threshold and are not filtered (hence, no colloid osmotic pressure exists in the filtrate). Red and white blood cells also are far too large to be filtered from the glomerular capillaries into the Bowman's capsules.

For each liter of blood that circulates through the kidneys of an adult male, about 125 c.c. of filtrate are formed, and this filtrate contains a cross section of all substances in the blood having molecular weights below 30,000 daltons. If we consider the hematocrit of a normal male to be 0.45 (see chapter 8), then the 125 c.c. of filtrate formed each minute by the kidney tubules contains about 23% of all of the plasma's small molecules (i.e., each liter of blood contains 550 c.c. of plasma [1000 - (0.45 x 1000)], and 125 c.c. of filtrate is 23% of the 550 c.c. of filterable plasma).

Table 13.1 lists the quantities of some representative chemical substances that are present in the blood plasma and compares these with the amounts found in the filtrate of the Bowman's capsules and in the urine. Although only six substances are listed in the table, hundreds of different chemical substances are present in normal blood plasma. For brevity, small ions such as Na^+, K^+, and Cl^- are not listed in the table, but all of these ions are filterable and varying quantities are found in the filtrate (and in the urine). The fates of these ions are described in the next section. The six substances that are listed in

TABLE 13.1 COMPOSITION OF BLOOD PLASMA, FILTRATE, AND URINE

SUBSTANCE	AMOUNT IN PLASMA	AMOUNT IN FILTRATE	AMOUNT IN URINE
All substances	550 c.c.	125 c.c.	1-2 c.c.
Water	500 c.c.	123+ c.c.	1-2 c.c.
Protein	40 *g*	0 *g*	0 *g*
Sugar (glucose)	500 *mg*	115 *mg*	0 *mg*
Amino acids	200 *mg*	46 *mg*	0 *mg*
Creatinine	5 *mg*	1.1 *mg*	2.2 *mg*
Ammonia	0 *mg*	0 *mg*	5.0 *mg*

The quantities listed in each column are based on the passage of one liter (1,000 c.c.) of blood through the kidneys (i.e., from the renal arteries to the renal veins). The 450 c.c. of volume represented by blood cells (which cannot be filtered) are omitted.

the table are representative of major chemical subgroups. Water is included because it is the major component to be filtered and is the medium in which all of the other substances are dissolved. Plasma protein is representative of nonfilterable substances (i.e., the molecular weights of the plasma proteins are high enough to preclude their filtration into the Bowman's capsules). Sugars and amino acids represent molecules that are small enough to be filtered but which are useful to the body and ought not be lost in the urine. Creatinine and ammonia represent true wastes that must be excreted.

Especially striking are the values in the far right column of table 13.1. These values show that (1) some substances filtered into the Bowman's capsules do not end up in the urine (e.g., glucose and amino acids), (2) some substances are present in greater amounts in the urine than are initially filtered into the Bowman's capsules (e.g., creatinine), and (3) some substances excreted in the urine are not produced by filtration of the blood (e.g., ammonia). The reasons for these seemingly unexpected values are explained in the following sections.

TUBULAR REABSORPTION

The filtrate that is driven from the glomerular capillaries into the Bowman's capsule passes into the proximal convoluted tubule, while the blood leaving the glomerulus enters the peritubular capillary network (see figure 13.3). The blood and the filtrate are separated from each other by the single layer of cells forming the proximal convoluted tubule's wall, the fine endothelial cell layer that forms the capillary wall, and any interposed kidney tissue (called **interstitium**).

Blood pressure in the peritubular capillary network falls from about 30 mm Hg in that part of the network near the efferent arteriole to 15–20 mm Hg in the regions of the capillary network that surround the loop of Henle (i.e., the vasa recta). Whereas the blood's hydrostatic pressure falls, the blood's colloid osmotic pressure (and the colloid osmotic pressure of the interstitium) is maintained at (or even above) 30 mm Hg; this is because no plasma proteins pass from the blood into the filtrate. Within the kidney tubule itself, the hydrostatic pressure remains at about 10 mm Hg. Considering the forces that are at play, it should be apparent that there is now a net force directed from the filtrate back into the bloodstream.

For example, if the blood pressure and colloid osmotic pressure in the peritubular capillary are 30 mm Hg (values that would exactly cancel one another), and if the hydrostatic pressure in the proximal convoluted tubule is 10 mm Hg, then there is a net pressure gradient of 10 mm Hg acting to draw materials from the filtrate, across the walls of the tubule and capillary, into the bloodstream (i.e., 30 minus 30 minus 10). This pressure gradient acting in the direction of the bloodstream is responsible in part for the return of filtered water to the bloodstream.

A number of other substances filtered into the Bowman's capsules are returned to the blood of the peritubular capillaries. Included here are the sugars and amino acids that were lost to the filtrate. Sugars and amino acids are reabsorbed by **active transport**; that is, enzymes in the membranes of the cells that line the proximal convoluted tubule actively bind and remove sugars and amino acids from the lumen of the tubule and transfer them into the interstitium and the bloodstream. Therefore, under normal circumstances, no sugars or amino acids are lost from the body in the urine. (In the condition known as **diabetes mellitus** [discussed in detail in chapter 14], the blood glucose level rises far above its normal value, and the capacity for complete tubular reabsorption is exceeded. This results in the appearance of glucose in the urine, which is the principal diagnostic symptom in an untreated case of diabetes mellitus.)

> 13-5

Positive ions (especially Na$^+$) that were filtered from the blood of the glomerular capillaries are also actively transported out of the tubule, into the interstitium, and into the blood of the peritubular capillaries. This is achieved in the following way. The cells forming the walls of the proximal convoluted tubules face the tubular filtrate on one side (i.e., the *apical* surface of the cells) and face the interstitium and peritubular capillaries on the other (i.e., the *basal* surface of the cells; see figure 13.7). The basal surface of the cells contains *sodium/potassium pumps*, which actively extrude sodium ions from the cell, thereby dramatically lowering the intracellular Na$^+$ concentration. At the apical surface of the cells, there are no sodium/potassium pumps and the plasma membrane has pores (or channels) that are permeable to Na$^+$. Therefore, Na$^+$ extrusion through the basal surface of the cells is followed by the diffusion of Na$^+$ into the cell through the apical surface. The net effect is to transfer Na$^+$ from the tubular filtrate into the interstitium and into the blood.

In order to preserve the electrical neutrality of the filtrate and the blood, chloride ions passively accompany the reabsorption of the Na$^+$. The transport of filtered solutes back

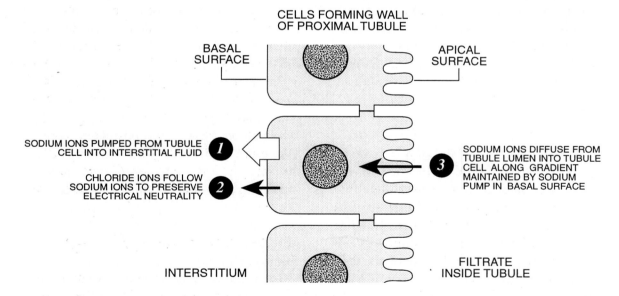

Figure 13.7
Active transport of sodium ions across the basal surface of the cells forming the wall of the proximal convoluted tubule lowers the intracellular Na$^+$ concentration and creates a gradient that causes Na$^+$ to diffuse into the cells from the filtrate. Chloride ions follow the movements of sodium in order to preserve electrical neutrality. In this way, 65% of the Na$^+$ and Cl$^-$ lost to the glomerular filtrate is regained by the interstitium and the blood of the peritubular capillaries.

into the bloodstream is accompanied by the return of water to the blood. Indeed, by the time the filtrate reaches the end of the proximal convoluted tubule and is about to enter the descending limb of the loop of Henle, 65% of its water content has been drawn back into the blood by the combined effects of the pressure gradient, active transport, and osmosis.

Because the passage of filtered solutes back into the blood of the peritubular capillaries is accompanied by the movement of filtered water, there is little change in the *concentration* of the filtrate as it passes through the proximal convoluted tubule. (That is to say, the *volume* of the filtrate diminishes considerably, but its *concentration* remains the same because of the balance between solute movement and water movement.)

Reabsorption of salts and water continues as the filtrate descends into the loop of Henle. Sodium ions are actively transported out of the tubule and into the interstitium and blood, and these positive ions are accompanied by the passive movements of chloride ions. The increasing tonicity (i.e., salt concentration) of the blood and the falling tonicity of the filtrate that result from the movements of these ions cause more water to return from the filtrate to the blood by osmosis. By the time the filtrate reaches the ascending limb of the loop of Henle, about 85% of all of the water originally lost from the blood by filtration has been reabsorbed. The return of water to the bloodstream from the proximal convoluted tubules and descending limbs of the loops of Henle is called **obligatory water reabsorption** ("obligatory" because it is in response to osmotic pressure gradients over which the body does not exercise much control).

Regulating the Solute Concentration of the Urine

The solute concentration of the urine is regulated by a hormone released from the posterior lobe of the **pituitary gland** (the pituitary is an endocrine gland located at the base of the brain; see also chapter 14). The hormone released by the posterior lobe of the pituitary is called **antidiuretic hormone** (abbreviated **ADH** and also known as **vasopressin**). ADH is produced by nerve cells in the brain's hypothalamus and is then conveyed to (and temporarily stored in) the pituitary's posterior lobe. When the solute concentration of the body fluids (blood, lymph, and tissue fluid) is too high, osmoreceptors in the hypothalamus sense the hypertonicity and bring about the secretion of additional ADH from the pituitary's posterior lobe. This causes the kidneys to conserve water and, as a result, produce a concentrated urine. On the other hand, if the solute concentration of the body fluids is too low, ADH secretion by the pituitary is reduced. The effect on the kidneys is to produce a dilute urine as large volumes of excess body water are allowed to accompany the excreted solutes. The reabsorption of filtered water controlled by the level of circulating ADH is called **facultative water reabsorption** and amounts to about 15% of the total reabsorbed water.

Producing a Dilute Urine

The solute concentration of a body fluid is measured in the units "milliosmoles per liter" (abbreviated mOsm). Ordinarily, the solute concentration of the glomerular filtrate is nearly the same as that of the blood plasma and equals about 300 mOsm. Because the reabsorption of salts from the proximal convoluted tubules is accompanied by the reabsorption of water, the filtrate reaching the loops of Henle still has a salt concentration of about 300 mOsm. To produce a dilute urine, more solute must be reabsorbed than water. This is achieved by the actions of the ascending limbs of the loops of Henle, the distal convoluted tubules, and the collecting ducts (fig. 13.8). In the absence of ADH, these parts of the nephron actively reabsorb Na⁺ but are impermeable to water. When

little ADH reaches the kidneys through the bloodstream, the solute concentration of the filtrate progressively falls, with the result that the filtrate passing from the collecting ducts into the kidney pelvis may have a solute concentration as low as 70 mOsm.

The reabsorption of filtered solute (but not water) from the ascending limbs of the loops of Henle, the distal convoluted tubules, and the collecting ducts raises the solute concentration of the interstitium and peritubular blood. As depicted in figure 13.8, some of the reabsorbed solute passes into the descending limb of the loop of Henle. This leads to a phenomenon called **countercurrent multiplication** in which a solute concentration gradient is created in the medulla of the kidney. In this gradient, the solute concentration of the medullary interstitium and bloodstream progressively increase with increasing proximity to the kidney pelvis. (Countercurrent multiplication of the solute concentration occurs whenever a solution flows in *opposite* directions within channels that are separated by permeable walls.) Even when the kidneys are producing a dilute urine, countercurrent multiplication creates a solute gradient extending from 300 mOsm (near the cortex-medulla boundary) to 700 mOsm (in the depths of the medulla).

It bears repeating that production of a dilute urine normally occurs when the ADH level of the bloodstream is low. Low levels of ADH reduce the water permeability of the distal convoluted tubules and the collecting ducts.

Figure 13.8
Producing a dilute urine. When the ADH level of the blood is low, active transport of Na⁺ (accompanied by Cl⁻) into the interstitium and peritubular bloodstream (depicted by solid arrows) proceeds without the accompaniment of water. As a result, the urine is more voluminous and contains a lower concentration of solute. PCT= proximal convoluted tubule; LH = loop of Henle; and DCT = distal convoluted tubule.

Producing a Concentrated Urine

A concentrated urine is produced in response to an insufficiency in the intake of water by the body (i.e., not drinking enough fluid). In

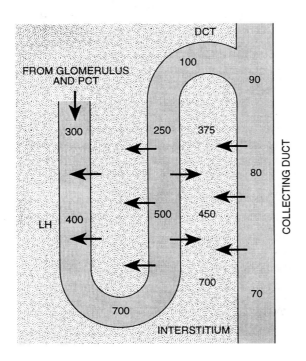

response to dehydration of the body fluids, the pituitary secretes large amounts of ADH into the bloodstream. When the ADH reaches the kidneys, it causes a dramatic increase in the water permeability of the distal convoluted tubules and collecting ducts. As a result, water is reabsorbed from the filtrate along with certain solutes. The countercurrent multiplication mechanism creates a steep solute concentration gradient in the medulla that extends from about 300 mOsm near the medulla's boundary with the cortex to about 1200 mOsm near the kidney pelvis. This further raises the osmotic gradient favoring the movement of water into the interstitium and blood (fig. 13.9). As water passes from the kidney tubules into the interstitial tissues and bloodstream, the filtrate becomes increasingly concentrated in salts and other solutes.

From the preceding discussion, it is clear that the composition of the glomerular filtrate is dramatically altered by tubular reabsorption as sugars, amino acids, sodium ions, and other substances of value are returned to the bloodstream. Not reabsorbed (or only minimally reabsorbed) are the metabolic wastes present in the filtrate. These wastes include creatinine, urea, and uric acid, which are nitrogen-containing waste products of tissue metabolism.

TUBULAR SECRETION

Since only 125 c.c. of filtrate are produced for each 1,000 c.c. of blood that passes through the kidneys, most of the waste products in the blood are not filtered into the Bowman's capsules. Instead, they pass with the blood from the glomerular capillaries into the peritubular capillaries. It is now clear that a good part of this waste is transferred into the filtrate of the nephron from the peritubular capillaries and interstitium. That is, waste molecules move in a direction *opposite to that of tubular reabsorption* by crossing the walls of the peritubular capillaries, the interstitium, and nephron and entering the filtrate within

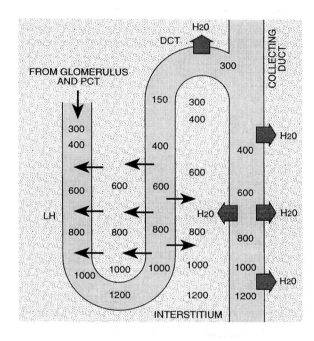

Figure 13.9
Producing a concentrated urine. A rise in the ADH level of the bloodstream acts to increase the water permeability of the walls of the distal convoluted tubules (DCT) and collecting ducts. As a result, the active transport of filtered solutes back into the inter-stitium and peritubular bloodstream (depicted by solid arrows) is accompanied by the osmotic flow of water (large grey arrows). Countercurrent multiplication steepens the solute concentration gradient in the kidney medulla, further promoting the reabsorption of water. The result is a urine whose concentration may exceed 1,200 mOsm. PCT = proximal convoluted tubule.

the tubules. This phenomenon is called **tubular secretion** and is based on active transport. The active transport of creatinine from the peritubular capillaries into the distal convoluted tubules explains why the creatinine concentration of the urine is so much greater than that of the original filtrate (see table 13.1). Other substances are also subjected to tubular secretion, including potassium ions and hydrogen ions. The secretion of H$^+$ into the filtrate explains why urine is so much more acidic than blood plasma.

13-5

Electrolyte Balance

Salt ions such as Na$^+$, K$^+$, and Cl$^-$ are also known as **electrolytes** and their balance in the blood, lymph, and tissues is maintained by the actions of the kidneys. Normal levels of these ions are maintained as a result of the kidneys excreting or retaining the ions in the appropriate quantities.

Most of the sodium and potassium ions that are lost to the filtrate from the glomerular capillaries are reabsorbed by active transport from the proximal convoluted tubule and loop of Henle. The final concentrations of these ions in the urine depends on what happens in the distal convoluted tubule and in the upper (cortical) section of the collecting duct.

The amounts of Na$^+$ and K$^+$ lost in the urine are regulated by a hormone called **aldosterone**. Aldosterone is produced in the outer portion (i.e., the cortex) of the adrenal glands (see also chapter 14). When the amount of aldosterone reaching the kidneys via the bloodstream is high, essentially all of the Na$^+$ that escaped active tubular reabsorption in the earlier regions of the nephron is actively reabsorbed from the distal convoluted tubule and collecting duct. As a result, the urine that is excreted from the body contains little or no Na$^+$. In contrast, when the amount of aldosterone reaching the kidneys is low, some sodium escapes reabsorption from the distal convoluted tubule and collecting duct and is excreted in the urine. Typically, the amount of Na$^+$ lost in the urine is about 2% of that which was originally filtered from the glomerular blood.

The effect of aldosterone on K$^+$ excretion and reabsorption is opposite to that of Na$^+$. Like sodium, most of the potassium lost by glomerular filtration is reabsorbed from the proximal convoluted tubule and loop of Henle. When the blood's aldosterone level is high, potassium ions are actively secreted from the peritubular blood into the distal convoluted tubule and cortical region of the collecting duct. As a result, the excreted urine contains K$^+$ (but lacks Na$^+$). When the amount of aldosterone reaching the kidneys is low, nearly all of the K$^+$ that reaches the distal convoluted tubule and collecting duct is reabsorbed into the blood.

Acid-Base Balance

The kidneys play an important role in **acid-base balance**–that is, maintaining the slightly alkaline status of the blood and tissue fluids. Normally, the pH of the blood is about 7.4 (recall that pH is a measure of a liquid's concentration of H$^+$; see chapter 2). If the pH falls appreciably below this (e.g., to pH 7.3 or lower), a state of **acidosis** is said to exist, whereas a rise in blood pH (e.g., to pH 7.5 or higher) produces **alkalosis**. The tendency to lower or raise the blood's pH hinges on the nature and quantities of the wastes and secretions emptied into the blood by the metabolism of the body's tissues.

Normally, significant changes in pH are prevented through the actions of the lungs and the kidneys.

The greatest threat to maintaining the proper pH of the blood and tissue fluids stems from the production of acidic end-products during tissue metabolism. There are three principal sources of these acids. One source is carbon dioxide (known as a **volatile acid** because it is gaseous), which by its reaction with water produces *carbonic acid* (i.e., H_2CO_3). Although carbonic acid is a weak acid, its partial dissociation produces large numbers of hydrogen ions (see chapter 10).

As you should recall from chapter 10, the carbon dioxide released into the blood by the tissues is ultimately excreted from the body by the actions of the lungs. During its journey from the tissues to the lungs, most of this carbon dioxide is converted into carbonic acid through the action of the enzyme *carbonic anhydrase* present inside of red blood cells. The dissociation of the carbonic acid within the red blood cells produces H^+ and bicarbonate ions (i.e., HCO_3^-).

A second source of acids are end-products of the degradation of tissue proteins and nucleic acids. The principal acids produced

in this way are sulfuric acid and phosphoric acid. These are known as **fixed acids** because they are not volatile. Finally, there are the **organic acids** produced during the metabolism of carbohydrates and lipids. The body's principal organic acid is lactic acid, produced for the most part during the metabolism of muscle tissue (see chapter 4).

Potentially harmful changes in the pH of the blood and tissue fluids are prevented through the actions of **buffers**. As described in chapter 2, the body's buffers are mixtures of either *a weak acid and the salt of that acid* or *a weak base and the salt of that base*. Together, these combinations are known as **buffer pairs**. The most important of the body's buffer pairs is the bicarbonate pair, which consists of carbonic acid and sodium bicarbonate (i.e., H_2CO_3 and $NaHCO_3$). As depicted in figure 2.13, when a base is added to a fluid containing this buffer pair, hydroxyl ions (i.e., OH^-) produced by the base's dissociation combine with H^+ to form water (thereby neutralizing the base). As a result, the pH of the fluid is not raised. On the other hand, when an acid is added to a fluid containing this buffer pair, hydrogen ions produced by the acid's dissociation combine with bicarbonate ions, thereby forming more

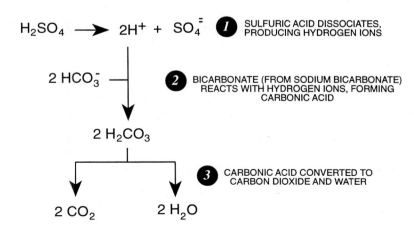

① SULFURIC ACID DISSOCIATES, PRODUCING HYDROGEN IONS

② BICARBONATE (FROM SODIUM BICARBONATE) REACTS WITH HYDROGEN IONS, FORMING CARBONIC ACID

③ CARBONIC ACID CONVERTED TO CARBON DIOXIDE AND WATER

Figure 13.10
How the bicarbonate buffer system prevents a lowering of the pH when fixed acids are produced. When sulfuric acid (produced by amino acid catabolism) dissociates, the resulting hydrogen ions combine with bicarbonate ions to form carbonic acid. The carbonic acid is then converted to carbon dioxide and water. There is, therefore, no net change in the amount of hydrogen ions present.

undissociated carbonic acid. Such action prevents the pH of the fluid from falling. Figure 13.10 illustrates the manner in which sulfuric acid (a fixed acid) is buffered by the action of bicarbonate. By regulating the amount of CO_2 present in the bloodstream, the lungs contribute to the control over the levels of carbonic acid and bicarbonate in the blood and tissue fluids.

The role of the kidneys in acid-base balance is to (1) regulate the loss of blood HCO_3^- in the urine after HCO_3^- is filtered from the glomerular capillaries and (2) generate additional HCO_3^- to compensate for bicarbonate losses resulting from the buffering of fixed and organic acids, and (3) transfer excess H^+ from the peritubular blood into the filtrate via tubular secretion. The actions of the kidneys in acid-base balance are illustrated in figure 13.11.

Bicarbonate ions are filtered from the blood into the Bowman's capsules as blood circulates through the glomerular capillaries. If this bicarbonate were lost in the urine, the body would quickly lose much of its capacity to prevent acidosis. The return of bicarbonate to the bloodstream is indirect and takes the following course. The cells that form the walls of the proximal and distal convoluted tubules contain the enzyme *carbonic anhydrase*, which combines CO_2 produced by the tubule cells (and CO_2 diffusing into the tubule from the bloodstream) with water to form carbonic acid. The H_2CO_3 produced in this way dissociates into H^+ and HCO_3^- and the H^+ is then secreted into the lumen of the tubule (i.e., into the filtrate), while the HCO_3^- is transferred into the bloodstream. Accordingly, for every H^+ secreted into the tubule, the systemic circulation gains a bicarbonate

Figure 13.11
Role of the proximal and distal convoluted tubules in recovering filtered bicarbonate and producing additional bicarbonate. P and D refer respectively to exchange mechanisms in the membranes of proximal (P) and distal (D) tubule cells.

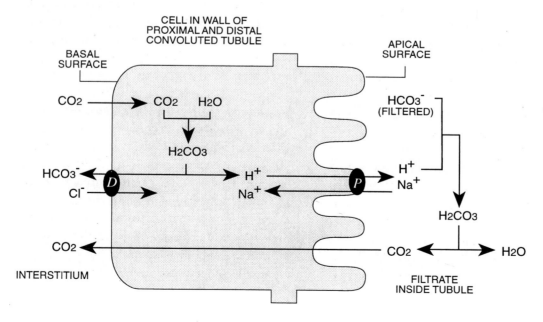

ion. As seen in figure 13.11, H^+ secretion into the tubule is coupled to the reabsorption of sodium ions. In the distal convoluted tubule, the transport of bicarbonate from the tubule cells into the interstitium is coupled to the movement of chloride ions into the tubule cells.

Most of the H^+ transferred to the tubular filtrate reacts with bicarbonate to form carbonic acid, and the carbonic acid then breaks down to form CO_2 and H_2O. The CO_2 then diffuses from the filtrate, through the tubular cells, and into the interstitium and bloodstream. Since most of the CO_2 in the bloodstream is converted to bicarbonate ions, the net effect is the return of filtered bicarbonate ions to the blood.

| 13-6 |

Ammonia

Although there is little or no ammonia (i.e., NH_3) present in the blood, urine does contain small amounts of ammonia (see table 13.1). The presence of a substance in urine that is not present in blood warrants some explanation. Ammonia is a waste product of the breakdown of nitrogen-containing materials in the body (primarily amino acids). Because most tissues are especially sensitive to ammonia and large quantities of this substance are toxic, waste ammonia is chemically modified prior to its release into the bloodstream from a tissue. Tissue cells combine their ammonia with carbon dioxide (also a waste), thereby forming *urea*. The urea, which is considerably less toxic than ammonia, is then released into the blood along with other wastes. Where then does the ammonia present in urine come from? The interstitial cells and the tubule cells that form the walls of the

nephrons also produce ammonia as a waste product of their metabolism. These cells, however, can transfer their waste directly into the glomerular filtrate (instead of the bloodstream). Consequently, the cells of the interstitium and nephron's walls do not convert ammonia to urea; instead, they release their ammonia directly into the filtrate.

Within the filtrate most of the ammonia reacts with hydrogen ions to form ammonium ions (i.e., NH_4^+). In effect, ammonia secreted by the kidney tubules acts as a buffer within the filtrate by combining with H^+; as a result, the urine is not as acidic as it otherwise would be.

UREA

As noted above, ammonia is a waste product of amino acid catabolism and most tissues of the body convert their ammonia to urea before emptying this waste into the bloodstream. Urea in the blood circulating through the kidneys is readily filtered into the Bowman's capsules of the nephrons and ends up in the urine. Some urea, however, is reabsorbed from the kidney tubules, and because the ascending limbs of the loops of Henle and the collecting ducts are readily permeable to urea, it is believed that the urea concentration of the ascending limb and neighboring interstitium undergoes *countercurrent multiplication* (see earlier). That is, urea diffusing out of the collecting ducts enters the nearby ascending limbs of the loops of Henle; the cycling of urea between the collecting ducts, interstitium, and ascending limbs serves to trap urea in the kidney medulla and is believed to make a significant contribution to the high osmolarity that characterizes the interstitium.

THE JUXTAGLOMERULAR APPARATUS

The wall of the distal convoluted tubule fuses with the wall of the afferent arteriole just before the arteriole's entry into the Bowman's capsule; this region of fusion is called the **juxtaglomerular apparatus** (figures 13.3 and 13.12). The juxtaglomerular apparatus plays an important role in the regulation of (1) blood pressure, (2) filtration into the Bowman's capsules, and (3) retention or excretion of sodium ions.

When the blood pressure in the systemic circulation falls, there is a corresponding reduction in the glomerular filtration rate (recall that it is the blood pressure that provides the force of filtration). Under these circumstances, the volume of filtrate proceeding through the various segments of the nephron is reduced. The reduced flow of filtrate in the distal convoluted tubule is sensed by a collection of cells in the wall of the distal tubule called the **macula densa** (fig. 13.12). It is believed that the macula densa causes the **granular cells** (in the walls of the afferent arteriole) to release the proteolytic enzyme *renin* into the bloodstream.

13-7

In the bloodstream, *renin* acts on a plasma protein called **angiotensinogen**, splitting off a short polypeptide called **angiotensin I**. As blood containing angiotensin I circulates

Figure 13.12
The walls of the distal convoluted tubule fuse with the wall of the afferent arteriole just prior to the arteriole's entry into the Bowman's capsule. This region of fusion, called the juxtaglomerular apparatus, plays an important role in regulating blood pressure, glomerular filtration, and salt balance. See text for details.

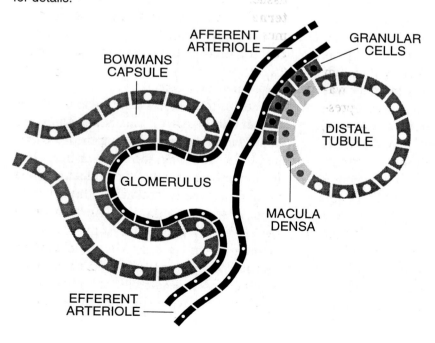

through the capillary beds of the lungs, angiotensin I is converted by ***angiotensin-converting enzyme*** to **angiotensin II**, a powerful vasoconstrictor. The generalized vasoconstriction that is promoted by angiotensin II acts to raise the blood pressure. Among the blood vessels constricted through the effects of angiotensin II are the efferent arterioles. Constriction of the efferent arterioles raises the blood pressure in the glomerular capillaries, and this increases the amount of filtration into the Bowman's capsules. (Angiotensin II may also promote the sensation of thirst, leading to greater ingestion of water.)

Among the other actions of angiotensin II is the stimulation of the pituitary gland to release additional quantities of antidiuretic hormone (ADH, see earlier); this promotes greater water reabsorption from the kidney's collecting ducts. The increased return of water to the bloodstream helps to raise the blood pressure.

Angiotensin II also acts on the body's adrenal glands, causing secretion of the hormone **aldosterone** (see also chapter 14). This hormone acts on the kidneys, promoting the reabsorption of sodium ions into the bloodstream from the glomerular filtrate. Sodium reentering the blood is accompanied by water, thereby further elevating the blood pressure. As the Na^+ concentration of the blood and filtrate rises, the macula densa acts to inhibit further secretion of ***renin***.

| 13-8 |

In response to an increase in blood volume, cardiocytes in the walls of the heart's atria secrete a hormone called **atrial natriuretic factor** (**ANF**). The effects of this hormone are opposite to those of angiotensin II. There is some evidence that ANF inhibits (1) ADH secretion by the pituitary, (2) aldosterone secretion by the adrenal glands, and (3) ***renin*** secretion by the kidneys. ANF also promotes salt and water excretion.

MICTURITION

Urine passes from the collecting ducts into the calyxes of the kidneys and then into the pelves. From each kidney pelvis, the urine is swept downward into the urinary bladder by peristaltic waves generated by smooth muscle tissue in the ureters' walls (fig. 13.1). Each minute, some 1-2 c.c. of urine enters the bladder, so that the chamber slowly fills.

The urinary bladder is emptied (voided) through a single tube, the **urethra**, which drains the bladder at its base. During the time that the bladder is filling, loss of urine through the urethra is prevented by the contractions of two **sphincter muscles** that encircle the urethra. The first of these (i.e., the sphincter closest to the bladder) is called the **internal sphincter** and is formed of smooth muscle tissue. The second sphincter, called the **external sphincter**, is formed of striated muscle, and is, therefore, under conscious control. Parasympathetic motor fibers reach the internal sphincter muscle through the **pelvic nerves**, which also contain parasympathetic motor fibers that innervate the smooth musculature of the bladder wall (i.e., the **detrusor muscle**). During the time that the bladder is filling, the detrusor muscle is maintained in an uncontracted state. The spinal cord communicates with the external urethral sphincter through the **pedundal nerves**.

Emptying of the urinary bladder is called **micturition** (or **urination**) and is controlled reflexively via the **micturition center** in the sacral region of the spinal cord. Although there is some variation according to a person's age and size, in the average adult male the urinary bladder can accommodate 200–300 c.c. of urine before its internal pressure begins to rise dramatically. Above a volume of 300 c.c., stretch receptors in the walls of the bladder are stimulated and initiate the flow of sensory impulses to the spinal cord's micturition center via the pelvic nerves. The micturition center responds by sending parasympathetic motor impulses to the detrusor muscle, causing the bladder wall to contract (while simultaneously relaxing the internal urethral sphincter). Despite the bladder's contraction, no urine exits the bladder unless the external urethral sphincter is consciously relaxed. If the emptying of the bladder at that moment is not socially convenient, the external sphincter remains contracted and after several minutes the bladder wall again relaxes (i.e., the sense of urgency subsides). This cycle repeats itself at intervals until either the external sphincter is consciously relaxed or the bladder's limiting volume is reached (at which time the emptying of the bladder can be prevented no longer).

Usually, the bladder is emptied voluntarily by consciously relaxing the external urethral sphincter at the same time that the micturition center relaxes the internal sphincter and contracts the bladder wall. Under these conditions, a sustained contraction of the bladder voids all but about 10 c.c. of urine. Once the urine has been voided, the external and internal sphincters contract, the bladder wall relaxes, and the urinary bladder begins to fill again. (During micturition, the contraction of the bladder wall squeezes shut the opening of the ureters at the rear of the bladder, temporarily halting the flow of urine from the kidneys into the bladder.)

In babies and young children, the reflexive filling and emptying of the urinary bladder is automatic. With increasing age, and through training, it soon becomes possible to control micturition by regulating the external urethral sphincter. In a similar regard, it is also possible to voluntarily initiate the voiding of urine even when the bladder is only partly filled (for example, when you are asked to provide a urine sample during a medical exam).

SELF TEST*

True/False Questions

1. The kidneys are the body's only excretory organs.

2. A volume of blood equal to all of the blood in the body circulates through the two kidneys every five or six minutes.

3. After circulating through the glomerular capillaries, systemic blood is immediately returned to the venous circulation.

---

* *The answers to these test questions are found in Appendix III at the back of the book.*

4. Gravity causes urine to drain from each of the kidneys into the urinary bladder. T

5. The nephron's net filtration pressure is equal to the glomerular blood pressure minus the colloid osmotic pressure minus the hydrostatic pressure exerted by the filtrate in the Bowman's capsule.

6. Ammonia is one of several nitrogenous wastes that are abundant in the blood plasma. T

7. By the time that the glomerular filtrate reaches the end of the ascending limb of the loop of Henle, about 85% of the filtered plasma water has been reabsorbed into the bloodstream. T

8. So-called "obligatory" reabsorption of water is under the control of antidiuretic hormone (ADH) secreted by the posterior lobe of the pituitary gland.

9. The kidneys play an important role in maintaining the slightly alkaline status of the blood and tissue fluids.

10. Unlike the internal urethral sphincter, the contracted state of the external urethral sphincter is under conscious control. T

Multiple Choice Questions

1. Which one of the following would not normally be said to play a role in excretion? (A) the kidneys, (B) the liver, (C) the spleen, (D) the lungs, (E) the skin. D

2. Blood leaving the glomerular capillaries passes first into the (A) interlobar veins, D

(B) interlobular veins, (C) afferent arterioles, (D) efferent arterioles, (E) renal veins.

3. The U-shaped segment of each nephron is known as (A) the vasa recta, (B) the loop of Henle, (C) Bowman's loop, (D) the peritubular capillary, (E) the distal tubule.

4. In a normally-functioning kidney, the net filtration pressure is about (A) 10 mm Hg, (B) 20 mm Hg, (C) 30 mm Hg, (D) 40 mm Hg, (E) 50 mm Hg..

5. Which of the following substances would not be filtered from the blood of the glomerular capillaries? (A) sugars, (B) salts, (C) amino acids, (D) vitamins, (E) plasma proteins.

6. Which of the following ions is actively transported across the basal surface of the cells forming the wall of the proximal convoluted tubule and into the interstitium? (A) potassium ions, (B) sodium ions, (C) chloride ions, (D) sulfate ions, (E) lactate ions.

7. Which of the following promotes the reabsorption of sodium into the blood from the kidney tubules? (A) antidiuretic hormone, (B) aldosterone, (C) insulin, (D) angiotensin I.

8. Which of the following tissues releases ammonia into the bloodstream? (A) kidney, (B) liver, (C) brain, (D) no tissue releases ammonia into the bloodstream.

9. The granular cells of the juxtaglomerular apparatus secrete (A) antidiuretic hormone, (B) aldosterone, (C) renin, (D) angiotensinogen, (E) angiotensin II.

CHAPTER 14

THE ENDOCRINE SYSTEM

In order for the functions and actions of the body to be fully integrated, one organ (or tissue) must be able to communicate with another. The most rapid way of achieving such communication involves the nervous system. As you learned in chapter 5, nerve fibers rapidly conduct impulses from one part of the body to another, and these impulses ultimately bring about physiological and/or mechanical changes in the target organ.

Interorgan communication may also be achieved in other ways. For example, certain tissues secrete "chemical messengers" into the bloodstream. These messenger molecules are then carried through the circulatory system to target organs. Upon reaching the target organ, the chemical messages bring about physiological changes. Chemical messengers circulating in the blood are more appropriately called **hormones**, and the tissues that release hormones are called **endocrine glands**. Collectively, the endocrine glands comprise the body's **endocrine system**.

ENDOCRINE VS. EXOCRINE GLANDS

Before we proceed any further, it is important to point out that there are two different classes of glands in the body–namely, **endocrine glands** and **exocrine glands** (fig. 14.1). Exocrine glands deliver their secretions into small ducts, which then convey the secretion onto an external or internal body surface such as the skin or the lumen of the digestive tract (e.g., oral cavity, stomach, and intestine). As a result, the secretion has an effect at (or near) the site of the gland. For example, water secreted by the sweat glands in the skin of the face cools the face; digestive enzymes secreted by the salivary glands of the mouth serve to digest food in the mouth; digestive enzymes secreted by the gastric pits of the stomach digest food in the stomach; and so on. In each case, the

action of the secretion is *near* the gland producing the secretion.

In contrast, the secretions of the endocrine glands may have effects at great distances from the location of the gland. This is because the glands' secretions enter the bloodstream and are carried throughout the body via the circulatory system. There is an added implication to secretion into the bloodstream; namely, that the secretions may affect several different and widely separated body parts. Because they lack ducts, the endocrine glands are sometimes referred to as the body's *ductless glands*.

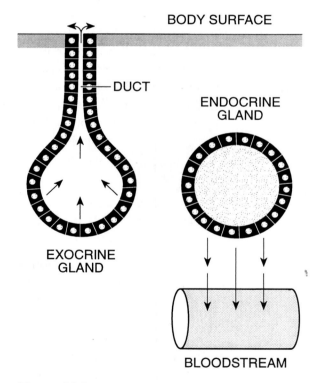

Figure 14.1
A comparison of exocrine and endocrine glands. The secretions of exocrine glands are conveyed by ducts to a body surface, whereas the secretions of endocrine glands enter the bloodstream.

HORMONES

Chemistry of Hormones

The hormones that are secreted by the endocrine glands take a variety of different chemical forms (fig. 14.2). Some hormones are relatively small molecules that are derived from **amino acids**. Included among the small hormone molecules are the thyroid hormones *thyroxine* and *triiodothyronine* and the adrenal **catecholamines** *epinephrine* and *norepinephrine*. Other hormones are **peptides** or **proteins**; that is, they consists of one or more chains of amino acids. Sometimes the chains of amino acids are quite short. For example, the pituitary hormone *antidiuretic hormone* is only nine amino acids long. In contrast, in other polypeptide hormones, chain lengths may be quite large, consisting of dozens of amino acids. Included here are the proteinaceous hormones *insulin* and *erythropoietin*. Finally, some hormones are a form of lipid called a **steroid** (e.g., *progesterone* and *cortisol*), derived principally from cholesterol.

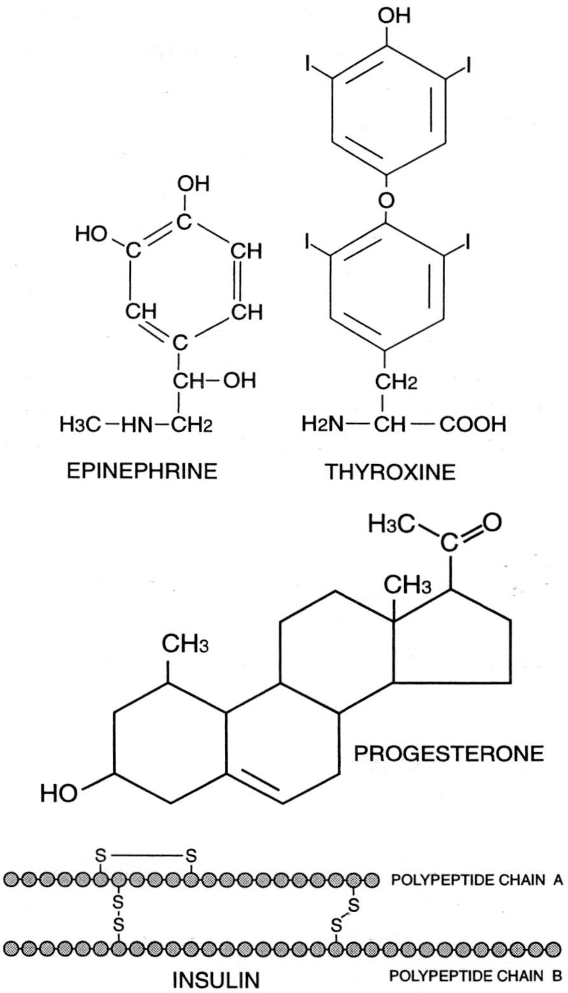

Actions of Hormones

Upon reaching a target tissue, the hormones either bind to or permeate the surfaces of the tissue cells and trigger processes that act to alter the metabolic activities of the cells. The changes that are produced may take the form of activating (or deactivating) enzymes that *already* exist in the cell, or the effect may be on *gene expression*, resulting in the production of additional enzymes.

The properties and physiological activities of cells are determined to the largest extent by the kinds of proteins that the cell

Figure 14.2
Varied chemistry of hormones. Some hormones (e.g., epinephrine and thyroxine [I = iodine atom], *top*) are derivatives of amino acids (compare their chemistry with that of tyrosine in figure 2.17). The steroid hormones (e.g., progesterone, *middle*) are derived from cholesterol. Protein hormones (e.g., insulin, *bottom*) are composed of one or more chains of amino acids (each amino acid is represented by a sphere). In insulin, there are two polypeptide chains (A and B). The two chains are linked together by disulfide bridges (i.e., -S-S-) between cysteine positions (again, see figure 2.17). As seen here, chain A also has an intrachain disulfide bridge.

possesses, especially the cell's enzymes. These proteins are the cell's functional machinery and are responsible for the specific actions that characterize a cell's (or tissue's) behavior. By affecting the target cell's enzymes, hormones can change the behavior of cells in several different ways. For the most part, the hormones bring about either (1) a change in the *nature* of these catalytic proteins, or (2) a change in the *amount* of enzyme that is present. To change the amount of enzyme, it is necessary to alter the rate of

enzyme synthesis, and this requires that the hormone somehow alter the expression of the cell's genetic material in the cell nucleus.

To fully appreciate the effects of hormones, it is necessary to understand the mechanism by which a cell's enzymes (and other proteins) are synthesized. The process is illustrated in simplified form in figure 14.3. The typical human cell contains thousands of different proteins. Among the things that make one protein different from another are (1) the *number* of amino acids in the

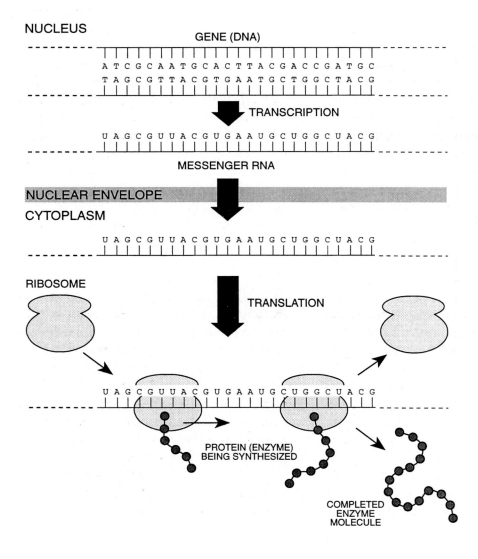

Figure 14.3
Transcription and translation of protein-encoding genes. A,T,G,C, and U represent the nitrogenous bases adenine, thymine, guanine, cytosine, and uracil (see text for details).

protein, (2) the *kinds* of amino acids that are present (recall that there are about 20 different amino acids that regularly occur in human proteins), and (3) the *order* in which the amino acids occur in the polypeptide chains that comprise the protein. For every human protein, these three characteristics are encoded in one or more **genes** present in the cell nucleus.

The genetic material in the cell nucleus is composed of DNA (i.e., deoxyribonucleic acid), the DNA taking the form of two intertwined polynucleotide chains that form a double helix (see chapter 2). For each gene that encodes a protein, it is the sequence of nitrogenous bases (one nitrogenous base in each nucleotide) along the polynucleotides that dictates the order of amino acids in the encoded protein. Whereas the genes that encode the cell's proteins are in the cell nucleus, the synthesis of proteins takes place in the extranuclear cytoplasm.

To produce a particular protein, the genes that encode the protein's structure must first be *transcribed*. During **transcription**, which takes place within the cell nucleus, the order of nitrogenous bases along one of the gene's two polynucleotide chains is used as a *template* for the synthesis of a related polynucleotide called **messenger ribonucleic acid** (or mRNA). In mRNA, the order of nitrogenous bases is *complementary* to that of the transcribed polynucleotide (again, see chapter 2).

Messenger RNA passes from the cell nucleus into the cytoplasm, where it combines with **ribosomes**. The ribosomes, together with a number of other factors contributed by the cytoplasm, *translate* the mRNA's nitrogenous base sequence into a particular amino acid sequence. This is done by "reading" the message as a sequence of "codons," each codon consisting of three successive nucleotides. Consequently, produced in the cytoplasm by the **translation** process are proteins whose chain lengths and amino acid sequences reflect the various mRNAs produced by gene transcription. Each mRNA molecule may be translated hundreds (or even thousands) of times, so that many copies of the final protein are produced from a single gene.

| 14-1 |

As noted earlier, some hormones initiate physiological changes in their target tissues by binding to the surfaces of the target cells. Included in this class of hormones are the protein hormones and the catecholamines epinephrine and norepinephrine. Other hormones, such as thyroxine and the steroid hormones, pass through the surface and into the interior of their target cells and affect gene expression. The mechanisms that bring about changes in the behavior of the target cells are described in the following sections.

Actions of Protein Hormones and Catecholamines

The surfaces (plasma membranes) of cells contain receptor molecules that bind chemical substances circulating in the bloodstream. The receptors are membrane proteins, and among the substances that they bind are circulating protein hormones and catecholamines (fig. 14.4). The binding of the hormone by the receptor triggers chemical reactions in the cell's plasma membrane that result in the production of a "second messenger" (the "first messenger" is the hormone molecule itself). The most common of the second messengers produced by a hormone's target cells is a small compound called **cyclic** adenosine monophosphate (cyclic AMP or cAMP).

The attachment of the hormone to the plasma membrane receptor serves to activate an enzyme in the membrane called *adenylate cyclase*. The active site of this enzyme faces the interior of the cell, where the enzyme converts cytoplasmic ATP to cAMP and pyrophosphate (PP$_i$; figure 14.4). The cAMP produced by the reaction diffuses through the cytoplasm and attaches to an inactive enzyme called *protein kinase*. *Protein kinase* molecules consist of two subunits (i.e., two polypeptide chains). One of the subunits is called the *catalytic* subunit, whereas the other subunit is an *inhibitory* subunit. So long as the two subunits are attached to one another, *protein kinase* possesses no catalytic

activity. However, when cAMP is released into the cytoplasm, it combines with the inhibitory subunit of *protein kinase*; this causes the two subunits to dissociate. Dissociation of the inhibitory subunit activates the catalytic subunit. The consequence of all this is that binding of the hormone at the cell's surface activates the cell's *protein kinase* enzyme molecules.

The function of activated *protein kinase* is to enzymatically attach phosphate groups to other cellular proteins. Some of the cellular proteins that are *phosphorylated* by *protein kinase* are themselves enzymes, and the addition of phosphate groups to these enzymes serves to activate them (for some enzymes, phosphorylation serves to inactivate the enzyme). Thus, a whole family of cellular

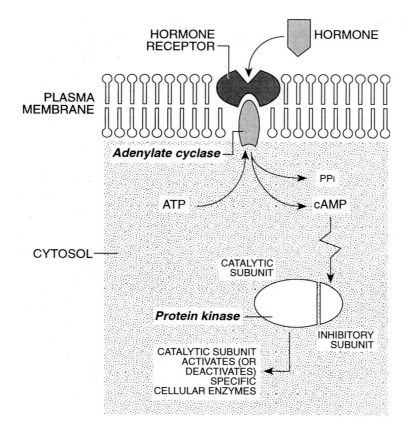

HORMONE
RECEPTOR

HORMONE

PLASMA
MEMBRANE

Adenylate cyclase

PPi

ATP

cAMP

CYTOSOL

CATALYTIC
SUBUNIT

Protein kinase

INHIBITORY
SUBUNIT

CATALYTIC SUBUNIT
ACTIVATES (OR
DEACTIVATES)
SPECIFIC
CELLULAR ENZYMES

Figure 14.4
Actions of protein and catecholamine hormones on their target cells.

enzymes is activated (or inactivated) as an indirect result of the initial binding of the hormone to the target cell's surface. The activation (or inactivation) of cellular enzymes alters the cell's metabolism and behavior (and this, of course, is the physiological goal of the hormone).

The cyclic AMP that is produced in response to hormone binding at the cell surface is degraded within the target cell by an enzyme called *phosphodiesterase*. Therefore, the sustained alteration of a cell's metabolism requires the continuous production of fresh cAMP at the plasma membrane. This, in turn, requires that additional hormone be bound by the plasma membrane's hormone receptors. | 14-2 |

In addition to cAMP, several other second messengers have been discovered. Included among them is **cyclic guanosine monophosphate** (or **cGMP**, a compound that is chemically quite similar to cAMP), Ca^{++} acting in concert with **calmodulin, inositol triphosphate (IP₃)**, and **diacylglycerol (DAG)**.

Some cells respond to hormone binding with plasma membrane changes that increase the membrane's permeability to calcium ions. Because calcium pumps keep the concentration of Ca^{++} much lower in the cytoplasm than it is outside the cell (or within the channels of the endoplasmic reticulum, see chapter 4), the increase in permeability is followed by a dramatic cytosolic influx of Ca^{++}. Within the cell, Ca^{++} combines with and activates the protein calmodulin. Activated Ca^{++}–calmodulin then activates specific cellular *protein kinases*.

In response to hormone binding to receptors in the plasma membrane, membrane phospholipids are enzymatically converted into IP₃ and DAG, both of which act as second messengers. IP₃ is released from the membrane and diffuses into the cytoplasm where it acts on the membranes of the endoplasmic reticulum. IP₃ opens channels in these intracellular membranes, permitting the influx of Ca^{++} into the cytoplasm. Ca^{++} combines with cytoplasmic calmodulin and the Ca^{++}-calmodulin complex then activates certain cellular *protein kinases*. DAG remains in the plasma membrane where it too serves to activate certain cellular *protein kinases*.

Actions of Steroid Hormones and Thyroxine

Like protein hormones, steroid hormones secreted into the blood by endocrine glands circulate in the bloodstream until they reach their target tissue. Because most steroids do not dissolve readily in water and are, therefore, poorly soluble in blood plasma, they are specifically transported within the plasma by carrier proteins.

Upon reaching its target cell, the steroid hormone separates from its carrier protein, permeates the target cell's plasma membrane, and enters the cytoplasm (fig. 14.5). Once within the target cell, the steroid hormone molecule combines with a cytoplasmic receptor molecule to form a **hormone-receptor complex**. The hormone-receptor complex then enters the nucleus of the cell where it attaches to specific genes. The effect on the gene(s) to which the hormone-receptor complex binds is to "turn the gene on;" that is, gene transcription begins in genes that were not being transcribed prior to binding the hormone-receptor complex. As a result, new mRNA is produced, which passes into the cytoplasm from the nucleus and is

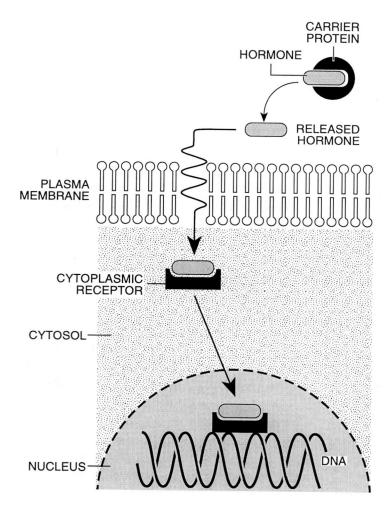

Figure 14.5
Mechanism by which steroid hormones affect their target cells
(see the text for discussion).

translated into protein. The new proteins appearing in the cell (usually enzymes) bring about a change in the cell's metabolism and behavior.

14-3

Thyroxine is the major hormonal product of the thyroid gland (see later). The manner in which thyroxine interacts with its target cells is similar—but not identical—to the action of a steroid hormone. Like a steroid hormone, thyroxine is carried in the blood bound to a carrier protein; and, following its dissociation from the carrier, passes into the cytoplasm of the target cell by permeating the cell's plasma membrane. Unlike a steroid hormone, however, thyroxine passes into the nucleus of the cell where it interacts with a nuclear receptor protein. The hormone-nuclear receptor complex then binds to certain genes, effecting their transcription and subsequent translation into additional cellular enzymes.

A SURVEY OF THE BODY'S MAJOR ENDOCRINE GLANDS

What follows is a survey of the body's major endocrine glands (and their hormonal secretions): (1) the **pituitary** gland, (2) the **hypothalamus**, (3) the **pineal** gland, (4) the **thyroid** gland, (5) the **parathyroid** glands, (6) the **thymus** gland, (7) the **heart**, (8) the **stomach**, (9) the **duodenum**, (10) the **pancreas**, (11) the **adrenal** glands, (12) the **kidneys**, and (13) the body's reproductive organs (i.e., the paired **ovaries** in females and the paired **testes** in males). For each of these glands, we will examine the hormones that they secrete and the effects of the hormones on their target tissues. Abnormalities in which a gland secretes either too much or too little hormone will also be considered. In this regard, excessive secretion of hormone is called *hypersecretion* or *hyperactivity* (i.e., "hyper"

= "over"); when too little hormone is secreted, this is called *hyposecretion* or *hypoactivity* (i.e., "hypo" = "under").

THE PITUITARY GLAND

The pituitary gland (also called the **hypophysis**) is sometimes referred to as the body's "master" endocrine gland because it influences so many body tissues and also regulates the behavior of several other glands. With this in mind, it may not be surprising to learn that the pituitary gland is located at the base of the brain (figures 5.14 and 14.6). The undersurface of the brain rests on a bony ledge formed by the skull. A small pocket in this ledge houses and protects the pituitary gland, which descends into the pocket from the overlying brain tissue. The stalk-like

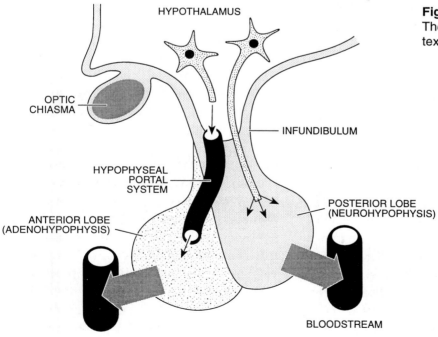

Figure 14.6
The pituitary gland (see the text for details).

connection of the gland with the overlying brain tissue is called the **infundibulum** and is rich in blood vessels and nerve fibers.

Anterior Lobe (Adenohypophysis)

The pituitary gland is divided into front and rear halves, each half, or *lobe*, secreting different families of hormones into the bloodstream. The front or **anterior lobe** (which is also called the **adenohypophysis**) secretes seven hormones: (1) **growth hormone** (abbreviated **GH**), which is also known as **somatotropic hormone** (abbreviated **STH**), (2) **thyrotropin** or **thyroid-stimulating hormone (TSH)**, (3) **adrenocorticotropic hormone (ACTH)**, (4) **prolactin (Pr)** (or **lactogenic hormone**), and the two *gonadotropic hormones* (5) **follicle-stimulating hormone (FSH)**, (6) **luteinizing hormone (LH**; also called **interstitial cell-stimulating hormone [ICSH]**), and (7) **melanocyte-stimulating hormone (MSH**, insignificant after birth and poorly understood in humans).

Relationship Between the Hypothalamus and the Anterior Lobe

The pituitary gland lies just below a region of the brain called the **hypothalamus**, which plays a major role in regulating the actions of the gland. Systemic blood that has circulated through the capillary networks of the hypothalamus does not return directly to the venous circulation; instead, this blood enters a *portal system* (the **hypophyseal portal system**) that carries it from the hypothalamus into the adenohypophysis. As illustrated in figure 14.6, neurosecretory cells in the hypo-

thalamus release a variety of hormones into the portal blood. When these hormones reach the adenohypophysis, they cause the cells of the adenohypophysis to secrete their hormones (some of the hypothalamic hormones act to inhibit pituitary hormone release).

The stimulating (and inhibiting) hormones released into the portal blood by the hypothalamus include **growth hormone-releasing hormone, growth hormone-inhibiting hormone, thyrotropin-releasing hormone, corticotropin-releasing hormone, prolactin-inhibiting hormone, gonadotropin-releasing hormone**, and **melanocyte-stimulating hormone-inhibiting factor**. Note that the names of the hypothalamic hormones reflect the names of the adenohypophyseal hormones whose secretion they promote or inhibit. For example, growth-hormone-releasing hormone promotes the secretion of growth hormone, thyrotropin-releasing hormone promotes the secretion of thyrotropin, and so on.

Hormones of the Adenohypophysis

Growth Hormone. Growth hormone (GH) influences tissue growth, and is especially important during the first 15-20 years of life, when most of the body growth takes place. The hormone's influences are especially marked on the growth of the skeleton. Hyposecretion (i.e., undersecretion) of GH during the normal growing years results in reduced growth. In its extreme form, hyposecretion of GH results in **pituitary dwarfism**, a condition in which the affected person retains childlike size and proportions. Although physical growth is diminished, mental development is entirely normal.

Hypersecretion of GH produces varied effects, depending upon when in life the excessive production and secretion of the hormone occurs. Overproduction of GH during the normal growing years results in a condition known as **gigantism** and is characterized by the affected person reaching extraordinary size (e.g., people who are 7 or 8 feet tall). When hyperactivity occurs after the normal growing years have ended, there is a different effect. Most bone growth occurs at the ends of the bones (i.e., at the **epiphyses**), and when skeletal growth is concluded, the epiphyses seal over. When there is excessive production of GH after the epiphyses seal over, the bones grow wider and become distended. The condition is known as **acromegaly** and is characterized by distortion of the facial bones and the bones of the arms and hands and the legs and feet.

Thyroid-Stimulating Hormone. Thyroid-stimulating hormone (abbreviated TSH and also known as **thyrotropin**) acts upon the thyroid gland (see below), causing the gland to produce and secrete the various thyroid hormones.

Adrenocorticotropic Hormone. As we will see later, the tissues of the adrenal glands are divided into an outer region called the **adrenal cortex** and an inner region called the **adrenal medulla**. The adrenal cortex and the adrenal medulla produce and secrete different sets of hormones. Whereas pituitary secretions have no effect on the medulla of the adrenal glands, adrenocorticotropic hormone (ACTH) released by the pituitary's anterior lobe acts to stimulate hormone production and secretion by the adrenal cortex.

Prolactin. Although prolactin is secreted by the anterior lobe of the pituitary gland in both males and females, it is only in females

that the hormone has a major impact. In females, prolactin acts to promote the production and release of milk (i.e., "lactation") from the mammary glands after the birth of a baby. Note that in this case, an endocrine gland (i.e., the pituitary) is acting (via hormonal secretion) to stimulate secretion by exocrine glands (i.e., the mammary glands).

Follicle-Stimulating Hormone. Follicle-stimulating hormone (FSH) is one of the two *gonadotropic hormones* secreted by the adenohypophysis. In females, FSH promotes the development of ovarian follicles (the chamber-like structures that contain and protect developing egg cells; see later and chapter 15). FSH also promotes the secretion of estrogen by the ovaries (again, see later). In males, FSH promotes the production of sperm cells in the testes.

Luteinizing Hormone. Luteinizing hormone (LH) is the other gonadotropic hormone produced by the adenohypophysis. In females, LH plays a role in inducing the monthly release of mature egg cells from the ovaries into the **Fallopian tubes** (see chapter 15). LH also promotes the secretion of progesterone by the ovaries and the conversion of ruptured ovarian follicles into **corpora lutea**. In males, LH acts on the **interstitial cells of Leydig** in the testes; these are the cells that produce and secrete the male sex hormones (i.e., the androgens, see later). LH is also called **interstitial cell-stimulating hormone (ICSH)**.

Melanocyte-Stimulating Hormone. The color (i.e., lightness or darkness) of skin depends on the amount of **melanin** pigment synthesized by cells called *melanocytes* found between the dermal and epidermal layers of the skin. In lower vertebrates , melanocyte-stimulating hormone promotes the synthesis of melanin in melanocytes, with resulting darkening of the skin. In normal humans, however, secretion is minimal and its effects remain uncertain.

Hormones of the Neurohypophysis

The cells that make up the posterior lobe of the pituitary gland do not actually manufacture the hormones that they secrete. Instead, the hormones are made by neurons in the overlying hypothalamus. The hormones are then carried by the axons of these hypothalamic neurons down through the infundibulum and into the pituitary's posterior lobe (fig. 14.6). The cells of the posterior lobe store the hormone molecules until instructed to secrete them. The hormones are secreted in response to nerve signals received from the hypothalamus.

Two hormones are secreted by the pituitary's posterior lobe: **antidiuretic hormone** (**ADH**; also known as **vasopressin**) and **oxytocin**. Both neurohypophyseal hormones are small peptides consisting of a sequence of nine amino acids. Only two of the nine amino acid positions distinguish the hormones chemically. From a physiological point of view, however, ADH and oxytocin are quite distinct.

Antidiuretic Hormone. ADH was discussed in connection with kidney function in chapter 13. This hormone acts on the kidneys' collecting ducts, promoting the return of water to the bloodstream during **facultative water reabsorption**. The greater the amount of ADH released into the bloodstream from the pituitary, the greater is the amount of facultative water reabsorption, and the more concentrated is the excreted urine. In contrast, when the amount of ADH released into the blood falls, there is less facultative water reabsorption and the excreted urine is greater in volume and more dilute. By influencing the amount of water that is returned to the blood from the kidney tubules, ADH also affects blood pressure (hence the alternate name, *vasopressin*).

Oxytocin. Although oxytocin is produced in both males and females, the hormone has no known function in males. In females, oxytocin has two effects. The first effect occurs during *parturition* (childbirth); at this time, oxytocin released into the blood acts to stimulate the forcible contractions of the smooth musculature of the wall of the uterus, thereby helping to expel the baby from the uterus. Oxytocin also causes constriction of the uterine blood vessels reducing the loss of blood. After childbirth, the continued release of oxytocin promotes contraction of smooth muscle tissue around the ducts of the mammary glands; this promotes **lactation** (i.e., the release of milk from the glands).

| 14-5 |

The hormonal secretions and actions of the pituitary gland are summarized in figure 14.7.

| 14-6 |

PINEAL GLAND

The **pineal gland** is a small gland located in the roof of the third ventricle of the brain (see figure 5.14). The hormone produced and secreted by this gland is **melatonin**, a derivative of the amino acid tryptophan (see figure 2.17). Although there is still considerable controversy about the production and

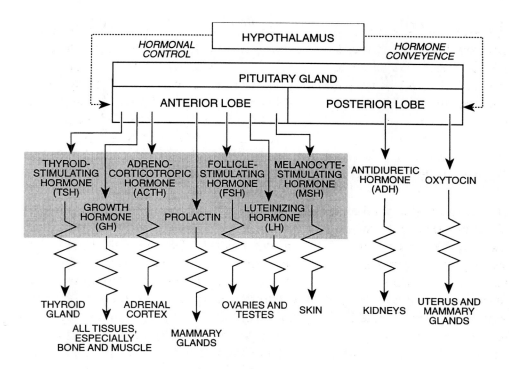

Figure 14.7
Hormonal secretions of the pituitary gland and the targets of these secretions. The adenohypohyseal hormones are in the shaded square.

action of melatonin in humans, it is believed that production of the hormone is influenced by light/dark cycles–more hormone being produced during darkness (i.e., nighttime) and less during daylight hours. Melatonin may have an effect upon that part of the brain that controls circadian rhythms (i.e., physiological events that are related to time of day, such as awakening). There is also evidence that melatonin may influence reproductive activity and capability.

THYROID GLAND

The thyroid gland consists of two lobes of tissue seated on the right and left sides of the trachea, just below the larynx (fig. 14.8). The two lobes of the thyroid are joined on the anterior (i.e., front) surface by a narrow strip of thyroid tissue called the **isthmus**. The major secretions of this gland are **thyroxine**, **triiodothyronine**, and **calcitonin**.

Thyroxine and Triiodothyronine

Thyroxine and triiodothyronine are unusual substances in that they contain the element **iodine**. As its name suggests, each molecule of triiodothyronine contains three atoms of iodine; thyroxine, however, contains four iodine atoms (and is also known as **tetraiodothyronine**). For simplicity, the two hormones are usually referred to as **T3** (triiodothyronine) and **T4** (tetraiodothyronine).

369

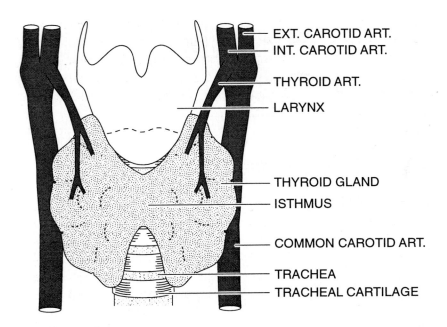

- EXT. CAROTID ART.
- INT. CAROTID ART.
- THYROID ART.
- LARYNX
- THYROID GLAND
- ISTHMUS
- COMMON CAROTID ART.
- TRACHEA
- TRACHEAL CARTILAGE

Figure 14.8
The thyroid gland.

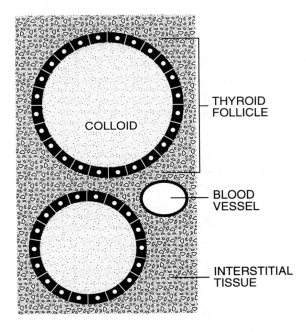

COLLOID

- THYROID FOLLICLE
- BLOOD VESSEL
- INTERSTITIAL TISSUE

Figure 14.9
Thyroid follicles and interstitial tissue.

The T3- and T4-producing cells of the thyroid are arranged to form hollow spheres called **follicles** (fig. 14.9). At the center of each follicle, there is a chamber containing a colloidal suspension of the protein **thyroglobulin**. Dietary iodine absorbed into the bloodstream is removed by the follicle cells as the blood flows through the thyroid tissue. The iodine is then transferred to the follicle chambers and temporarily stored in association with the thyroglobulin. Individual atoms of thyroglobulin-bound iodine are sequentially added to molecules of the amino acid **tyrosine**, first forming **monoiodotyrosine** and then **diodotyrosine**. Monoiodotyrosine and diiodotyrosine are then removed from the colloid by the follicle cells and used to synthesize T3 and T4. T4 accounts for about 90% of the hormone produced by the follicular cells, while the remaining 10% is T3.

T3 and T4 are secreted into the bloodstream where they combine with plasma proteins, forming a complex called **protein-bound iodine** (PBI). Upon reaching the target tissue, the hormones are released from PBI and are taken up by the target cells. Once inside the cell, T4 is converted into T3, which then enters the cell nucleus. Within the cell nucleus, T3 is bound by receptor proteins that affect gene expression.

T3 and T4 bring about an increase in cell metabolism that is reflected by increases in oxygen consumption and the production of heat by the target cells. The effect is believed to be due to stimulation of the synthesis of electron carrier proteins of the mitochondrial membranes. The effect of T3 and T4 in liver is to promote the breakdown of stored glycogen into glucose (called **glycogenolysis**, see chapter 12). The glycogenolytic effect is due to both direct action of the hormones and to the stimulatory effect that the T3 and T4 have on other hormones that promote glycogenolysis (e.g., *epinephrine*; see later).

T3 and T4 also promote the uptake of glucose into the blood from the small intestine. *Insulin*, a hormone that promotes the removal of sugar from the blood by the liver (see later), is inhibited by T3 and T4. These effects, together with the glycogenolytic effect described above, explain why T3 and T4 act to bring about a rapid increase in the amount of sugar circulating in the blood.

Relatively common are disorders in which the amounts of T3 and T4 secreted into the bloodstream are either too low (this is known as *hypothyroidism*) or too high (known as *hyperthyroidism*). The proper levels of T3 and T4 secretion that characterize normal individuals are the result of a delicate feedback mechanism in which the T3 and T4 levels of blood reaching the hypothalamus affect the release of thyrotropin-releasing hormone from this region of the brain. In turn, thyrotropin-releasing hormone regulates the amount of thyroid-stimulating hormone that is produced and released by the adenohypophysis.

Hypothyroidism

Hypothyroidism is an abnormality in which the thyroid gland fails to secrete normal amounts of T3 and T4. When hypothyroidism is the result of a dietary iodine insufficiency, one of the symptoms is the abnormal growth of the thyroid gland, producing a massive enlargement of the throat known as an **endemic goiter**. The goiter is the result of disturbing the feedback loop that controls the thyroid's size. Because the thyroid is not secreting enough T3 and T4, the T3 and T4 levels reaching the hypothalamus are abnormally low. In an effort to raise the T3 and T4 levels, the hypothalamus responds by releasing more thyrotropin-releasing hormone. This causes the anterior lobe of the pituitary to release more thyroid-stimulating hormone. The elevated level of thyroid-stimulating hormone causes excessive growth of the thyroid gland.

Hypothyroidism in a baby is difficult to detect because during pregnancy T3 and T4 are transferred across the placenta from the mother's blood to the baby's blood. As a result, a hypothyroid baby appears normal. However, hypothyroidism during infancy leads to a condition called **cretinism**, characterized by retarded physical and mental growth, abnormal bone development, low body temperature, and a general lethargy. If diagnosed sufficiently early, these effects can be reversed by administering T3 and T4.

Hypothyroidism in adults is characterized by a condition known as **myxedema** in which an excessive amount of tissue fluid fills the spaces between tissue cells and the bloodstream. The skin, especially the facial skin, appears abnormally puffy. Other symptoms include low body temperature, weight gain, and a general lethargy.

Hyperthyroidism

Hyperthyroidism is an abnormality in which the thyroid gland secretes abnormally high amounts of T3 and T4. The disease is rare in infants, but in adults is accompanied by high heart rate and blood pressure, a general irritability (i.e., excessive "nervous energy"), and loss of weight. Two especially obvious physical symptoms are **exophthalmos** and **toxic goiter** (also called **Grave's disease**). Exophthalmos is a bulging of the eyes resulting from swelling of the eye muscles and tissues behind the eyes. In instances of toxic goiter, it is the enlarged thyroid that is the source of the excessive amounts of T3 and T4 that are secreted. Toxic goiters usually are not as large as endemic goiters.

Calcitonin

Blood contains small amounts of calcium derived by absorption from the digestive tract or by decalcification of bone. **Calcitonin**, together with parathyroid hormone (see below), affects the level of calcium in the blood and other body fluids. Calcitonin acts to lower the blood's calcium level by reducing calcium absorption and increasing calcium excretion via the kidneys. Calcitonin also influences the amount of phosphate excreted from the body. The hormone is a short

polypeptide and is produced by the *interstitial cells* of the thyroid. These cells, also known as **C-cells**, lie between neighboring thyroid follicles (fig. 14.9).

PARATHYROID GLANDS

The parathyroid glands (fig. 14.10) consist of four small masses of tissue buried in the rear surface of the thyroid gland (two nodules in each lobe of the thyroid). The major hormonal secretion of this gland is a polypeptide called **parathormone** (also called **parathyroid hormone** or **PTH**). Parathormone's principal effect is its influence on the level of calcium (and to a lesser extent phosphate) that is circulating in the blood. An increase in the amount of parathormone being secreted into the bloodstream is followed by an increase in the blood's calcium (and phosphate) levels. Normally, the increase in blood calcium is the result of increased absorption through the digestive tract (the major source of calcium being milk). Since PTH increases the

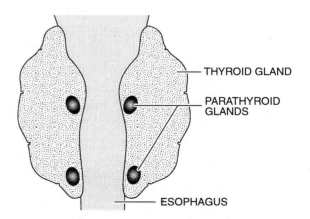

Figure 14.10
The four parathyroid glands are buried in the rear surface of the thyroid gland.

level of calcium in the blood, its effects are *antagonistic* to those of calcitonin. The blood's calcium level can also be raised by withdrawing calcium from the body's skeleton; the latter event poses a danger for a mother during pregnancy.

14-7

THYMUS GLAND

The thymus gland is located between the sternum and aorta and is just inferior to (i.e., below) the thyroid gland. Relative to the size of the body, the thymus is a large organ in babies and children but begins to diminish in size after puberty and is believed to have only minor activity in adults. The principal hormonal secretion of the thymus is **thymosin** (or **thymopoietin**). As described in chapter 9, thymosin influences the development of T-lymphocytes, first in the thymus gland itself and later in the organs to which the T-cells migrate (i.e., the spleen, lymph nodes and bone marrow). Proliferative activity of T-cells in these lymphoid tissues is also promoted by thymosin.

HEART

Upon being stretched by a greater or more forcible influx of blood, **cardiocytes** in the walls of the atria secrete the hormone **atrial natriuretic factor** (**ANF**) into the bloodstream. ANF is a small protein that acts to increase the excretion of salt ions and water through the kidneys. The enhanced loss of water serves to reduce the blood volume and, as a result, lower the blood pressure.

Evidence exists that ANF also inhibits the secretion of the enzyme **renin** by the kidneys

(see chapter 13) and the secretion of the hormone aldosterone (see later) by the adrenal glands. The eventual consequences of these actions is to further lower the blood volume (by reducing retention and reabsorption of filtered ions and water) and blood pressure. The actions of and relationships among renin, aldosterone, and angiotensin were discussed in chapter 13.

STOMACH

The presence of food in the stomach stimulates **G-cells** in the gastric pits (see chapter 11) to release the hormone **gastrin** into the bloodstream. Gastrin then acts to promote the production and secretion of hydrochloric acid (by the pits' oxyntic cells) and pepsinogen (by the chief cells). Gastrin also has a modest stimulatory effect on the pancreas promoting the release of pancreatic enzymes.

DUODENUM

When partially digested food enters the duodenum, duodenal cells release two hormones into the bloodstream; these are **secretin** and **cholecystokinin** (CCK). Secretin's principal effect is on the pancreas, where it promotes the release of large quantities of bicarbonate-rich pancreatic fluid. This fluid passes from the pancreas into the duodenum where it acts to neutralize the acidic chyme that has just arrived from the stomach. CCK also acts on the pancreas, promoting the synthesis and secretion of the pancreatic enzymes (see chapter 11). Vigorous contractions of the walls of the gallbladder that cause the emptying of bile into the duodenum also are promoted by CCK.

PANCREAS

We considered the pancreas previously in connection with the digestive system and learned then that the pancreas is a major source of digestive enzymes. In addition to its role as a digestive organ, the pancreas is an important endocrine gland. The pancreas produces two major hormones: **insulin** and **glucagon**. Both hormones are small proteins and are involved in the regulation of sugar metabolism in the body. Like calcitonin and parathormone, insulin and glucagon are antagonistic (i.e., they have opposite effects).

The pancreatic digestive enzymes and the pancreatic hormones are produced and secreted by different types of cells. The enzyme-producing cells, called **acinar cells**, make up most of the pancreatic tissue; this is *exocrine* tissue, since its secretory products exit the pancreas through ducts (which convey the enzymes toward the duodenum). The pancreas' endocrine tissue consists of clusters of cells scattered through the acinar tissue. The cell clusters, called **islets of Langerhans** (fig. 14.11), are surrounded by capillaries into which the hormonal secretions are emptied.

Insulin and glucagon are secreted by two different types of cells in the islets. The so-called **beta cells** (which make up about 70% of the cells in each islet) secrete insulin, whereas the **alpha cells** (approximately 30% of the islet cells) secrete glucagon.

Insulin secretion by the beta cells occurs in response to a rise in the blood's sugar level. The secreted insulin acts to lower the amount of sugar circulating in the blood by promoting the uptake of sugar by the liver and by adipose (i.e., fat) tissue scattered through the body. Within the liver, insulin promotes the conversion of sugar to glycogen (glycogen is

a high molecular weight polymer of glucose and is the principal form in which sugar is stored in the body; see chapters 2 and 4). In adipose tissue, insulin acts to promote the conversion of sugar to fat. During exercise, insulin promotes the transfer of glucose into muscle cells.

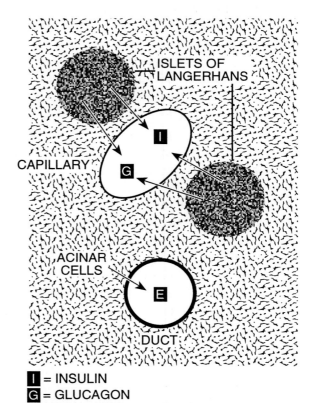

I = INSULIN
G = GLUCAGON
E = DIGESTIVE ENZYMES

Figure 14.11
Pancreas tissue consists principally of digestive enzyme-producing acinar cells, whose secretions enter the pancreatic duct. Scattered through the acinar tissue are the islets of Langerhans, whose alpha and beta cells produce the hormones glucagon and insulin. Enzymes produced by the acinar cells pass into pancreatic ducts, whereas the hormones enter the tissue's capillaries.

In addition to its effects on sugar metabolism, insulin promotes the uptake of amino acids from the blood and the incorporation of these amino acids into cellular proteins. Consequently, together with human growth hormone, insulin is required for proper tissue growth.

Glucagon is antagonistic to insulin, promoting the breakdown of glycogen into glucose and the release of this sugar into the blood. The pancreas secretes insulin and glucagon in response to the amount of sugar circulating in the bloodstream. Consequently, a delicate interplay between these two hormones ensures that the blood sugar level is neither too high nor too low.

Diabetes Mellitus

There are two relatively common physiological abnormalities involving the secretion and actions of the pancreatic hormones; these are **diabetes mellitus type I** (about 10% of all cases) and **diabetes mellitus type II** (about 90% of all cases). In diabetes mellitus type I (also known as "insulin-dependent diabetes" and "juvenile-onset diabetes"), there is an insulin deficiency (that is, *hyposecretion* of insulin) due to malfunction of the beta cells of the islets of Langerhans. In diabetes mellitus type II (also known as "noninsulin-dependent diabetes"), insulin secretion is approximately normal, but the target tissues do not respond with increased sugar uptake from the blood. The diseases' most obvious clinical symptoms are abnormally high plasma glucose levels following a meal (a condition called **hyperglycaemia**) and the appearance of sugar in the urine. The origin of these symptoms may be explained as follows.

In a normal person, the digestion of carbohydrate and the absorption of the resulting sugars begins to raise the blood sugar level. However, the rise in blood sugar is paralleled by increased secretion of insulin into the bloodstream. As a result, sugar uptake by the liver and adipose tissue is stimulated. Recalling the discussion of the digestive system presented in chapter 11, sugars absorbed into the bloodstream following digestion of carbohydrate enter the **hepatic portal system**. The portal system diverts venous systemic blood through the narrow sinuses of the liver before the blood enters the inferior vena cava and returns to the right side of the heart. Sugar removed from the portal blood by the liver is stored as glycogen. Small amounts of the sugar are allowed to reenter the blood as sugar is needed to fuel tissue metabolism between meals. Now, in diabetes mellitus types I and II, the amount of sugar removed from the blood (following digestion of carbohydrate and absorption of sugar) is diminished. Reduced sugar removal by the liver causes the blood sugar level to rise above normal. Recalling the discussion in chapter 13 of the physiology of the kidneys, sugar molecules are sufficiently small to be filtered from the blood of the glomerular capillary networks into the Bowman's capsules. Whereas in a normal person the small amount of filtered sugar is not lost in the urine because the sugar molecules undergo tubular reabsorption, in a diabetic much of the filtered sugar is lost in the urine. This is because the amount of sugar that is filtered greatly exceeds that which can be reabsorbed into the bloodstream.

The loss of sugar in the urine is not, in itself, especially harmful. What is harmful is the variety of additional physiological consequences of elevated blood sugar levels.

For example, the increased sugar level of the blood and the body dehydration that results from excessive urination increase the blood's viscosity, and this places an additional burden on the heart. Since insulin also promotes the uptake of amino acids and synthesis of protein by the body's tissues, protein metabolism is adversely affected. In diabetes mellitus type I, the insulin deficiency slows the rate of fat synthesis in adipose tissue. As a result, fatty acids may be released into the blood and converted in the liver to ketones. This leads to an elevated ketone level of the blood called **ketosis**. For poorly understood reasons, the hyperglycaemia that characterizes untreated type I diabetes leads to major circulatory problems, including atherosclerosis ("hardening of the arteries").

Diabetes mellitus type I appears to be an **autoimmune disease** (see chapter 9) and is influenced by hereditary factors. In persons suffering with this type of diabetes, the body's immune system produces antibodies that attack the beta cells of the islets of Langerhans, progressively destroying them and leading to greatly diminished insulin production and secretion. Diabetes mellitus type I can be treated by administering exogenous insulin to the diabetic. Until quite recently, the only sources of insulin that were available were samples purified from the pancreases of pigs and cows. Although *ovine* (pig) and *bovine* (cow) insulins are proteins that are chemically similar to human insulin, they are not identical and, as a result, they are not entirely effective. Now, however, it is possible to produce large amounts of human insulin using genetic engineering methodologies.

Like diabetes mellitus type I, diabetes mellitus type II is hereditary. However, the symptoms of diabetes mellitus type II develop more slowly (they rarely appear before

reaching the age of 40) and usually occur in people who are obese. As noted earlier, insulin secretion may be normal but the target tissues fail to respond properly to the hormone. In some individuals suffering from diabetes mellitus type II, ingestion and digestion of carbohydrate may be followed by an excessive secretion of insulin. This, in turn, can cause too much sugar to be removed from the blood, thereby producing what is called **reactive hypoglycaemia** (i.e., too little sugar in the blood). When the amount of sugar circulating in the blood is too low, there usually are symptoms of hunger, weakness, impaired mental function, and blurred vision.

$\boxed{\text{14-8}}$

THE ADRENAL GLANDS

Seated on the upper surface of each kidney is an **adrenal gland** (fig. 14.12). Each adrenal gland is divided structurally and functionally into two regions. The outer portion of each gland is called the **adrenal cortex**; this region accounts for about 70% of the gland's mass. The central portion of each gland is called the **adrenal medulla** and accounts for the remaining 30% of the gland's mass. The adrenal cortex and adrenal medulla secrete different families of hormones.

Secretions of the Adrenal Cortex

In response to pituitary ACTH, the cortexes of the adrenal glands secrete two classes of hormones known as **mineralocorticoids** and **glucocorticoids**. All are steroids that are derived from cholesterol.

The principal mineralocorticoid is **aldosterone**. As discussed in chapter 13,

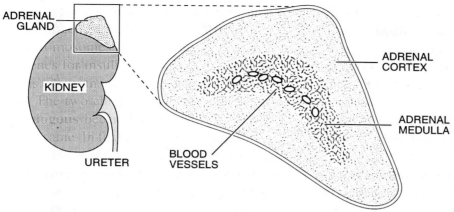

Figure 14.12
The organization of an adrenal gland. On the left side of the drawing, the right adrenal gland is shown seated on the upper surface of the right kidney. The enlarged section view to the right shows that internally the adrenal gland is divided into two distinct zones: outer (cortical) and inner (medullary). Each zone produces and secretes its own hormones.

aldosterone plays an important role in maintaining the body's salt balance. Aldosterone promotes the reabsorption into the bloodstream of Na^+ filtered into the kidney tubules, while promoting the excretion of K^+ through the kidneys.

The main glucocorticoid is **cortisol**. This hormone has a variety of influences on metabolism. It promotes the synthesis of sugars from amino acids and promotes the storage of glycogen. Cortisol also promotes the release of fatty acids from adipose tissue and the use of fatty acids as energy sources during strenuous activity. Because of this, cortisol is said to have a "glucose-sparing effect." Cortisol also has anti-inflammatory effects by reducing white blood cell migration into sites of injury. The production and secretion of aldosterone and cortisol by the adrenals is regulated by ACTH released by the anterior lobe of the pituitary.

Although the ovaries in females and testes in males are the major sources of the sex hormones (see later), the adrenal cortex produces small quantities of androgens and estrogens. The significance of this is unclear.

Secretions of the Adrenal Medulla

In response to sympathetic nerve stimulation, the adrenal medulla secretes the **catecholamines epinephrine** (also called **adrenaline**) and **norepinephrine** (also called **noradrenaline**). Epinephrine accounts for 75-80% of the total catecholamine secretion, while norepinephrine accounts for the remaining 20-25%. Both of these hormones are **sympathomimetic**; that is, their effects mimic the actions of the sympathetic division of the autonomic nervous system. Epinephrine promotes an increase in blood sugar by mobilizing sugar that has been stored in the liver as glycogen (norepinephrine has a similar but less dramatic effect on glycogen mobilization). Epinephrine also promotes the breakdown of stored fats into fatty acids and the release of these fatty acids into the bloodstream. Epinephrine increases heart rate and blood pressure and acts as a vasodilator in skeletal muscle. Norepinephrine is a strong vasoconstrictor and, like epinephrine, raises the blood pressure.

THE KIDNEYS

In addition to being the body's most important excretory organs, the kidneys play a secondary role as endocrine glands. The kidneys produce and secrete the hormone **erythropoietin**. This hormone regulates the production of red blood cells by the bone marrow. The discovery of this hormone is relatively recent and quite interesting. When an animal suffers a blood loss and becomes anemic, the bone marrow responds by increasing its output of new red blood cells. That the increased erythropoietic activity is the consequence of hormonal stimulation was recognized for the first time when plasma taken from an anemic animal was injected into the bloodstream of a normal animal. The normal animal quickly developed **polycythaemia**; that is, within a few days of receiving the "anemic plasma," the number of circulating red blood cells increased to a level well above normal. Naturally, it was suspected that something in the injected plasma was the cause of the elevated erythropoietic activity. The plasma factor was eventually isolated and identified as a glycoprotein and named "erythropoietin" because

of its stimulatory effects on erythropoiesis. Not long after the discovery of erythropoietin, the source of this hormone was identified as the kidneys.

Erythropoietin acts on the bone marrow's stem cells, causing their more rapid and more frequent development into red blood cell progenitors (see chapter 8). Erythropoietin may also cause earlier release of developing red cells from the bone marrow. For example, several days following a severe blood loss, it is not unusual to find larger numbers of *reticulocytes* or even *normoblasts* (again, see chapter 8) in the peripheral blood.

The fact that the kidneys are the source of erythopoietin explains why people who suffer kidney losses and are placed on dialysis must also receive periodic transfusions of blood. In the absence of fully functional kidneys, the lack of erythropoietin secretion causes reduced bone marrow erythropoietic activity, and this results in anemia.

THE REPRODUCTIVE ORGANS

The reproductive tissues (i.e., the **ovaries** in females and the **testes** in males) serve as endocrine glands in addition to their roles in the production of eggs and sperm.

The Ovaries

In order to appreciate the functions of the ovaries as endocrine glands, it is necessary to be familiar with the **ovarian cycle** and the role of the ovaries as the sources of egg cells. At birth, the ovaries of a female contain hundreds of thousands of egg cells (also called **ova** [singular = **ovum**] or **oocytes**). The egg cells are surrounded by a layer of non-reproductive cells called **follicle cells**. Collectively, each ovum and its surrounding follicle cells comprise a **primary follicle** (fig. 14.13, stage 1). The primary follicles remain dormant in the ovaries until the onset of puberty (12 to 15 years of age), when the ovarian cycle or **menstrual cycle** begins.

In the average female, each ovarian cycle lasts about 28 days, day number 1 being the first day of **menstruation** (i.e., loss of blood from the uterus), as the endometrial lining of the uterus breaks down and is discharged.

Beginning on day 1 of each cycle, about 20 primary follicles begin to undergo additional growth and development; that is, the number of follicle cells increases and the size of the egg cell increases. These maturing follicles are called **secondary follicles** (fig. 14.13, stage 2). The growth of one secondary follicle (in one of the two ovaries) dominates the others, and by about day 5 of the cycle, all of the secondary follicles except the dominant follicle cease development and begin to break down and disappear; these degenerating

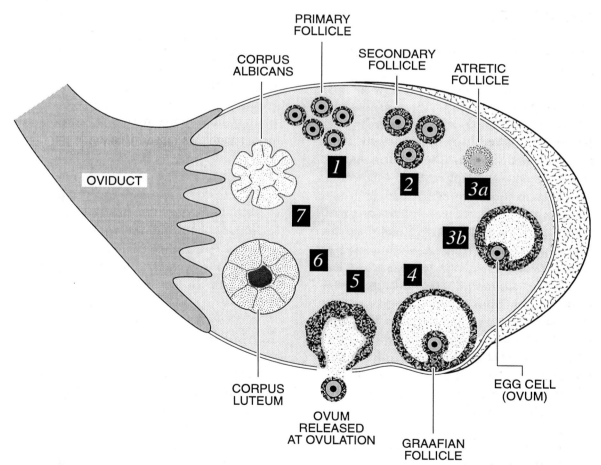

Figure 14.13
Sequence of stages (numbered 1 through 7) of follicle development in an ovary during the ovarian (menstrual) cycle.

structures are called **atretic follicles** (fig. 14.13, stage 3a). The follicle that continues to grow is called a **Graafian follicle** and is characterized by several layers of follicle cells surrounding the egg cell (fig. 14.13, stages 3b and 4).

The development of secondary follicles and the Graafian follicle in particular is promoted by follicle-stimulating hormone (FSH) that is being released by the anterior lobe of the pituitary gland. As the Graafian follicle grows, the follicle cells secrete **estrogen**, one of the two major ovarian hormones (actually, estrogen is a mixture of three chemically similar steroids called **estradiol**, **estriol**, and **estrone**). As the Graafian follicle grows, it creates a bulge in the surface of the ovary.

On (or about) day 14 of the menstrual cycle, the mature Graafian follicle fuses with the surface of the ovary, ruptures, and releases the ovum (together with some of the follicle cells) into the abdominal cavity (fig. 14.13, stage 5). This process, called **ovulation**, is believed to be triggered by a dramatic rise (6 to 10-fold) in the level of luteinizing hormone (LH) released by the pituitary's anterior lobe, together with a more modest rise (about two-fold) in FSH (fig. 14.14). The increase in pituitary FSH and LH secretion occurs in the 16 hours that precede ovulation.

After its release from the ovary, the ovum is guided into the **oviduct** (or **Fallopian tube**). The cells that line this tube are ciliated, and the beating of the cilia acts to convey the ovum toward the uterus. Meanwhile, the center of the ruptured Graafian follicle fills with a blood clot and is then transformed into a **corpus luteum** (fig. 14.13, stage 6). The cells of the corpus luteum (i.e., former Graafian follicle cells) continue the secretion of estrogen but in greater quantities. Also released by the corpus luteum are large amounts of a second ovarian hormone called **progesterone**. If the egg cell released at ovulation is not fertilized and implanted in the wall of the uterus (the usual case), then beginning at about day 21 the corpus luteum begins to degenerate, and the secretion of estrogen and progesterone declines. The decline in estrogen and progesterone secretion is mediated through a feedback mechanism involving the hypothalamus and pituitary gland. The corpus luteum becomes a vestigial body called a **corpus albicans** (fig. 14.13, stage 7). The changes in the levels of FSH, LH, estrogen, and progesterone that characterize the ovarian cycle are summarized in figure 14.14.

14-9

The ovarian cycle is accompanied by cyclic changes in the lining of the uterus that prepare the uterus for reception and implantation of a fertilized egg. If no fertilized egg reaches the uterus and the corpus luteum begins to degenerate, the secretion of FSH and LH falls. The drop in FSH and LH leads to the breakdown of the uterine lining and a loss of blood and tissue from the uterus. This discharge is called **menstruation**. As noted above, the first day of bleeding is taken as the first day of the 28-day ovarian cycle.

Actions of the Ovarian Hormones

As we have seen, the secretion of female (and male, see later) sex hormones by the reproductive organs (or "gonads") is under the control of FSH and LH that are released by the anterior lobe of the pituitary gland. The release of these pituitary hormones remains low until puberty, at which time the hypothalamus increases its secretion of gonadotropin-releasing hormone. Gonadotropin-

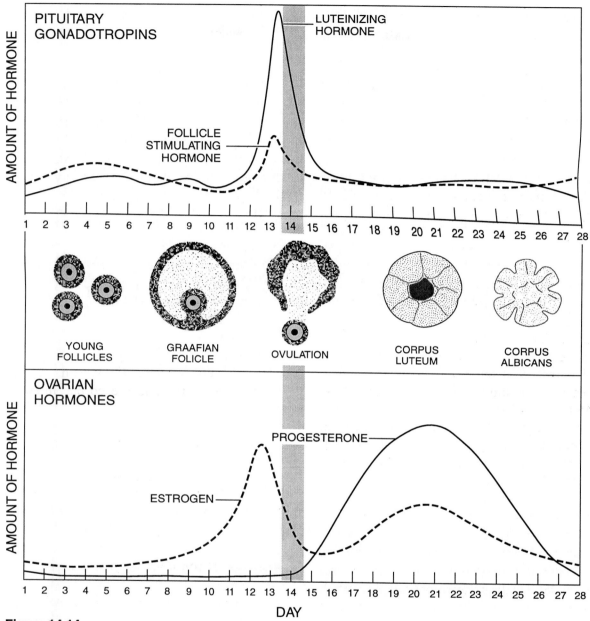

Figure 14.14
Changes in the levels of gonadotropic hormones (i.e., follicle-stimulating hormone and luteinizing hormone) and ovarian hormones (i.e., estrogen and progesterone) during the ovarian cycle.

releasing hormone acts on the pituitary, and the elevated release of FSH and LH that results from this promotes the release of the sex hormones. The rising levels of sex hormones in the blood at puberty are responsible for dramatic changes in the shape and appearance of the body. These changes are accompanied by rapid growth (which occurs somewhat later in males than in females).

Estrogens. As noted earlier, estrogen is really a mixture of three steroid hormones, among which estradiol is the most abundant and most important. Estrogen has a variety of important influences on the body's functions and actions. From a metabolic point of view, the principal effects of estrogen are on protein synthesis and tissue growth. The most direct targets of estrogen action are the so-called *accessory reproductive organs*. At puberty, estrogen promotes the growth of these accessory structures, and in the absence of estrogen, none of the changes associated with puberty transpire. Accordingly, estrogen stimulates the growth of the uterus, the growth and development of the external genitalia, and the growth of the mammary glands. During puberty, estrogen promotes rapid growth of the skeleton, then abruptly halts growth in the later teenage years by inducing the sealing of the growing ends (i.e., the *epiphyses*) of the bones. In adult females, estrogen is necessary for the maintenance of the accessory reproductive structures and the initiation and continuation of the menstrual cycle.

Estrogen also influences electrolyte balance (i.e., the excretion and retention of water and salts) and acts to maintain the female *secondary sex characteristics*, such as (1) the distribution of fat to the hips, abdomen, and mammary glands, (2) the elevated pitch of the voice, and (3) the pattern of hair distribution on the body.

At the time of ovulation, estrogen acts on the cervical glands so that they produce a secretion that favors the survival of sperm and facilitates sperm entry into the uterus from the vagina. Estrogen also increases the ciliary beating of the oviduct cells so that the ovum is efficiently swept from the ovary to the uterus.

Progesterone. Prior to ovulation, the amount of progesterone circulating in the blood is very small. However, following ovulation the level of progesterone rises rapidly, as this hormone and estrogen are produced and secreted in large amounts by the corpus luteum. Working in concert with estrogen, progesterone promotes the development of the wall of the uterus (i.e., the **endometrium**), so that the wall can accept and adequately nourish a fertilized egg and developing embryo (see also chapter 15). The endometrium becomes thick and spongy and also highly vascular (i.e., rich in tiny blood vessels). If the egg cell released at ovulation is not fertilized, the secretion of progesterone declines. The fall in progesterone is followed by the progressive death of the tissues of the endometrium as the blood supply to the tissues is shut down. The dying (i.e., *necrotic*) tissue, along with small amounts of blood, is sloughed off the surface of the uterus and gives rise to the menstrual discharge marking the onset of a new ovarian cycle.

In the event that the egg released at ovulation is fertilized and is implanted in the endometrial lining of the uterus, the corpus luteum remains intact in the ovary and continues to secrete progesterone (i.e., the corpus luteum is not transformed into a corpus albicans). This is due to the secretion of LH by the placenta (see chapter 15). During the pregnancy, follicle development and ovulation are halted as FSH and LH are no longer secreted by the pituitary. Progesterone continues to be secreted by the corpus luteum for the first five weeks of pregnancy and promotes additional and more complex developmental changes in the uterine wall, including the development of the placenta (see chapter 15). After this time, however, the placenta takes over the role of progesterone (and estrogen) secretion, as the corpus luteum degenerates and disappears from the ovary.

Birth control pills that are prescribed for women who wish to avoid pregnancy contain a mixture of synthetic estrogens and progesterones. By acting either on the hypothalamus or anterior lobe of the pituitary, these pills are believed to prevent the dramatic surge in luteinizing hormone that precedes ovulation. As a result, no ovulation occurs (i.e., no egg is transferred to the oviduct).

The Testes

The gamete-producing organs of males are the **testes**, which produce **sperm cells** (or **spermatozoa**). In addition to producing

sperm, the testes produce and secrete male sex hormones called **androgens**.

The organization of a testis is illustrated in figure 14.15. The sperm-producing tissue is arranged to form large numbers of tiny, coiled tubules called **seminiferous tubules**. The seminiferous tubules lead into the **rete testis**, which empties into a coiled channel called the **epididymis**. In turn, the epididymis empties into the **ductus deferens** (or **vas deferens**), which conveys sperm out of the testes and into the **urethra** of the **penis** (see also chapter 15).

Beginning at puberty, sperm cell production in the seminiferous tubules proceeds on a continuous basis. (There is no cycle of sperm production or release comparable to the ovarian cycle of the female.) The production and maturation of each sperm cell, a process called **spermatogenesis**, begins near the outer wall of each seminiferous tubule and proceeds inwardly, toward the tubule's lumen (figures. 14.16 and 14.17). Develop-

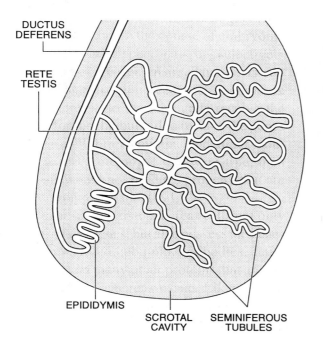

DUCTUS
DEFERENS

RETE
TESTIS

EPIDIDYMIS

SCROTAL
CAVITY

SEMINIFEROUS
TUBULES

Figure 14.15
Organization of one of the two testes. Sperm cells are produced in the seminiferous tubules and move from there into the rete testis, epididymis, and ductus deferens.

ment begins with cells called **spermatogonia**. These cells give rise to **primary spermatocytes** which then form **secondary spermatocytes**. In the final stages of maturation, the secondary spermatocytes become **spermatids**, which lose much of their cytoplasm and develop the tail-like locomotor appendage, called a **flagellum**, which characterizes the mature sperm cell.

The maturation of sperm is assisted by large cells in the wall of the seminiferous tubules called **Sertoli cells** (maturing sperm are actually embedded in these cells, figures 14.16 and 14.17). These cells supply nutrients to the developing gametes, remove improperly forming sperm cells by phagocytosis, and assist in the withdrawal of cytoplasm from the sperm during the final stages of sperm maturation.

Hormone Production and Secretion by the Testes

Maturing sperm and Sertoli cells do not produce or secrete hormones. Rather, the

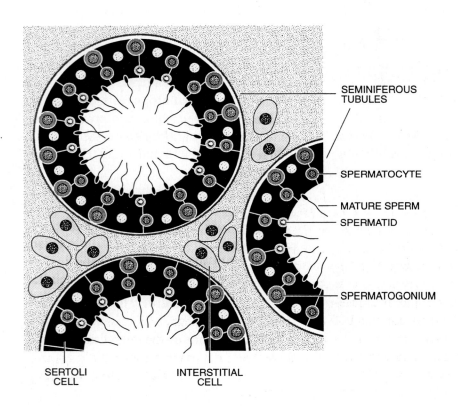

Figure 14.16
Sperm cells are produced in the testes' seminiferous tubules (shown here in cross-section). Development of sperm begins at the wall of the tubule with stem cells called spermatgonia. Successively mature cells are seen closer to each tubule's lumen and mature sperm cells line the lumen. The male sex hormones (androgens) are produced and secreted by the interstitial cells (cells of Leydig).

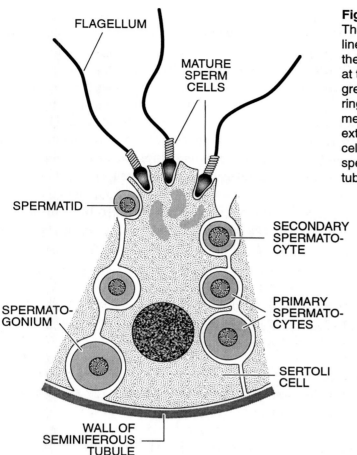

FLAGELLUM

MATURE
SPERM
CELLS

SPERMATID

SECONDARY
SPERMATO-
CYTE

SPERMATO-
GONIUM

PRIMARY
SPERMATO-
CYTES

SERTOLI
CELL

WALL OF
SEMINIFEROUS
TUBULE

Figure 14.17
The walls of the seminiferous tubules are lined by Sertoli cells. These cells assist the development of sperm, which begins at the base of the Sertoli cells and progresses toward the apical surface. During the late stages of sperm development, Sertoli cells remove much of the extranuclear cytoplasm from the sperm cells. The tails (or flagella) of mature sperm (top of figure) project into the tubule's lumen.

male sex hormones, called *androgens*, are produced by the **cells of Leydig** in the **interstitial tissue** that separates neighboring seminiferous tubules (fig. 14.16). Like the female sex hormones, the androgens are steroid molecules. By far, the most abundant of the androgens is **testosterone**.

The production of sperm and the secretion of androgens by the testes are regulated by the same hormonal secretions of the anterior lobe of the pituitary gland that affect the ovarian cycle. Follicle-stimulating hormone promotes sperm production in the seminiferous tubules, whereas luteinizing hormone promotes the secretion of androgens by the cells of Leydig found in the interstitial tissue.

Because sperm development in the seminiferous tubules does not proceed beyond the primary spermatocyte stage unless testosterone is present, LH (which promotes testosterone secretion) is said also to influence sperm development. It is the effects of luteinizing hormone on the interstitial tissue of the testes that prompt the hormone's alternative name–**interstitial cell-stimulating hormone (ICSH)**.

Androgens. Testosterone is the most abundant of the androgens, accounting for about 90% of all male sex hormone secreted by the testes. Like the estrogens of the female, the male sex hormones have a diversity of effects. For example, the androgens promote the growth and development of the accessory reproductive organs of the male (i.e., the penis, seminal vesicle, prostate gland, and scrotum; see also chapter 15). During the teen years, androgens stimulate muscle development and rapid growth in height (i.e., growth of the skeleton). In the late teen years, the androgens promote the sealing of the epiphyses at the end of the bones, thereby bringing skeletal growth to a conclusion.

14-10

The androgens promote the development and maintenance of the male *secondary sex characteristics*. For example, the androgens promote growth of the larynx, which acts to lower the pitch of the voice. Androgens also influence the pattern of hair distribution on the body (e.g., the facial, chest, and back hair that is so much greater in the male than the female). Interestingly, small quantities of androgens (and estrogens) are produced in the adrenal glands in both males and females. Normally, however, the masculinizing effect of adrenal androgens in a female is slight and so too are the feminizing effects of adrenal estrogens that are produced in males.

SELF TEST[*]

True/False Questions

1. Exocrine glands release their secretions into ducts, whereas endocrine gland secretions are released into the bloodstream.

2. The catecholamines are a special group of steroid hormones.

3. Whereas transcription occurs within the cell nucleus, translation takes place in the cytoplasm.

4. Steroid hormones can permeate the cell membranes (plasma membranes) of a tissue, whereas protein hormones do not enter their target cells; instead, protein hormones bind to surface receptors, triggering the release of intracellular second messengers.

5. Whereas the anterior lobe of the pituitary gland manufactures and secretes hormones, the hormones secreted by the pituitary's posterior lobe are manufactured in the hypothalamus.

6. The thyroid hormones T3 and T4 readily dissolve in the water of the blood plasma.

7. In babies and children, the thymus gland plays an important role in the development of the immune system.

8. Follicle-stimulating hormone plays an important physiological role in females that have attained puberty but has no physiological effects in normal, adult males.

·9. The androgens (male hormones) are secreted by the interstitial cells of the

---

[*] *The answers to these test questions are found in Appendix III at the back of the book.*

testes, not by maturing sperm cells.

10. Secretin and cholecystokinin are hormones secreted into the blood by the pancreas.

Multiple Choice Questions

1. Which of the following does not have an endocrine function? (A) thyroid gland, (B) pancreas, (C) heart, (D) stomach, (E) all of these have an endocrine function.

2. So-called "second messengers" are produced by cells in response to (A) nerve stimulation directly from the brain, (B) the action of steroid hormones, (C) the action of protein hormones, (D) excess sugar (i.e., glucose) in the bloodstream.

3. Most of the iodine in the body is used in the manufacture of (A) thyroid-stimulating hormone, (B) thyroxine, (C) body starch, (D) DNA, (E) eye pigments.

4. Which one of the following hormones would you expect to have the greatest influence on the volume of the blood circulating in the body? (A) parathormone, (B) insulin, (C) FSH, (D) ADH, (E) none of these hormones affects the volume of the circulating blood.

5. Which one of the following is not a "second messenger?" (A) cAMP, (B) cGMP, (C) calcium ions, (D) all of these play roles as second messengers.

6. The pineal gland secretes (A) melatonin, (B) thymosin, (C) melanocyte-stimulating hormone, (D) oxytocin, (E) parathormone.

7. Which one of the following conditions is symptomatic of hyperthyroidism? (A) cretinism, (B) endemic goiter, (C) exophthalmos, (D) hypoglycaemia, (E) acromegaly.

8. "Juvenile-onset diabetes" is an alternative name for (A) diabetes mellitus type I, (B) diabetes mellitus type II, (C) diabetes insipidus.

9. Epinephrine and norepinephrine are (A) mineralocorticoids, (B) glucocorticoids, (C) catecholamines, (D) membrane receptors.

10. The so-called female "secondary sex characteristics" are (A) regulated by estrogen, (B) regulated by progesterone, (C) regulated by aldosterone, (D) essential to survival.

CHAPTER **15**

THE REPRODUCTIVE SYSTEM AND INHERITANCE

It has been estimated that the various tissues and organs of an adult human being contain more than 40,000,000,000,000 (that's forty thousand billion!) cells. All of these cells are derived from a single cell called a **zygote** (i.e., the fertilized egg cell) through cell growth and cell division. In many tissues of the body (e.g., the epithelial lining of the digestive organs and the blood-forming tissues), cells continue to grow and divide for most of a person's life. However, in some tissues, such as muscle and nerve, cell division ceases some time after birth, and subsequent tissue growth results from individual cell growth without division.

The ongoing cell growth and division that characterizes most of the body's tissues accounts for the growth of the organism as a whole and also the replacement of dying cells. With the exceptions of such tissues as muscle and nerve, replacement of old tissue cells with new ones results in a complete turnover of the body's cellular composition every few years. (A portion of the body's total mass is represented by noncellular materials that are secreted by cells; for example, most of the mass of bone and cartilage is represented by secreted calcium salts and proteins.)

In this chapter, we will explore the processes by which the zygote is formed from sperm and egg cells and the mechanisms by which the cells and tissues of the body are formed. Finally, we will briefly examine some examples of human inheritance that illustrate the processes by which human traits are passed from generation to generation.

FERTILIZATION

Human life begins when a **sperm cell** from the male parent (i.e., the father) fuses with an egg cell (or **ovum**) from the female parent (i.e., the mother). Fusion of egg and sperm is called **fertilization**, and the new cell that is formed in the process is a human *zygote* (fig. 15.1). In order to fertilize an egg, 200,000,000 - 500,000,000 sperm cells must be deposited in the vagina of the female reproductive tract (see later). At first, the sperm cells are incapable of fertilizing an egg. However, secretions from the female tract induce **capacitation**–a process by which the sperm

acquire the ability to fertilize an egg.

Because there is a very high mortality rate of sperm within the female tract, only a few hundred sperm successfully reach the egg cell as it travels from the ovary toward the uterus (fig. 15.1, stage 1). Collectively, the surviving sperm employ enzymes that are concentrated in the sperm's **acrosome** (a tiny cap at the forward tip of the sperm head) to progressively digest a path through the **corona radiata** (an encapsulating layer of follicle cells) and the **zona pellucida** which surround the ovum (fig. 15.1, stage 2). One of the enzymes is *hyaluronidase* (also called *acronase*), which digests polysaccharide; another is a trypsin-like enzyme that digests protein. As soon as one sperm cell makes contact with the surface of the egg cell, their plasma membranes fuse (fig. 15.1, stage 3) and other sperm are barred from entering the egg cell. The nucleus of the sperm fertilizing the egg is drawn into the egg cell's cytoplasm (fig. 15.1, stage 4). The nucleus of the ovum (whose development has been arrested at *metaphase II* of meiosis, see later) completes its division and fuses with the sperm cell nucleus. The resulting combined nucleus contains genetic information contributed by each parent. This combination of genetic information will determine the characteristics and properties of the zygote and the billions of cells that are to be derived from the zygote during embryonic and fetal development.

15-1

CHROMOSOMES AND GENES

Inheritable information takes the form of individual units called **genes** comprised of the chemical substance **deoxyribonucleic acid** (i.e., DNA; see chapter 2). It is esti-

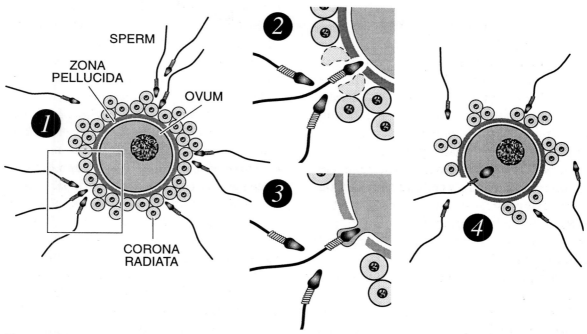

Figure 15.1

Fertilization. Human life begins when a sperm cell from the male parent fuses with an egg cell from the female parent. The many sperm that surround the egg (stage 1) secrete enzymes that digest a path through the corona radiata and zona pellucida. This permits a single sperm to contact the surface of the egg (stage 2). The plasma membranes of the sperm and egg fuse (stage 3) and the sperm cell nucleus is drawn into the egg cell's cytoplasm (stage 4). The two nuclei move toward one another and fuse to form the single nucleus of the zygote. Stages 2 and 3 are magnifications of the rectangular region outlined in stage 1.

mated that the nuclei of human cells contain hundreds of thousands of genes. The genes of a cell nucleus are linked together in linear arrays called **chromosomes**. Chromosomes are not entirely DNA but also contain large amounts of protein. Although the number of chromosomes in the cell nucleus varies considerably among different animal and plant species, each species has a specific **chromosome number**. In humans, the chromosome number is 46; that is, the hundreds of thousands of genes are apportioned among 46 chromosomes. The chromosome number of a species has no special significance. In goldfish, which are considered simpler or-

ganisms than humans, the chromosome number is 96. In chimpanzees, which are very closely related to humans, the chromosome number is 48, but then the chromosome number in the tobacco plant is also 48.

Within a species, each chromosome exhibits a specific and characteristic shape during the metaphase (see later) of cell division; therefore, each human chromosome can be individually distinguished. For purposes of identification and study, the human chromosomes are assigned specific numbers. (The numbers are 1 through 23, not 1 through 46; the reason for this will be explained later.)

The functional machinery of a cell is its proteins. That is to say, the structure, organization, and functions of cells rest upon the kinds of proteins that are present in the cell. What distinguishes one protein from another (chemically and functionally) is the number and order (i.e., sequence) of the various amino acids that comprise the protein's polypeptide chains. It is these two properties of proteins that are encoded in the genes that comprise a chromosome. Thus, for simplicity, a chromosome may be thought of as a sequence of blueprints, each blueprint spelling out the specific structure of all (or part) of a particular cell protein.

This concept may be illustrated using an example that is already familiar to you. Each of the hemoglobin molecules present in a red blood cell consists (in part) of four polypeptide chains: a pair of so-called *alpha* chains and a pair of *beta* chains (see chapter 8, figure 8.4). Each of these two types of polypeptide chains is encoded by different genes. The genes are called the *alpha chain genes* and the *beta chain genes*. The alpha chain genes of the cells of the body are found on chromosome number 16, whereas the beta chain genes are found on chromosome number 11. The chromosomal locations of thousands of other human genes are also known.

From the preceding discussion, it is clear that a chromosome may be viewed as a linear sequence of blueprints for cellular proteins. There are, therefore, genes for the alpha and beta chains of hemoglobin, genes for the protein hormone *insulin*, genes for the plasma protein *albumin*, and so on.

Homologous Chromosomes

To further your understanding of the nature

of human chromosomes (and the chromosomes of other living things), let's suppose that the genes that encode insulin, alpha and beta globin chains, albumin, and a number of other proteins that we'll simply call protein "X," protein "Y," and protein "Z" are all on the same chromosome. (*This is not actually true because you already know that alpha globin chain genes are on chromosome number 16 and beta chain genes are found on chromosome number 11; but making such a supposition will help to illustrate several important points.*) Such a hypothetical chromosome is illustrated in figure 15.2.

If, having identified our hypothetical chromosome containing the genes for insulin, alpha and beta globin chains, albumin, and proteins "X," "Y," and "Z," we were now to explore the gene sequences of the other 45 chromosomes in the nucleus, we

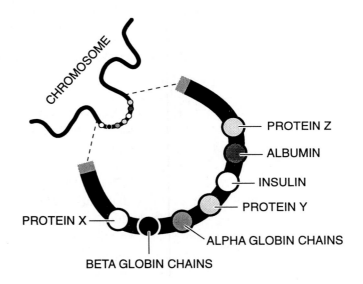

Figure 15.2
A hypothetical human chromosome. The chromosome consists of a series of thousands of genes, each gene encoding the chemical structure of all (or part) of a protein.

would find a second chromosome that had the same gene sequence. That is, there are *two* chromosomes in the nucleus containing the genes for insulin, alpha and beta globin chains, albumin, and proteins "X," "Y," and "Z." The two chromosomes are said to be **homologous** because their gene sequences are the same. In fact, for every chromosome in the nucleus, there is an homologous partner whose sequence of genes (which may number in the thousands) is the same (fig. 15.3). Therefore, human cells are said to contain *23 pairs of homologous chromosomes*. Because the two *homologues* of a pair are the same, there are only 23 *different* chromosomes in a human cell. (This is why the chromosomes are numbered 1 through 23 and not 1 through 46.) The number of different chromosomes present in a cell is given by the symbol *n*. Therefore, for humans, $n = 23$.

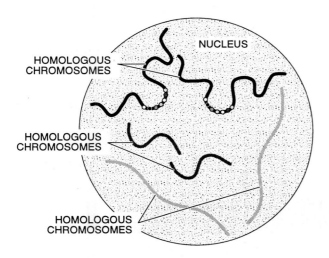

Figure 15.3
A diploid cell nucleus contains pairs of homologous chromosomes. Depicted here are three of the 23 pairs of homologous human chromosomes.

Because every chromosome has a homologous partner, this implies that the genes occur in pairs. For example, there are two insulin genes—one on each of two homologous chromosomes; there are two albumin genes, two beta globin chain genes, and so on. Organisms (or individual cells) that contain pairs of homologous chromosomes are said to be **diploid**, whereas organisms (or individual cells) that have only one copy of each chromosome (e.g., bacteria, certain other microorganisms, and [as we will see] sperm and egg cells) are said to be **haploid**. Thus, haploid organisms (or individual cells) have *n* chromosomes, whereas diploid organisms (or individual cells) have *2n* chromosomes.

Homozygous and Heterozygous Genes

It is usually the case that the two genes occupying equivalent sites on homologous chromosomes are identical; that is, they encode polypeptides with precisely the same number and order of amino acids. Such genes are said to be **homozygous**. However, for a small percentage of gene pairs, there may be small differences between the polypeptide encoded by one gene and the polypeptide encoded by the equivalent gene on the homologous partner chromosome. When these small differences exist, the two genes are said to be **heterozygous**. This notion may be illustrated using an example that is already familiar to you.

In chapter 9, you learned that there are four human blood types in the ABO series; namely, type A, type B, type AB, and type O. These blood types are based upon the presence (or absence) of specific antigens in the membranes of red blood cells. The antigens themselves are not proteins; rather they are a combination of lipid and carbohydrate called a *glycosphingolipid*. All (normal) humans produce identical core glycosphingolipids to which specific terminal sugars are added enzymatically, thereby completing the antigen molecule. The nature of the terminal sugar that is added to the core determines the final structure of the antigen, and this depends upon the enzyme (called a *glycosyltransferase*) that is available to catalyze the sugar addition. When the enzyme that makes the transfer is the one that adds *acetylgalactosamine*, the result is antigen A; on the other hand, when the enzyme that makes the transfer is the one that adds *galactose,* the result is antigen B. *It is the enzymes catalyzing these sugar transfers that are encoded by genes.*

A person with type A blood has only the A antigen in the membranes of his (or her) red blood cells because he (or she) has the acetylgalactosamine-transferring enzyme (but not the galactose-transferring enzyme). A person with type B blood has only the B antigen in the membranes of his (or her) red blood cells because he (or she) has the galactose-transferring enzyme (but not the acetylgalactosamine-transferring enzyme). A person with type AB blood has *both* the A antigens and the B antigens in the membranes of his (or her) red blood cells

because both enzymes are present; and, finally, a person with type O blood has *neither* the A antigens nor the B antigens in the membranes of his (or her) red blood cells because neither enzyme is present.

The *glycosyltransferase* enzymes are encoded by genes that occupy equivalent sites on homologous chromosomes. Three gene variations exist: (1) gene "A," which encodes the enzyme *acetylgalactosamine transferase*, (2) gene "B," which encodes the enzyme *galactosyltransferase*; and (3) gene "O" which is not expressed in the form of a detectable antigen, because the enzyme encoded by the gene is unable to function in terminal sugar transfer. There are, therefore, six different **genotypes** corresponding to the four ABO blood types; these are AA, AO, BB, BO, AB, and OO (fig. 15.4). Every human being has one of these six genotypes, determined at birth when genes contributed by the male and female parents are combined in the zygote.

If a person's genotype is AA (i.e., the person inherited an A gene from each parent), he is said to be *homozygous* and has type A blood. In contrast, a person whose genotype is AO (i.e., the person inherited an A gene from one parent and an O gene from the other parent) is said to be *heterozygous*, although he also has type A blood. During a

Figure 15.4
The human ABO series blood types and their genetic origins.

BLOOD TYPE	GENOTYPE	CONDITION
A	AA	Homozygous
A	AO	Heterozygous
B	BB	Homozygous
B	BO	Heterozygous
AB	AB	Heterozygous
O	OO	Homozygous

blood test, both individuals would be found to have type A blood, but genetically one is homozygous and the other is heterozygous. Note that all people with type O blood are homozygous and that all people with type AB blood are heterozygous (see figure 15.4). Therefore, because there are three alternate forms of the genes occupying the two ABO sites of homologous chromosomes, there is blood type variation in the human population. Likewise, it is the alternate forms of thousands of other human genes that serve as the source of all other variations in the human population. (Another way of saying this is that it is the potentially heterozygous gene pairs that make each one of us different, whereas the genes that are always homozygous are the sources of uniformity among all humans.)

THE CELL CYCLE AND MITOSIS

The Cell Cycle

During fertilization, the sperm and egg cell nuclei fuse, so that chromosomes contributed by the male and female parents are combined in a single nucleus. About 24 hours after fertilization, the zygote begins the first of many cell divisions that will give rise to the billions of cells of the developing embryo and fetus. That is, the zygote divides to form two cells; each of these cells divides, thereby giving rise to four cells; the four cells become eight cells; the eight become sixteen, and so on. The period of time between the completion of one round of division and the completion of the next round is called a **cell cycle**. As illustrated in figure 15.5, the cell cycle is divided into two major phases: the **interphase** and the **mitotic phase.** Each of

these phases is further subdivided into shorter intervals. In interphase, the chromosomes are diffusely spread through the nucleus and cannot be distinguished as individual bodies. During what is called the **S period** ("S" for "synthesis") of the interphase, every gene of every chromosome is replicated, so that the genetic complement of the cell is temporarily doubled. A period of cell growth (called the **G1 period**) precedes the S period, and a second period of cell growth (called the **G2 period**) follows the S period. In the mitotic phase, the chromosomes condense into visible bodies, and through a process described more fully later, become apportioned between the two daughter cells produced by nuclear and cytoplasmic division. As a result of mitosis, each daughter cell receives one copy of every replicated gene. Therefore, the resulting daughter cells are genetically identical.

For the series of cell divisions that immediately follows fertilization of an egg cell, the duration of the cell cycle is quite short. Indeed, the periods that precede and follow the S period (i.e., G1 and G2) are almost nonexistent. However, later in embryonic and fetal development, the length of the cell cycle increases, as more cell growth precedes and follows the S period. There is no cell growth during the mitotic phase.

Mitosis

When the zygote divides, the two daughter cells that are formed are genetically identical. When these two cells divide to form four cells, all four progeny are genetically identical. Indeed, virtually all cell divisions produce genetically identical progeny. The mechanism that acts to ensure that identical genetic

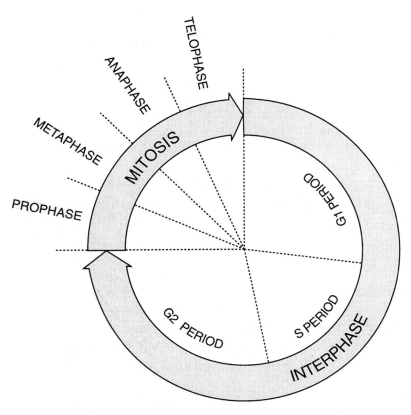

Figure 15.5
The cell cycle. All growing and dividing cells exhibit a cell cycle. The relative lengths of the interphase and mitotic phase vary according to the type of tissue or cell. In this illustration, the length of the mitotic phase is exaggerated for clarity; usually, the mitotic phase accounts for less than 10% of the cell cycle. During the S period, the nuclear genes (DNA) are replicated. The growth periods that immediately precede and follow the S period, called the G1 and G2 periods, vary in length. For the series of cell cycles that immediately follow fertilization, the G1 and G2 periods are very small.

complements are transferred to the two daughter cells produced by division is a process called **mitosis.**

During mitosis, there is a progressive change in the structure and appearance of the chromosomes. Although mitosis is a continuous process, for convenience it may be divided into four major stages called **prophase, metaphase, anaphase,** and **te-** lophase (figs. 15.5 and 15.6). The first stage (i.e., prophase) is characterized by the condensation of the dispersed interphase **chromatin** into visible bodies, namely the chromosomes (figure 15.6, stages 1 and 2), the disappearance of the nuclear envelope, and the formation of **spindle fibers** that extend between the separating **centrosomes**. The dramatic nature of the condensation of

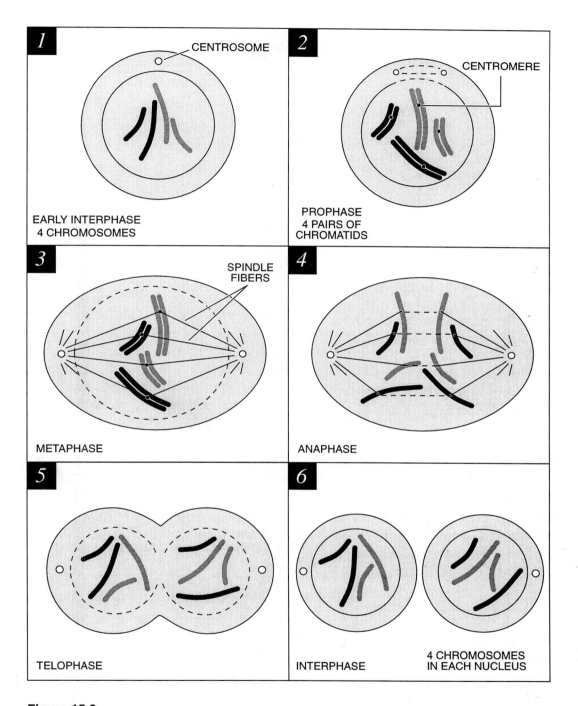

Figure 15.6
Mitosis. Mitosis marks the completion of one cell cycle and the beginning of the interphase of the next cell cycle. Although mitosis is a continuous process, for convenience it may be divided into four phases: prophase, metaphase, anaphase, and telophase. Shown here are the fates of two of the 23 pairs of homologous chromosomes at specific stages of mitosis. For simplicity, interphase chromosomes are shown in their condensed state.

chromatin into visible chromosomes can be appreciated when one considers that in a human cell, the DNA is 10-15 feet long in its dispersed state but condenses to form chromosomes whose combined length is less than 1/25 of an inch.

The chromosomes are distinguishable by light microscopy and are seen to be composed of two sister **chromatids** held together at the **centromere**. The sister chromatids are the products of the replication of chromosomal DNA during the S period of the interphase of the cell cycle. By the end of prophase, the centrosomes have migrated toward diametrically opposite poles of the cell, with spindle fibers extending between the two poles. In metaphase, the chromosomes migrate toward the center of the spindle apparatus (fig. 15.6, stage 3). The centromeres are duplicated, so that each chromatid becomes an independent chromosome and is attached to a spindle fiber connected to one of the two poles.

The onset of anaphase is characterized by the movement of the chromosomes toward opposite poles of the mitotic spindle (fig. 15.6, stage 4). In the final phase of mitosis, the telophase, the chromosomes reach the poles of the spindle and begin to undergo decondensation. During the telophase, a new nuclear envelope is formed enclosing the chromosomes (fig. 15.6, stage 5).

During telophase, a process called **cytokinesis** occurs (fig. 15.6, stage 6) and divides the cell into two daughter cells, thereby physically separating the two complements of chromosomes. Cytokinesis is a process that is distinct from but synchronized with nuclear division.

As a result of mitosis and cytokinesis, one cell (i.e., the "parental cell") has formed two daughter cells that are genetically identical to one another and also are genetically identical to the parental cell (figure 15.6, compare stages 6 and 1).

| 15-2 |

Noncycling Cells

Mitosis and cytokinesis produce the billions of cells of the embryo and fetus. Even after birth, these processes continue at a high rate and provide for the growth of the baby and child. As noted earlier in the chapter, the cells of some tissues continue to grow and divide throughout life in order to replace lost and dying cells. For example, skin cells, the epithelial lining of the digestive and respiratory tracts, and blood cells undergo continuous growth and proliferation by mitosis. Other tissues, such as muscle and nerve, lose their capacity for mitosis within a short time after birth and are not replaced if lost by injury or disease. Finally, the cells of some tissues (e.g., liver tissue cells) may exist in a nongrowing, nondividing state for a long period of time. Such cells are said to be temporarily *noncycling* and have been arrested in the interphase of the cell cycle. Noncycling or arrested cells may ultimately resume growth and division if some of the tissue is lost. For example, the surgical removal of a small portion of the liver is followed by renewed growth and division of the remaining liver tissue, replacing the tissue that had been excised.

TISSUE SPECIALIZATION (SELECTIVE GENE EXPRESSION)

Virtually all of the many billions of cells that comprise the organs of the body are produced by mitosis and cytokinesis. Since these processes produce cells that are genetically identical, it is fair to ask why the cells in different organs of the body may appear and function differently? Indeed, what is it that distinguishes one type of cell from another?

Fundamentally, the functional and structural specificity of a cell is based on the types of *proteins* that it produces. For example, erythrocytes are unique in that they are the only cells of the body to produce the protein *hemoglobin*. The islet cells of the pancreas are unique in that they are the only cells of the body to produce the protein *insulin*. B-lymphocytes are unique in that they are the only cells of the body that produce (and secrete) *antibodies*. Only the liver produces and secretes *albumin* into the blood plasma. Only muscle cells produce and organize vast amounts of the proteins *actin* and *myosin* into the thick and thin filaments that form a sarcomere. Each of these specialized proteins (i.e., hemoglobin, insulin, antibodies, albumin, actin, myosin, etc.) is encoded by a pair (or several pairs) of nuclear genes on homologous chromosomes. We might therefore ask whether in the course of embryonic and fetal growth, specific tissues selectively acquire specific gene pairs. That is, as cell growth and division occur, for example in the pancreas, do the islet cells receive the genes for insulin but fail to receive the genes for hemoglobin, albumin, and other non-pancreatic proteins? If this were true, then mitosis would not produce progeny that are truly genetically identical.

It is now clear that pancreas cells not only have insulin genes, but also possess the genes for hemoglobin, albumin, antibodies, actin, myosin, and many other proteins that are not produced by pancreas tissue. Likewise developing blood cells in the bone marrow

(e.g., erythroblasts) possess the genes for insulin as well as the genes for hemoglobin. Indeed, all cells produced by mitosis possess a full complement of human genes. However, pancreas cells *express* their insulin genes (i.e., they produce insulin) but do not express their hemoglobin genes. Erythroblasts express their hemoglobin genes but do not express their insulin genes. B-lymphocytes express their antibody genes but do not express their hemoglobin genes. Thus, the key to cell and tissue specialization is the **selective expression** of certain genes; that is, the proteins encoded by some of the nuclear genes are manufactured, while other proteins are not produced despite the presence of the genes that encode them.

MEIOSIS

The two cells produced at the conclusion of mitosis are genetically identical to one another, and they are also genetically identical to the parental cell from which they were derived. The divisions of the zygote that produce the billions of cells of the embryo are mitotic divisions, so that all progeny have the same 46 (23 pairs of) human chromosomes. Since the zygote is formed by the fusion of a sperm cell from the male parent and an egg cell from the female parent, it is clear that neither the egg nor the sperm that fertilizes it can possess 46 chromosomes. If the egg and sperm each contained 46 chromosomes, then fertilization would produce a zygote with 92 chromosomes; moreover, all of the cells derived by mitosis from this zygote would contain 92 chromosomes. Such a state would imply that the numbers of chromosomes in the nuclei of human cells ought to double with each generation of human beings. Clearly, this is not the case. Rather, the chromosome number is preserved at 46 from one generation to another.

The human chromosome number remains at 46 because sperm and egg cells are **haploid**. That is, these cells contain only 23 chromosomes (half the number of chromosomes found in other cells). The 23 chromosomes present in sperm and egg cells are not a random half of the normal genetic complement of 46 chromosomes. Rather, every egg and sperm cell contains *one member of each homologous pair of chromosomes*. Consequently, when the nuclei of sperm and egg fuse at fertilization, the resulting zygote acquires 23 *pairs* of homologous chromosomes. In each homologous pair, one chromosome is derived from the male parent and the other chromosome is derived from the female parent.

Thus, sperm cells and egg cells are unique in the nature of their genetic complements. These cells are produced only in the reproductive tissues (ovaries of females and testes of males). Whereas all other cells of the body are produced by mitosis, sperm and egg cells are produced by a related process called **meiosis**.

An in-depth discussion of meiosis on a cellular as well as a genetic basis is beyond the scope of this book and is normally treated at length in textbooks of genetics. Therefore, we will limit our discussion of meiosis to a description of some of the major meiotic events and their implications.

The goal of meiosis is to produce cells (*gametes*) that are haploid. This is achieved because the paired chromatids of the cell embarking on this process are apportioned among *four* daughter cell nuclei, each nucleus acquiring half the number of chromosomes of the diploid parental cell. Although the resulting cell nuclei contain only half the diploid number of chromosomes, the chromosome set is genetically complete, *because each nucleus acquires one member of each pair of homologous chromosomes*. The two sequential divisions producing first 2 and then 4 cells are called **meiotic division I** and **meiotic division II**. Like mitosis, the meiotic divisions are further subdivided into 4 phases. For meiotic division I, these are **prophase I**, **metaphase I**, **anaphase I**, and **telophase I**; for meiotic division II, they are **prophase II**, **metaphase II**, **anaphase II**, and **telophase II**. The events that take place in several of these phases are depicted in figure 15.7. The homologous chromosomes are *assorted randomly* during anaphase I (see below), and this accounts in part for the genetic variation that characterizes the human population. Additional genetic variation occurs during metaphase I as a result of a phenomenon called **crossing-over**; during crossing-over, genes occupying equivalent sites are exchanged between homologous pairs of chromatids. The genetic implications of random assortment and crossing-over are principal subjects of genetics courses. For simplicity, crossing-over is not depicted in figure 15.7.

Figure 15.7

Meiosis (opposite page). During the two rounds of nuclear division that characterize meiosis, a diploid cell gives rise to four haploid cells. Depicted here are the fates of 4 (2 homologous pairs) of the 46 chromosomes present in the parental cell. In the first round of division, homologous chromosomes are separated and drawn into separate nuclei. In the second round of division, paired chromatids are separated and drawn into separate nuclei. (For simplicity, interphase chromosomes are shown in the condensed state.)

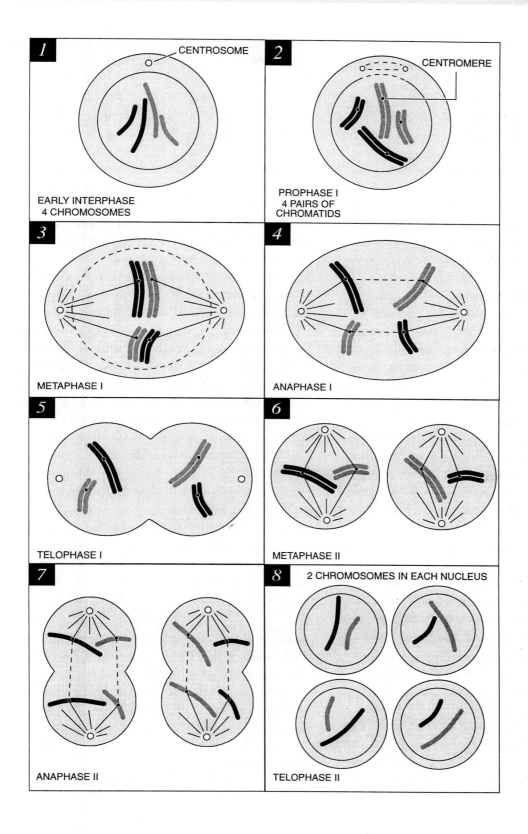

1 CENTROSOME

EARLY INTERPHASE
4 CHROMOSOMES

2 CENTROMERE

PROPHASE I
4 PAIRS OF
CHROMATIDS

3 METAPHASE I

4 ANAPHASE I

5 TELOPHASE I

6 METAPHASE II

7 ANAPHASE II

8 2 CHROMOSOMES IN EACH NUCLEUS

TELOPHASE II

As noted earlier, meiosis involves two successive rounds of nuclear division, each round subdivided into its own prophase, metaphase, anaphase, and telophase. In prophase I (figure 15.7, stage 2), the chromosomes become visible as condensation of the chromatin begins. As in mitosis, each chromosome can be seen to consist of two chromatids (resulting from replication of all genes during the S period of the interphase that precedes the first round of division). In contrast with the events that characterize mitosis, homologous chromosomes (i.e., homologous pairs of chromatids) become aligned side-by-side so that genes encoding products of similar or identical function are situated adjacent to one another. This phenomenon is called **synapsis**. At this point, genes may be exchanged between homologous segments of adjacent chromatids via the mechanism called *crossing-over*. When crossing-over involves a pair of homozygous genes, no change in the genetic makeup of either chromatid results (i.e., identical genes have been exchanged). However, when crossing-over involves a heterozygous gene locus, the gene sequences of both participating chromatids are altered. Consequently, crossing-over creates new combinations of genes in homologous chromosomes. (Crossing-over is *not* depicted in figure 15.7!)

In meiotic metaphase I (fig. 15.7, stage 3), the nuclear envelope disappears and the spindle apparatus forms, much as it does in mitosis. The paired chromatids attach via their centromere to the spindle fibers arising from opposite poles of the cell.

In meiotic anaphase I , (fig. 15.7, stage 4), homologous chromosomes (but not sister chromatids) separate from each other and move along the spindle fibers to opposite poles of the spindle. (In the metaphase of mitosis, *sister chromatids are separated* and move to opposite poles; this is an important distinction between meiosis and mitosis.)

Meiotic telophase I (fig. 15.7, stage 5) brings the first round of meiotic division to a conclusion as the chromosomes aggregate at their respective poles. The period between the end of telophase I and the onset of meiotic prophase II is usually quite short. In contrast to the events characterizing mitotic interphase, the interphase between the two rounds of meiosis is not accompanied by the replication of the genetic material.

The events characterizing meiotic prophase II are similar to mitotic prophase, although each cell nucleus has only half the number of chromosomes as does a cell in mitotic prophase. Each chromosome remains composed of sister chromatids formed in the interphase that preceded meiotic prophase I.

In meiotic metaphase II (fig. 15.7, stage 6), the paired chromatids migrate to the center of the newly forming spindle apparatus. Then, in meiotic anaphase II (fig. 15.7, stage 7), sister chromatids separate from one another and move along the spindle to opposite poles of the cell. Finally, in meiotic telophase II, the separated chromosome groups are enclosed in a newly developing nuclear envelope (fig. 15.7, stage 8) and begin decondensation. Meiosis ends with cytokinesis, which yields 4 separate cells.

15-3

What has taken place during meiosis can be summarized thusly: (1) a single parental cell has given rise to 4 daughter cells; (2) each daughter cell has one-half the number of chromosomes as the parental cell; (3) each daughter cell has one copy of every gene locus; and (4) as a result of random assortment and crossing over, all of the daughter cells have different genetic compositions.

SPERMATOGENESIS AND OOGENESIS

In both males and females, gene replication followed by the two division rounds of meiosis converts a single diploid cell into four haploid cells. In males, the process occurs in the testes and is called **spermatogenesis**; in females, the process occurs in the ovaries and is called **oogenesis**.

Spermatogenesis

Spermatogenesis occurs in the **seminiferous tubules** of the testes by the meiotic division of cells that are part of the tubules' walls. These cells are called **spermatogonia** (see figures 14.17 and 15.8). Prior to meiotic prophase I, the genes and chromosomes of the spermatogonia are replicated, thereby forming cells that are called **primary spermatocytes**. Each chromosome of a primary spermatocyte consists of a pair of chromatids. Meiotic division I converts each primary spermatocyte into two **secondary spermatocytes**. Meiotic division II then converts each secondary spermatocyte into two haploid **spermatids**. Therefore, four spermatids are produced by the meiotic divisions of one spermatogonium. Each spermatid undergoes a series of morphological changes in which most of the cytoplasm is lost and the cell develops a tail-like **flagellum** that will act to propel the cell once it is activated inside of the female reproductive tract. The mature gamete is called a **spermatozoan** or **sperm cell**. The meiotic changes of the spermatogonium and the maturation of spermatids into sperm cells are assisted by the actions of another type of cell in the walls of the seminiferous tubules; these cells are called **Sertoli cells** (see figure 14.17).

The important stages of spermatogenesis are depicted in figure 15.8. For simplicity only four of the spermatogonium's 46 chromosomes are followed through meiosis. To distinguish the members of a homologous pair, one partner is depicted in black and the other in grey. Prior to the first meiotic division of spermatogenesis, genes present in one of the two sister chromatids comprising each chromosome are exchanged with corresponding genes in one of the sister chromatids comprising the homologous chromosome (i.e., crossing-over occurs). Exchanges involving heterozygous gene pairs are a source of genetic variety among the spermatids that are formed. Additional genetic variation results from random assortment of homologous chromosomes among the daughter nuclei. Thus, the four sperm cells produced by meiosis of one spermatogonium are genetically different.

All four of the sperm cells produced by meiotic division are **viable**; that is to say, all are potentially capable of fertilizing an egg cell. In contrast (see later), meiosis in the female produces a single viable gamete (the egg cell) and two (or three) nonviable (nonfunctional) cells called **polar bodies**.

In males the conversion of spermatogonia into primary spermatocytes begins as early as during embryonic development. However, meiosis proceeds no further until the male attains puberty. The initial stages of spermatogenesis appear to be promoted by the hormone testosterone, whereas later stages of spermatogenesis are promoted by follicle-stimulating hormone (see chapter 14). Once spermatogenesis begins in the male at puberty, it takes place continuously for the next 50 or more years.

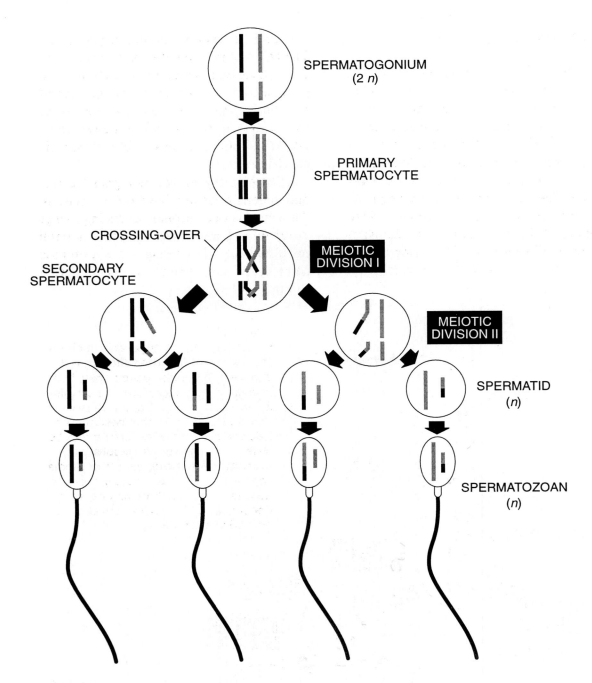

Figure 15.8
Spermatogenesis. During spermatogenesis, meiosis produces 4 haploid, viable sperm cells from a single diploid spermatogonium. For simplicity, the fates of only 2 of the 23 pairs of homologous human chromosomes are followed here. Notice that during meiotic division I, crossing-over results in an exchange of homologous chromosome segments. Although only one crossover point is depicted here, there may be many crossovers. Notice also that all 4 sperm cells are genetically different. For simplicity, the nuclear envelope is omitted in this drawing.

Oogenesis

Oogenesis, the production by meiosis of haploid egg cells or **ova** in the ovary, is depicted in figure 15.9. Although the process is fundamentally similar to spermatogenesis, there are some important differences. Oogenesis begins with diploid cells called **oogonia** that are first converted into **primary oocytes**. The primary oocytes then begin the first meiotic division. This division is said to be "unequal" in that one of the two daughter cells receives nearly all of the cytoplasm of the parental cell, while the other daughter

cell receives very little cytoplasm (fig. 15.9). The cell that receives the bulk of the cytoplasm is called a **secondary oocyte** and will give rise at the completion of the second meiotic division to a functional egg cell or ovum. The other cell, called a **first polar body**, will not produce a viable ovum (it is functionless).

The secondary oocyte undergoes the second meiotic division; like the first division, the second is also unequal, producing a large daughter cell and a nonfunctional **second polar body**. The large daughter cell, which is haploid, has the capacity to mature into a

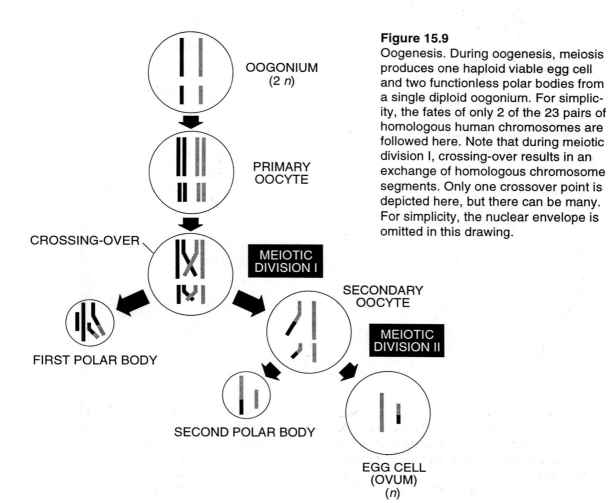

Figure 15.9
Oogenesis. During oogenesis, meiosis produces one haploid viable egg cell and two functionless polar bodies from a single diploid oogonium. For simplicity, the fates of only 2 of the 23 pairs of homologous human chromosomes are followed here. Note that during meiotic division I, crossing-over results in an exchange of homologous chromosome segments. Only one crossover point is depicted here, but there can be many. For simplicity, the nuclear envelope is omitted in this drawing.

OOGONIUM
(2 *n*)

PRIMARY
OOCYTE

CROSSING-OVER

MEIOTIC
DIVISION I

FIRST POLAR BODY

SECONDARY
OOCYTE

MEIOTIC
DIVISION II

SECOND POLAR BODY

EGG CELL
(OVUM)
(*n*)

viable (potentially fertilizable) egg cell. Therefore, whereas meiosis in males produces four viable gametes, meiosis in females produces one viable gamete and two functionless polar bodies. (The first polar body is capable of completing a second round of division; if this occurs, there will be one viable gamete and *three* polar bodies.)

During the first meiotic division of oogenesis, crossing-over occurs between one of the two sister chromatids comprising each chromosome and one of the two chromatids comprising the homologous chromosome. Pairs of sister chromatids and whole chromosomes are also randomly assorted among the two progeny of each division. Thus, like sperm, the egg cell and polar bodies are genetically different.

Oogenesis gets underway during the embryonic development of a female, so that at birth the ovaries of the average female contain hundreds of thousands of primary oocytes. Each primary oocyte is enclosed by a layer of supportive cells, thereby forming a **primary follicle** (see chapter 14). No further development of follicles or egg cells occurs until puberty.

At puberty, under the influence of pituitary hormones, the female begins her succession of ovarian (or menstrual) cycles. During each ovarian cycle, meiosis is resumed by the prospective ovum of the **Graafian follicle** (see chapter 14), and by the time of ovulation meiosis has proceeded to the secondary oocyte stage. Following ovulation, meiosis progresses to meiotic metaphase II but is arrested at this stage. The second round of meiotic division and cytokinesis are completed only if the ovum is fertilized by a sperm cell. If the egg cell is not fertilized, it is absorbed and digested by the lining of the oviduct (see later).

ORGANIZATION OF THE FEMALE AND MALE REPRODUCTIVE TRACTS

Female Reproductive Tract

In females, the organs of the reproductive system remain in an infantile condition until the onset of puberty, which occurs around age 13. The anatomical and chemical changes that occur during puberty are believed to be initiated by events taking place in the brain, although the precise nature of these events remains obscure. It is apparent that well before the onset of puberty, the ovaries are already in a state capable of releasing egg cells. This has been amply demonstrated through experiments in which injections of pituitary gonadotropic hormones into immature laboratory animals trigger the early onset of puberty, the initiation of the ovarian cycle, and the secretion of the ovarian hormones. Whereas, the pituitary glands of immature animals contain these gonadotropins, under normal conditions the release of these hormones is somehow deferred.

In females, the succession of ovarian cycles continues on a regular basis until the age of 45 to 55 years, when the cycle starts to become irregular and eventually ceases altogether. This is known as the **menopause** and may extend over a period of several years. (In males, there is no dramatic change in reproductive function; instead, testicular function slowly declines.) During the menopause, sensitivity of the ovaries to pituitary gonadotropins disappears, and there is no further development of ovarian follicles. The halt to follicle development is accompanied by a strong reduction in the secretion of estrogen.

The organization of the reproductive tract

of an adult female is depicted in figure 15.10 (see also figure 15.12). The major organs of the tract are the **vagina**, **uterus**, **Fallopian tubes** (or **oviducts**) and the **ovaries**. The vagina is a passageway that extends upwardly from the body surface into the thick-walled and muscular uterus. At its junction with the uterus, the vagina forms the **cervix**. Entering the uterus on either side from above are the Fallopian tubes (or oviducts). Each of these narrow channels begins near the outer edge of an ovary, curves over the ovary's upper surface and then descends toward the uterus.

The vagina acts to receive sperm from the male during **intercourse** (also called **copulation** or **coitus**). From the vagina, the sperm make their way through the uterus and into the Fallopian tubes. If the tubes contain an egg cell recently released from the ovary (see below), and if the sperm successfully fertilize the egg, the resulting zygote is swept toward the uterus and after several days in the uterus, becomes implanted in the uterine wall. The growth and development of the embryo and fetus take place within the uterus.

15-4

Male Reproductive Tract

The organization of the male reproductive system is depicted in figure 15.11. Sperm are carried out of each of the **testes** by narrow

Figure 15.10
Side view of the organization of the female reproductive tract. Associated with each ovary (only the right ovary is depicted here) is an oviduct (or Fallopian tube), which conducts ova (egg cells) from the ovary toward the uterus. The vagina leads from the uterus to the body surface.

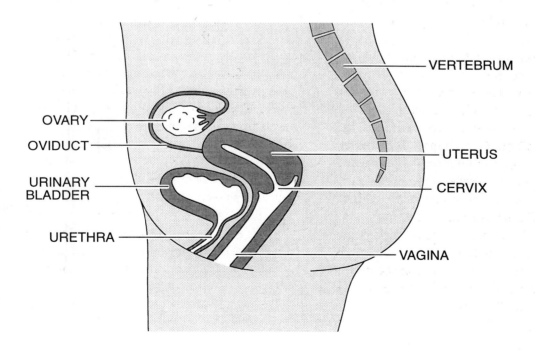

tubes called **vas deferens** (it is these tubes that are either tied or cut when a male is sterilized). The vas deferens arch upward and around the urinary bladder. Just beneath and behind the bladder, the two vas deferens merge with channels exiting the paired **seminal vesicles**. The two (left and right) common channels that are formed in this way are called **ejaculatory ducts**. The ejaculatory ducts enter the **prostate gland** and merge with the **urethra** as it exits from the undersurface of the **urinary bladder**. The common channel that is formed by the merging of these tubes retains the name *urethra* Below the prostate gland lies a pair of small glands called **bulbourethral glands**; their secretions also enter the urethra. Collectively, the sperm cells that are released from

the testes and the secretions of the seminal vesicles, prostate gland, and bulbourethral glands create the sperm suspension called **semen**. During **ejaculation**, semen is carried through and out of the penis by the **urethra**.

The secretions of the seminal vesicles (i.e., *seminal fluid*) contain a rich supply of the sugar *fructose*. It is believed that the fructose serves a nutrient role for the sperm cells following ejaculation. More than half the volume of semen released during ejaculation is represented by seminal fluid. The prostate gland produces an alkaline fluid (called *prostatic fluid*) that acts to neutralize the acidic sperm suspension that is released through the vas deferens and also neutralizes acidic vaginal secretions of the female.

Figure 15.11
Side view of the organization of the male reproductive tract.
See the text for details.

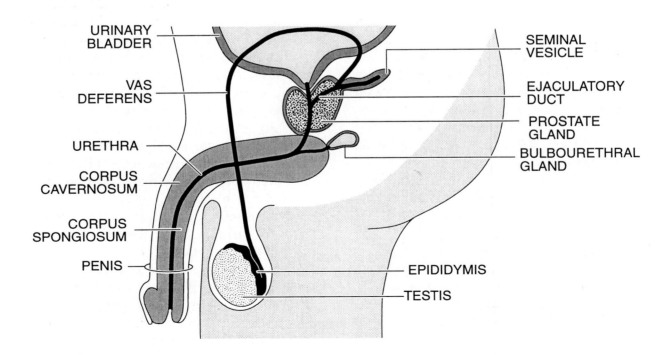

In a normal male, the influx of blood into the **corpus cavernosum** (on the dorsal side of the penis) and the **corpus spongiosum** (on the ventral side of the penis) cause the penis to become *erect* and this facilitates intercourse. During intercourse, forcible contractions of smooth muscle tissue in the walls of the vas deferens push sperm into the ejaculatory duct. Seminal fluid driven from the seminal vesicles by contractions of the vesicles' walls and similar contractions of the prostate gland flush the semen through the urethra, out of the penis, and into the vagina.

15-5

In the average adult male, the ejaculated semen contains about 120,000,000 sperm cells per c.c. Since the ejaculate is about 3.5 c.c. in volume, this means that altogether about 420,000,000 sperm cells are released during a single *emission*. Because the mortality rate of the sperm once they are within the female reproductive tract is so high, such large numbers of sperm cells are required for fertilization of an egg cell to be likely (even though only one sperm cell fertilizes an egg). (It is for this reason that males whose sperm counts are below 20,000,000 per c.c. are likely to be infertile.) Of the millions of sperm cells that may enter the female reproductive tract during intercourse, only a few hundred survive to reach an egg cell in an oviduct.

Sperm cells are stored in the epididymis and vas deferens and can remain alive within the male reproductive tract for several weeks. However, once emitted in the semen, their life span is about three days at body temperature (i.e., at the temperature of the female reproductive tract). Sperm may be kept alive for months or even years in semen that has been frozen and stored at very low temperatures.

FERTILIZATION AND IMPLANTATION

The egg cell that is released from the ovary on day 14 of the ovarian cycle is drawn by fluid currents into a Fallopian tube. These currents are created by the beating of cilia that cover the finger-like projections (**fimbriae**) that exist at the mouth of the Fallopian tube (fig. 15.12, stage 1). The egg cell then begins a 3-day journey along the Fallopian tube toward the uterus (figure 15.12, stages 2, 3, and 4). Since the egg cell has no independent means of locomotion, it relies on the beating cilia of the cells that line the oviduct in order to be moved along.

15-6

Following ovulation, the egg cell remains viable for 12 to 24 hours. Since the journey to the uterus requires about three days, this implies that the egg must be fertilized somewhere between the mouth of the oviduct and approximately the first one-third of the oviduct's length (see figure 15.12). If the egg cell is not fertilized, it is absorbed and destroyed by the cells lining the oviduct.

When sperm are deposited in the vagina during intercourse, they propel themselves into the uterus and into the Fallopian tubes. Because sperm cells are motile and can move relatively quickly (about 5 inches per hour), they can reach the mouth of each of the Fallopian tubes within a few hours. (There is some evidence that sperm may reach the end of the oviduct within 30 minutes. This is believed to result from contractions of the uterus or "backward" currents created in the oviducts by ciliary beating.)

After being deposited in the female reproductive tract, sperm remain viable for about three days. This implies that pregnancy is possible during an interval of about four days during each ovarian cycle. In the average

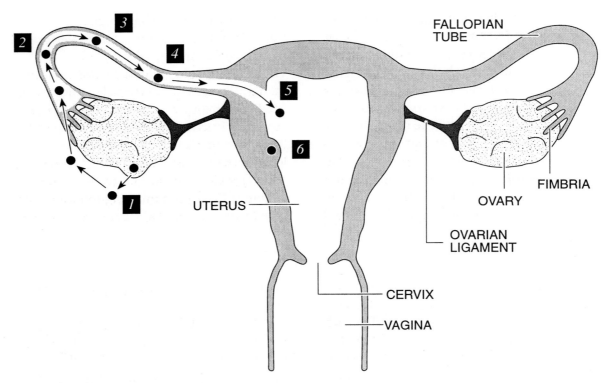

Figure 15.12
Front view of the female reproductive tract. 1, 2, 3, 4, 5, and 6 show
the egg cell (zygote) as it progresses toward (and is eventually
implanted in the wall of) the uterus (see also figure 14.13).

female, this "fertile interval" extends from about day 12 until day 15. For example, if sperm are deposited in the female reproductive tract on day 12, they will quickly reach the ends of the oviducts, where they will remain viable (capable of fertilizing an egg cell) until ovulation occurs on day 14. In such an instance, the egg cell will be fertilized almost immediately following ovulation. On the other hand, if sperm enter the female reproductive tract on day 15, they may reach and fertilize the egg cell while it is in the lingering hours of its viability.

During fertilization, the sperm cell penetrates the corona radiata and zona pellucida of the egg and enters the egg cell's cytoplasm

(see figures 15.1 and 15.13, stages 1 and 2). Fertilization triggers the completion of the egg cell's second round of meiosis, thereby producing a second polar body (only the first polar body is shown in figure 15.13). The sperm cell nucleus and egg cell nucleus then fuse to form a single nucleus. Because the nuclei of the two gametes are haploid, their fusion creates a diploid nucleus. The 23 chromosomes provided by the sperm cell represent one of each of the 23 homologous pairs originally present in the spermatogonium. By the same token, the 23 chromosomes provided by the egg cell nucleus represent one of each of the 23 homologous pairs originally present in the oogonium.

Therefore, the fusion of the sperm and egg cell nuclei at fertilization produces a **zygote** whose nucleus contains 23 *pairs* of homologous chromosomes.

Soon after the two nuclei have fused, the chromosomes are replicated, and after about 24 hours, the first mitotic division is initiated. This division produces two diploid cells (fig. 15.13, stage 3). A second round of chromosome replication quickly follows, and about 10 hours later, a second mitotic division. As a result, there are now four diploid cells. The rounds of chromosome replication and mitosis continue as the fertilized egg slowly proceeds toward the uterus.

Because the dividing zygote does not grow between successive divisions, the cells produced by these rounds of mitosis are consecutively smaller and smaller. (Growth in size does not begin until implantation in the wall of the uterus, where the rich supply of blood provides the raw materials needed for growth.) Continued mitosis produces a ball of cells, called a **morula**, that is no greater in

Figure 15.13
Developmental stages of the egg cell that follow ovulation. The six numbered stages that are represented here correspond to the six stages shown in figure 15.12.

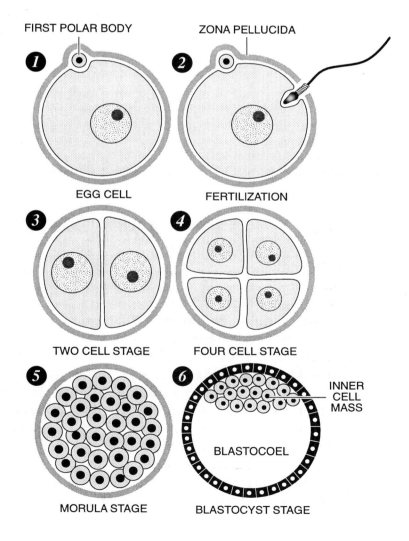

FIRST POLAR BODY

ZONA PELLUCIDA

EGG CELL

FERTILIZATION

TWO CELL STAGE

FOUR CELL STAGE

MORULA STAGE

BLASTOCYST STAGE

INNER CELL MASS

BLASTOCOEL

size than the original zygote (fig. 15.13, stage 5). By the time the uterus is reached, the morula is transformed through rearrangements of the positions of the cells into a **blastocyst**. The organization of a blastocyst is depicted in figure 15.13, stage 6. As seen in the figure, the blastocyst consists of an outer coat of cells called **trophoblast** cells and an **inner cell mass** facing a fluid-filled chamber (i.e., the **blastocoel**). It is the inner cell mass that will give rise to the embryo proper. The blastocyst remains in this state within the uterus for several days before **implantation** occurs. On about the seventh day fol-

lowing ovulation (corresponding to day 21 of the ovarian cycle), the blastocyst attaches to the wall of the uterus.

15-7

EARLY EMBRYONIC AND FETAL DEVELOPMENT

Enzymes secreted by the trophoblast cells digest and liquefy the **endometrial lining** of the uterus, and the blastocyst buries itself in the uterine wall (fig. 15.14). By about the 12th day (following ovulation), the blasto-

Figure 15.14
Formation of the three primary germ layers from the inner cell mass during the early days of embryonic development.

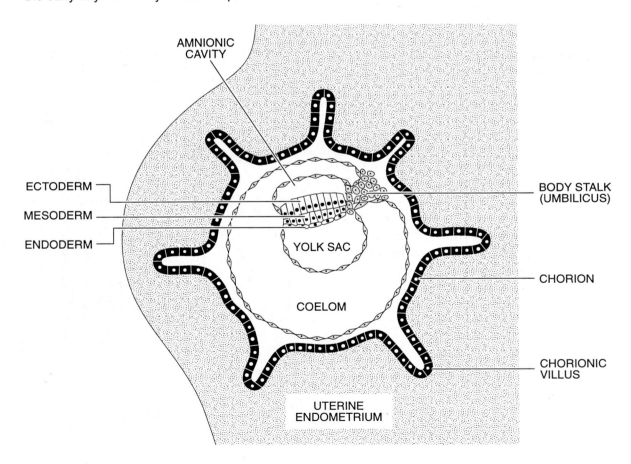

cyst is fully embedded in the endometrium. The trophoblast cells now form the **chorion**, which sends out fingerlike projections, called **chorionic villi**, into the endometrium (fig. 15.14). Nutrients are acquired from the solubilized endometrium by the villi and provide the raw materials needed for growth and development of the early embryo. Eventually, the chorionic villi facing the lumen of the uterus will disappear, whereas those of the opposite face will further differentiate, eventually contributing to the **placenta** (the interface between the embryonic and maternal circulations and through which nutrients and wastes are exchanged between the two bloodstreams).

The inner cell mass gives rise to the three **primary germ layers** from which all body parts are derived; these are the **ectoderm**, **mesoderm**, and **endoderm** (fig. 15.14). The ectoderm will give rise to the skin, hair, nails, nervous system, receptor cells of the sense organs and the lining of the mouth and anus. The mesoderm will give rise to the skeleton, muscles, blood, blood vessels, lymphatics, and the epithelium of the kidneys. The endoderm will give rise to the epithelial lining of the digestive tract and its associated structures, the lining of the urinary bladder, and the lining of the trachea and lungs.

Also formed in the early embryo are the **yolk sac** and the **amnion** (fig. 15.14). The yolk sac is a yolk-filled pouch derived from endoderm. In many animal species (e.g., birds), this structure plays an important role as a source of nourishment for the developing embryo. In humans, nourishment is derived from the maternal blood, and so the yolk sac has little function and undergoes little further growth. Eventually, the yolk sac is included in the **umbilical cord**.

The amnion is a fluid-filled sac of modest size in the early embryo (fig. 15.14) but which progressively grows to completely enclose the embryo (and fetus; see figure 15.15). By acting as a cushion, the amnion plays a protective role for the developing embryo. It also serves as a reservoir into which metabolic wastes are discharged.

The term "embryo" is usually used when referring to the future baby's first eight weeks of intrauterine development; after this time, and for the remainder of the pregnancy, the prospective baby is referred to as a **fetus**. Small numbers of embryonic and fetal cells are sloughed into the amnionic fluid. Because these cells are genetically identical to those of the embryo, much can be learned about the future child from an examination of the cells. For example, from the chromosomal com-

Figure 15.15
The embryo at about six weeks of development. Note that the amnion has grown to completely encompass the embryo.

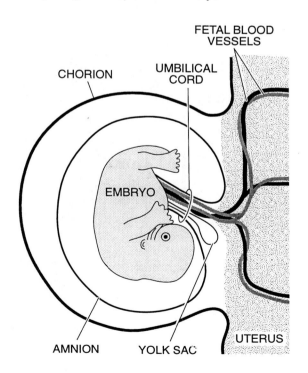

plement of the cells, it is possible to tell the baby's sex. Certain potential genetic abnormalities (e.g., Down's syndrome) can also be predicted by such analyses. The procedure in which a small amount of amnionic fluid containing fetal cells is withdrawn from the mother for such analyses is called **amniocentesis**. Normally, amniocentesis is not performed until the 14th week of pregnancy.

LATE FETAL DEVELOPMENT AND PARTURITION

Figure 15.16 shows the state of fetal development by about the 16th week. By this time, there has been considerable growth of the fetus. The size of the uterus also increases in order to accommodate the growing fetus. Note that the amnion and chorion merge, forming a single, thicker membrane. Arterial blood leaves the fetus through the **umbilical arteries** and enters capillary beds in the placenta's chorionic villi . As the fetal blood flows through these capillaries, there is an exchange of fetal metabolic wastes and carbon dioxide for nutrients and oxygen in the mother's blood. Nutrient-rich and oxygen-rich blood then returns to the fetus through the **umbilical vein**. It is important to recognize that the fetal blood and mother's blood do not mix. Rather, the two bloodstreams are separated by capillary walls. The exchanges that take place between the two bloodstreams occur primarily by diffusion.

15-8

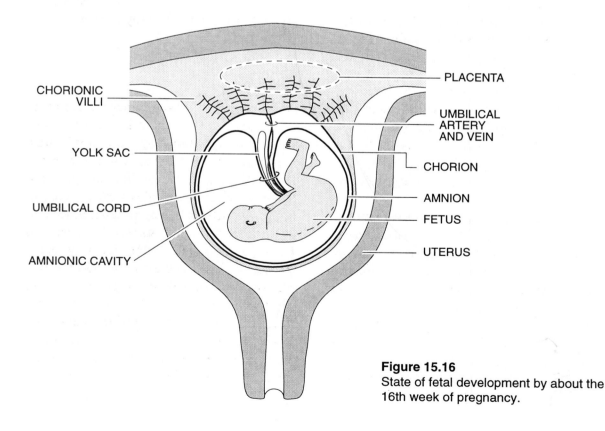

Figure 15.16
State of fetal development by about the 16th week of pregnancy.

CHORIONIC VILLI

YOLK SAC

UMBILICAL CORD

AMNIONIC CAVITY

PLACENTA

UMBILICAL ARTERY AND VEIN

CHORION

AMNION

FETUS

UTERUS

Counting from day 1 of the ovarian cycle in which fertilization occurred, pregnancy lasts about 39 weeks (9 months). As noted earlier, the first eight weeks constitute the period of "embryonic development," whereas the last 31 weeks constitute the period of "fetal development." At the end of the 39 weeks of *gestation*, several changes occur in the hormone levels of the mother's bloodstream: (1) there is an increase in the **estrogen:progesterone** ratio, (2) **oxytocin** is secreted by the posterior lobe of the pituitary gland, and (3) the uterus releases **prostaglandins**. These changes induce vigorous contractions of the uterus' muscular walls, and these contractions herald the onset of **parturition**–the birth of the baby.

As the end of gestation approaches, there is a dramatic rise in the number of oxytocin receptors in the surfaces of the muscle cells of the uterus. This change is reflected by an increase in the uterus' contractility, amplifying its responsiveness to oxytocin. The increased contractility of the uterus may also be attributed to the mechanical stretching of the uterus that accompanies fetal growth in the final weeks of pregnancy and to the increasing estrogen:progesterone ratio of the mother's bloodstream (estrogen is known to increase the contractility of the uterus, whereas progesterone reduces uterine contractility).

At the blastocyst stage of development, the trophoblast cells (which form the chorion of the early embryo) secrete a hormone that prevents menstruation and favors implantation of the blastocyst in the uterine wall and development of the early embryo. This hormone, called **human chorionic gonadotropin**, acts in the same manner as luteinizing hormone (see chapter 14), in that it promotes the growth and maintenance of the ovary's corpus luteum (which is the source of estrogen and, more importantly at this time, progesterone).

Despite its importance as a source of progesterone during the early weeks of pregnancy, the corpus luteum begins to break down within the ovary and by the 6th week ceases hormone secretion. However, at the same time that hormone secretion by the corpus luteum declines, the developing placenta begins to secrete estrogen and progesterone. Indeed, the placenta secretes ever-increasing amounts of these hormones as the pregnancy evolves (amounts far greater than those that were provided by the corpus luteum).

Just prior to parturition, placental progesterone secretion falls, thereby increasing the estrogen:progesterone ratio of the mother's blood. This change, coupled with the release of oxytocin and prostaglandin noted earlier, promote *labor* and the baby's birth.

TWINS

About one in every 90 pregnancies results in the birth of twins. There are two types of twins: **fraternal twins** and **identical twins**. Fraternal twins result from the rare occurrence in which two mature eggs are released from the ovaries during a single ovarian cycle and both are fertilized by sperm cells. Since two zygotes are formed, fraternal twins are also known as **dizygotic twins**. Each zygote is transformed into a blastocyst, and the two blastocysts independently implant in the wall of the uterus. Each of the twins becomes encapsulated in its own amnionic cavity and has its own placenta. Because they are formed from separate zygotes, fraternal twins are genetically no more alike than are any other family siblings. Accordingly, they may or may not be of the same sex.

Identical twins are genetically identical (which of course implies that they are the same sex). Identical twins arise when the inner cell mass of the blastocyst separates into two separate masses of cells. Each cell mass develops into a complete embryo and fetus. Depending upon *when* the separation of the two masses of cells occurs, the two fetuses may share an amnionic cavity and placenta or they may have separate amnionic cavities and placentas (the later the separation, the more likely they will share a single amnion and placenta). Because identical twins develop from a single zygote, they are also called **monozygotic twins**.

Natural multiple births (triplets, quadruplets, etc.) and those resulting from the use of fertility drugs are rarely monozygotic. It is more likely that several mature ova are released from the ovaries during a single ovarian cycle.

SOME FUNDAMENTAL PRINCIPLES OF INHERITANCE

Inheritance of ABO Blood Type

As noted earlier in this chapter and also in chapter 9, there are four different human blood types corresponding to the ABO antigen series. These are blood types A, B, AB, and O. Which one of these blood types a given individual has is determined genetically and is called the **phenotype**. A person's phenotype is readily identified by a clinical blood test. We also saw earlier in this chapter that there are three different forms of the genes that encode the antigen-forming *glycosyltransferase* enzymes. Gene A en-

codes the enzyme *acetylgalactosamine transferase* which produces antigen A; gene B encodes *galactosyltransferase,* which produces antigen B; and gene O encodes an inactive enzyme that produces no clinically detectable product. Because there is more than one form of the *glycosyltransferase* gene, the ABO series genes belong in the class of potentially *heterozygous* genes. Since there are three variations of the gene, there are six possible **genotype** variations corresponding to the four phenotypes (see figure 15.4). That is, a person whose is phenotypically blood type A may be genotypically AA (homozygous) or genotypically AO (heterozygous). A person whose is phenotypically blood type B may be genotypically BB (homozygous) or genotypically BO (heterozygous). All persons with blood type AB are heterozygous, and all persons with blood type O are homozygous (i.e., they are OO).

Now, consider a male who has type AB blood. In all of his diploid cells (including the spermatogonia in his testes), the homologous pair of chromosomes carrying the ABO series genes would reveal the A gene on one member of the pair and the B gene on the other member. When sperm cells are produced by spermatogenesis, only one member of each homologous pair of chromosomes ends up in each sperm cell. Therefore, half of the sperm that he produces will have the A gene-containing homologue, and the other half of his sperm will have the B gene-containing homologue. Consequently, depending on which of the two types of sperm fertilizes an egg cell, he will either pass on the A gene to the zygote or will pass on the B gene. The same rules apply to a female who has type AB blood. Oogenesis will produce two types of egg cells with respect to the ABO series genes: A gene-containing egg cells and B gene-containing egg cells.

To illustrate the fundamental rules of inheritance, let's explore the various ABO blood types that are possible in the children born to parents who have type AB blood. The chances that a zygote is formed from an A gene-containing sperm are 50:50 (i.e., the odds are 0.5), and the chances that the zygote is formed from an A gene-containing egg are also 50:50. Therefore, the chances of the zygote getting an A gene from both parents are 1:4 (i.e., 0.5 X 0.5 = 0.25). The same odds favor formation of a zygote with B genes from both parents. However, the odds of a zygote getting an A gene from one parent and a B gene from the other parent are 1 in 2 (i.e., 50:50). This is because there are two different ways in which this can occur; namely, (1) if the zygote receives an A gene from the sperm and a B gene from the egg, and (2) if the zygote receives a B gene from the sperm and an A gene from the egg. Thus, from a statistical point of view, one-half of the children born to type AB parents will have AB blood, one-fourth will have type A blood, and the remaining one-fourth will have type B blood.

All of these conclusions are more easily drawn and understood, if we use what is called a **Punnett square** to explore the potential genotypes of the zygotes that are formed at fertilization. Such a Punnett square is shown in figure 15.17. The male phenotype and genotype are shown to the left of the square and in the neighboring "subsquares" are placed the letters corresponding to the two types of sperm cells that are produced. Above the Punnett square are identified the phenotype and genotype of the female parent. In the subsquares below are placed the letters corresponding to the two types of egg

cells that are produced. Finally, in each of the remaining four "offspring squares" are placed two letters corresponding to the gene donated by the egg (above) and the gene donated by the sperm (to the left). For our example in which both parents have type AB blood, one of the four offspring squares has the genotype AA, two have the genotype AB, and one has BB. | 15-9 |

Figure 15.18 is a Punnett square showing all of the possibilities that exist among the offspring of an heterozygous type A father and an heterozygous type B mother. | 15-10 |

Finally, figure 15.19 is a Punnett square showing all of the possibilities that exist among the offspring of a type AB father and a type O | 15-11 |

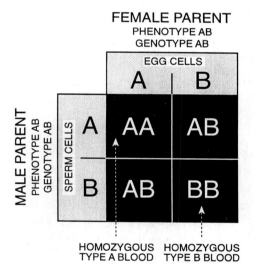

FEMALE PARENT
PHENOTYPE AB
GENOTYPE AB

Figure 15.17
A Punnett square showing the genotypes that are possible among the zygotes formed by parents with type AB blood. See text for explanation of how to use a Punnett square.

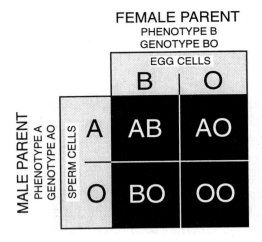

FEMALE PARENT
PHENOTYPE B
GENOTYPE BO

Figure 15.18
Punnett square showing the genotypes possible among the zygotes formed by parents with heterozygous type A blood and heterozygous type B blood.

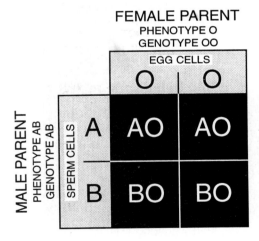

FEMALE PARENT
PHENOTYPE O
GENOTYPE OO

Figure 15.19
Punnett square showing the genotypes possible among the zygotes formed by parents with type AB blood and type O blood. Notice that none of the children will have the same blood type as either parent.

mother. Notice that in this case, none of the zygotes (and, therefore, none of the children) will have the same blood type as their parents.

Rhesus Blood Type (Rhesus Factor)

The **Rhesus factor** is another example of an inherited red blood cell antigen. The Rhesus antigen is named after the *Rhesus* monkey (in which the antigen was first discovered). Every person is either *Rhesus positive* (usually symbolized **Rh⁺**) or *Rhesus negative* (usually symbolized **rh⁻**). Positivity and negativity are based on the presence or absence of the Rhesus antigen genes and their antigenic products. Although there are several different forms of these genes, the most common (and clinically the most important) is the one that encodes the *type D* Rhesus antigen. Accordingly a person whose red blood cells are found to contain the type D antigen is classified as Rh⁺, whereas a person whose red blood cells lack this antigen is rh⁻.

For simplicity, we will call any gene conferring Rhesus positivity an "Rh⁺ gene" and call any gene that does *not* confer Rhesus positivity an "rh⁻ gene." There are, therefore, three possible human genotypes (see figure 9.19); namely, Rh⁺Rh⁺ (an Rh⁺ gene inherited from each parent); Rh⁺rh⁻ (an Rh⁺ gene inherited from one parent and an rh⁻ gene inherited from the other), and rh⁻rh⁻ (rh⁻ genes inherited from both parents). So long as at least one Rh⁺ gene is inherited, the person's blood cells will contain the Rhesus antigen and be classified as Rhesus positive. (The **dominance** of the Rh⁺ gene over the rh⁻ gene is the reason that an upper case "R" is used in "Rh⁺" and a lower case "r" in "rh⁻"). In the United States, about 85% of all

Caucasians, 95% of all African-Americans and 99% of all Asian Americans are Rhesus positive. If no Rh⁺ gene is present, the person is Rhesus negative.

As discussed in chapter 9, the Rhesus factor takes on special significance during pregnancies in which the mother is rh⁻ and the developing fetus is Rh⁺. Under these circumstances, the mother may become sensitized to Rhesus antigens that enter her bloodstream from the fetus either during the pregnancy (rare) or during parturition (more likely). The mother's immune response results in the production of antibodies against Rhesus antigens. These antibodies may lead to the destruction of fetal blood cells either in the current pregnancy or in a similar ensuing pregnancy. The damage done to the fetal circulation, called **Erythroblastosis fetalis** (or **hemolytic disease of the newborn [HDN]**), can be fatal. Knowing one's Rhesus blood type is, therefore, of importance.

To illustrate the inheritance of the Rhesus factor, let's consider the various possibilities among the children of a female who is rh⁻ and a male who is Rh⁺. Because the female is rh⁻, we automatically know that her genotype is rh⁻rh⁻). However, the Rh⁺ male may be either homozygous (i.e., Rh⁺Rh⁺) or heterozygous (i.e., Rh⁺rh⁻). For purposes of this illustration, we'll suppose that he is homozygous. The Punnett square of figure 15.20 shows the results. Because all of the eggs produced by the female will contain the rh⁻ gene, and because all of the sperm produced by this male will contain the Rh⁺ gene, this means that the genotypes of all of the children will be Rh⁺rh⁻. Therefore, all of the children will have Rhesus positive blood and each pregnancy should be monitored for the possibility of Erythroblastosis fetalis. (Note also that all of the children of these parents

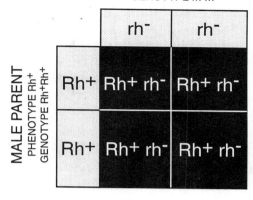

Figure 15.20
A Punnett square showing the genotypes that are possible among the zygotes that are formed by parents with Rhesus positive (father) and Rhesus negative (mother) blood.

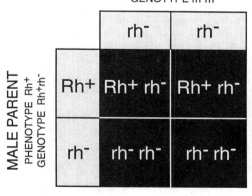

Figure 15.21
A Punnett square showing the genotypes that are possible among the zygotes that are formed by a father who is heterozygous Rhesus positive and a mother who is Rhesus negative .

419

will have genotypes that are different than the parents.)

As shown in the Punnett square of figure 15.21, if the male parent is heterozygous, then (on a statistical basis) only half of the children would have Rhesus positive blood (the other half would have Rhesus negative blood).

| 15-12 |

DETERMINATION OF SEX AND SEX-LINKED INHERITANCE

Sex Chromosomes

The two chromosomes that constitute an homologous pair are readily identified within the metaphase nucleus of a cell, because these chromosomes are the same size and shape. However, in the diploid cells of normal males, the homologous partners of only 22 of the 23 different chromosomes can be identified on the basis of their similar shape and size. That is, there are two chromosomes left over after you have matched up the other 22 pairs of homologues on the basis of their physical appearances. It turns out that the two "odd" chromosomes are the **sex chromosomes**; that is, they are the chromosomes containing genes that contribute to the determination of the individual's **genetic sex** (i.e., the sex as defined by the chromosome content of the cells). The other 22 pairs of chromosomes are called **autosomes** (or autosomal chromosomes).

In males, one of the two sex chromosomes is much larger than the other and contains many more genes. This chromosome is called an **X-chromosome**. The other, smaller sex chromosome is called a **Y-chromosome**. Thus, the 46 chromosomes present in the diploid tissue cells of normal males consist of 22 pairs of autosomal chromosomes, one X-chromosome, and one Y-chromosome.

In females, the situation is different. In the diploid tissue cells of females, there are 22 pairs of autosomal chromosomes and *two* X-chromosomes. That is, normal female cells do not contain a Y-chromosome; rather, they contain two X-chromosomes. With regard to their sex chromosomes, males may be described as **XY**, whereas females are **XX**.

Sex Determination

When spermatogenesis in males produces haploid sperm, each sperm cell receives either an X-chromosome or a Y-chromosome from the diploid spermatogonium. In females, oogenesis produces haploid egg cells containing one of the two X-chromosomes originally present in the oogonium. The Punnett square shown in figure 15.22 shows that the genetic sex of the zygote formed when a normal egg cell is fertilized by a normal sperm cell depends upon whether the sperm contains a Y-chromosome or an X-chromosome. If the sperm contains a Y-chromosome, then the zygote is genetically male. This is because the zygote contains one X-chromosome (donated by the egg cell) and one Y-chromosome (donated by the sperm cell). If the sperm cell that fertilizes the egg cell contains an X-chromosome, then the zygote is female (because the zygote contains two X-chromosomes, one donated by the egg cell and one donated by the sperm cell). Because the genetic sex of the zygote depends upon which sex chromosome is present in the sperm cell, it is said that "the father determines the genetic sex of the child." Figure 15.22 also suggests that from a

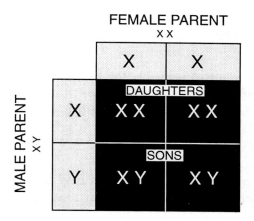

FEMALE PARENT
X X

MALE PARENT
X Y

	X	X
X	DAUGHTERS X X	X X
Y	SONS X Y	X Y

Figure 15.22
Punnett square depicting the determination of the genetic sex of the zygote. In this instance, "X" and "Y" represent whole chromosomes (i.e., the sex chromosomes), not individual genes. Note that from a statistical point of view, half of the zygotes will be male (XY) and half will be female (XX).

statistical point of view, there should be equal numbers of male and female children. That is, the odds of an egg cell being fertilized by a Y-chromosome-containing sperm cell are 50:50 (and the same odds apply to X-chromosome-containing sperm cells). Yet, in actual fact, there are slightly more male babies born than female babies.

Because male cells contain a single X-chromosome, whereas female cells contain two X-chromosomes, female cells have twice as many X-chromosome-associated genes. As a result, this might be expected to affect gene expression. For example, it might be expected that female cells would produce twice as much gene product (e.g., enzymes and other proteins). However, this does not occur. Instead, very early in embryonic development, one of the two X-chromosomes in female cells becomes inactive (i.e., its genes are not expressed). Which of the two X-chromosomes is inactivated appears to be random, so that in some cells the X-chromosome derived from the mother is the one that will be inactivated, whereas in other cells it will be the X-chromosome derived from the father that is inactivated. The upshot of this is that X-chromosome-associated genes are equally expressed in females and males. The inactivated X-chromosome in female cells undergoes *condensation* and is readily visible when the cell nucleus is examined by microscopy. The small lump of condensed chromatin is called a **Barr body**.

In the preceding discussion, care was taken to use the term "genetic sex" rather than the solitary word "sex" when referring to the gender of a zygote (or a prospective child). The reason for this is that it is possible (though relatively rare) for an XY zygote to develop into a *female* and for an XX zygote to develop into a *male*.

SRY Gene

The Y-chromosome contains a number of different genes. Among these is a single *control gene* called **SRY** (which stands for "sex-determining region of the Y-chromosome"). So long as the SRY gene is functioning properly, the hormone **testosterone** (see chapter 14) is produced. As testosterone spreads through the embryo, it promotes development into a male having normal male anatomy and function.

While it is beyond the scope of this book to explain all the ways in which chromosomal aberrations can occur, simply stated, it is possible for an XX zygote to acquire the SRY gene (i.e., the sperm cell that fertilized the egg contained an X-chromosome bearing an SRY gene). Though genetically female (i.e., XX), as embryonic development proceeds (and testosterone floods through the tissues), this embryo becomes male. Apparently, every zygote and embryo is capable of becoming male or female (i.e., possesses all the necessary genes for "maleness" and "femaleness") and which of these developmental paths is taken depends upon the presence (or absence) of the SRY gene. Another way of saying this is that all embryos are female unless they acquire the SRY gene. It is also possible for an XY zygote to have lost its SRY gene. Although genetically male, in the absence of testosterone, this embryo develops into a female.

Sex-Linked Inheritance

As already noted, the sex chromosomes contain genes whose expression guides the embryo's development into a male or female. X-chromosomes, however, also contain genes that encode characteristics unrelated to sex; Y-chromosomes lack these genes. For example, the genes that encode blood coagulation factors VIII (antihemophilic factor) and IX (Christmas factor) are located on the X-chromosome (see also chapter 8). Genes that are present on X-chromosomes but absent from Y-chromosomes are called **sex-linked genes** (about 200 sex-linked genes have been identified so far in humans). As we will see, these genes give rise to the seemingly peculiar patterns of **sex-linked inheritance**.

The fact that certain genes are present on X-chromosomes but are absent from Y-chromosomes creates certain peculiarities with regard to inheritance patterns. Consider, for example, the gene that encodes blood coagulation factor number VIII (i.e., see chapter 8). The tissue cells of females have two copies of this gene, one on each X-chromosome. However, the tissue cells of males have only one copy of the gene, because they have only one X-chromosome (remember that most genes found on X-chromosomes are not present on Y-chromosomes).

From a practical point of view, it doesn't make much difference if a person (male or female) has one copy of the factor VIII gene or two copies of the gene. Either copy ensures that normal blood coagulation factor is produced. However, consider a case in which a gene may be "defective" (by *defective* is meant that the gene's expression produces a protein that fails to function in the normal manner). Such defective genes arise through mutations that occur either spontaneously (i.e., naturally) or which are induced by chemical or physical forces (such as exposure to mutagenic chemicals, radiation, etc.). In males, a single defective sex-linked gene implies that there will be no correct product of that gene's expression. In contrast, in females, when the sex-linked gene of one X-chromosome is defective, the normal gene present on the other X-chromosome ensures that the correct protein encoded by the gene is produced. The male would exhibit the symptoms of the presence of the defective sex-linked gene, whereas the female would not.

Abnormalities that are associated with sex-linked genes are called **sex-linked diseases**. **Hemophilia A** (failure to produce adequate amounts of coagulation factor VIII) and **hemophilia B** (failure to produce adequate amounts of coagulation factor IX; again, see chapter 8) are examples of sex-linked diseases. These diseases are much more common in males than in females because they can stem from the presence of a single defective gene. The only way that a female exhibits a sex-linked disease (for example, hemophilia A) is if the genes on both of her X-chromosomes are defective.

The Punnett squares of figures 15.23, 15.24, and 15.25 illustrate patterns of inheritance of hemophilia. In these figures, the normal coagulation factor gene is represented as X^H and the defective gene by X^h. The letter Y is used to show Y-chromosomes, which, as you have learned, do not carry either the normal or the abnormal coagulation factor gene.

Figure 15.23 shows the progeny of a male who has hemophilia and a perfectly normal female. Note that the contents of each sub-square reveal both the blood coagulation characteristics of the offspring and also the offspring's sex. Because they inherit the Y-chromosome from their father (and not the defect-carrying X-chromosome), *all sons of this hemophilic father will have blood that coagulates normally*. In contrast, all daugh-

ters will inherit from their father an X-chromosome that contains the defective blood coagulation gene. Since these daughters will also inherit a normal X-chromosome from their mother, their blood will coagulate properly. These daughters are called "carriers" because they carry the defective sex-linked gene on one X-chromosome and may pass it on to one (or more) of their children.

The Punnett square of figure 15.24 illustrates the possible progeny of a normal male and a female who is a carrier of the hemophilia gene. Note that on a statistical basis, half of the sons of this normal male will be hemophiliacs and half of the daughters of this carrier mother will be normal.

In neither of the two previous illustrations does the possibility exist that a daughter will be a hemophiliac. Indeed, sex-linked diseases in females are very rare. In order to suffer from a sex-linked disease, a female must inherit a defective gene-containing X-chromosome from her mother *and* a defective gene-containing X-chromosome from her father. With regard to hemophilia, this

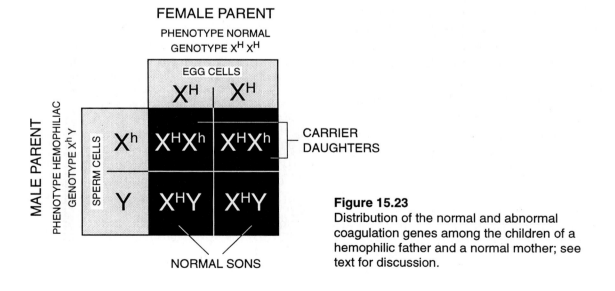

Figure 15.23
Distribution of the normal and abnormal coagulation genes among the children of a hemophilic father and a normal mother; see text for discussion.

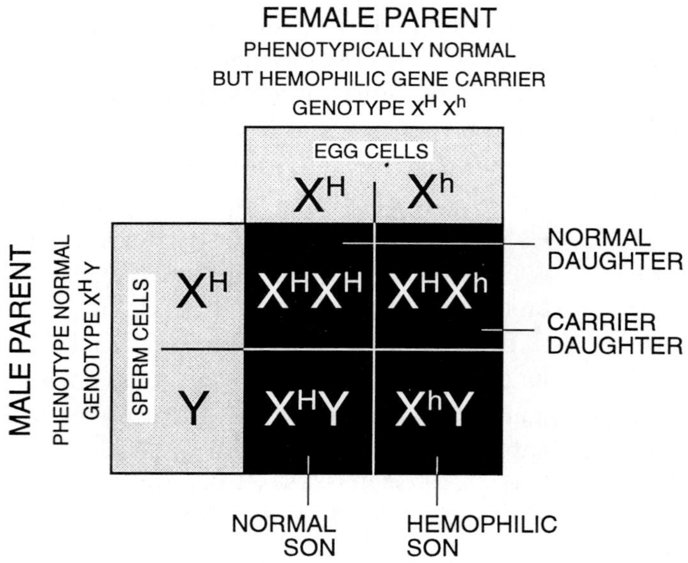

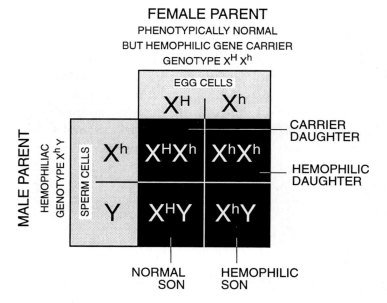

FEMALE PARENT
PHENOTYPICALLY NORMAL
BUT HEMOPHILIC GENE CARRIER
GENOTYPE $X^H X^h$

Figure 15.24
Distribution of the normal and abnormal coagulation genes among the children of a normal father and a carrier mother; see text for discussion.

FEMALE PARENT
PHENOTYPICALLY NORMAL
BUT HEMOPHILIC GENE CARRIER
GENOTYPE $X^H X^h$

Figure 15.25
Distribution of the normal and abnormal blood coagulation genes among the children of a hemophiliac male and a female who is carrying a defective sex-linked blood coagulation gene.

can occur only if (1) the father is a hemophiliac and the mother is a carrier, or (2) both the father and mother are hemophiliacs. The former possibility is illustrated in the Punnett square of figure 15.25.

Figures 15.23, 15.24, and 15.25 show the peculiar patterns of inheritance that characterize sex-linked genes. In addition to hemo-

philia, a number of other human diseases are believed to result from sex-linkage. These include (1) defective green color vision (one of several different forms of "color-blindness"), (2) juvenile muscular dystrophy, and (3) Hunter's syndrome (characterized by mental retardation and abnormal physical development).

The hemophilia gene arises spontaneously in families as a result of mutation of the normal gene. The presence of hemophilia in the royal families of Europe can be traced to its appearance in Queen Victoria of England. As is shown in the family pedigree of figure 15.26, Victoria's defective gene was not transmitted to descendents in the royal fam- ily of England but was passed through two daughters and one son to other European royal families. In the case of the royal family of Russia, the occurrence of the disease in Czarevitch Alexis, son of Czar Nicholas II and Czarina Alexandra, indirectly lead to events that changed the history of the world … but that's another story.

Figure 15.26
Pedigree showing the transmission of the sex-linked hemophilic gene through the royal families of Europe. The gene arose in Queen Victoria of England and spread through two of her daughters and one of her sons to other royal families. The path leading to Czarevitch Alexis of Russia is shaded. Victoria also had six normal children who married into other European royal families.

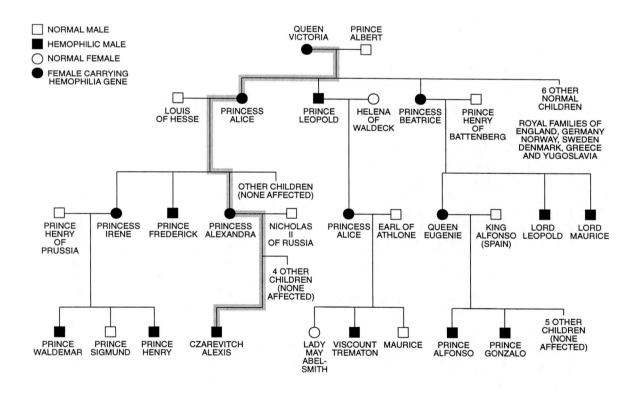

SELF TEST*

True/False Questions

1. In general, organisms that are smaller and simpler than human beings have a lower "chromosome number" than humans.

2. Daily exercise can increase the size of a muscle; exercise does this by stimulating mitosis, thereby increasing the number of muscle fibers in the muscle.

3. Homozygous genes encode identical products, whereas heterozygous genes encode different products.

4. You would not expect to find the genes that encode the protein insulin in the nuclei of cells other than those of the pancreas.

5. The G1, S, and G2 periods comprise the so-called interphase of the cell cycle.

6. In most tissues of the body, the telophase of mitosis ends with cytokinesis.

7. During anaphase I of meiosis, sister chromatids are separated from one another and enter different cells.

8. In females, meiosis does not begin in the tissues of the ovaries until the age of puberty.

9. Under normal circumstances, X-chromosomes are present only in the nuclei of the cells of females (and not normally in the cells of males).

10. Fraternal twin sisters are the result of the separate fertilizations of two different egg cells (i.e., the twins are formed from two different zygotes).

11. Hemophilia is a sex-linked disease that occurs more frequently in males than in females; this is because in females both X-chromosomes have to carry a defective coagulation factor gene in order for the disease to occur. In contrast, in males only one defective coagulation factor gene is required.

Multiple Choice Questions

1. In humans, the so-called "chromosome number" is (A) 12, (B) 23, (C) 36, (D) 46, (E) 52.

2. When the genes occupying equivalent sites on homologous chromosomes encode identical gene products, the genes are said to be (A) homozygous, (B) heterozygous, (C) polyzygous, (D) dizygotic.

3. The chromosomes of a cell are replicated during (A) mitosis, (B) cytokinesis, (C) the G1 period, (D) the G2 period, (E) the S period.

4. Which one of the following correctly lists the order of phases of mitosis? (A) metaphase, anaphase, prophase, telophase, (B) anaphase, metaphase, prophase, telophase, (C) prophase, anaphase, metaphase, telophase, (D) prophase, metaphase, anaphase, telophase.

5. Fertilization of an egg cell by a sperm cell normally takes place in (A) the uterus, (B) the oviduct, (C) the ovary, (D) the vagina, (E) the morning.

6. How long is an egg cell viable after it has

The answers to these test questions are found in Appendix III at the back of the book.

been released from an ovary at ovulation? (A) usually less than one day, (B) about 2 to 3 days, (C) 14 days, (D) 28 days, (E) egg cells are no longer viable once they have been released from an ovary.

7. By the time that a fertilized egg reaches the uterus it has been transformed into a (A) morula, (B) zygote, (C) embryo, (D) fetus.

8. Which one of the following "sperm counts" falls within the normal range? (A) 1,000 per c.c., (B) 10,000 per c.c., (C) 100,000 per c.c., (D) 120,000,000 per c.c.

9. Which one of the following primary germ layers gives rise to the brain and spinal cord? (A) ectoderm, (B) mesoderm, (C) endoderm, (D) neuroderm.

10. Identical twins occur when (A) a single egg cell is fertilized by two sperm cells, (B) two egg cells are fertilized by a single sperm cell, (C) each of two egg cells is fertilized by a separate sperm cell, (D) a single egg cell is fertilized by a single sperm cell, (E) a single egg cell separates (divides) into two cells during ovulation.

11. The body's "sex-linked" genes (A) are found only in the reproductive organs, (B) are found only on Y-chromosomes, (C) are found only on X-chromosomes.

USING THE
HYPERCARD SOFTWARE

This human physiology textbook is accompanied by a high-density 3.5 inch floppy computer diskette containing software that may be used in conjunction with the reading. The software is a "HyperCard* stack" titled "EHP© " that can be used with a Macintosh* computer ("EHP" stands for *Essentials of Human Physiology*). The EHP© stack assists in the illustration of concepts presented in the chapters using interactive demonstrations and simple animations; the stack also tests you on your understanding of the reading material. The "cards" of the EHP© stack are keyed to the text using small pictures of rectangular buttons similar to the buttons used in the stack itself. Each time that you encounter one of these buttons while reading the text, you are referred to one or more specific cards in the EHP© stack. For example, the button illustrated to the right directs you to the fourth card in the series of cards associated with the concepts that are presented in Chapter 7.

7-4

The remainder of this appendix is devoted to instructions for the installation and use of the EHP© software on your Macintosh computer. If you are an experienced Macintosh user, are familiar with the HyperCard environment, and have run HyperCard stacks before (especially stacks created with HyperCard version 2.0 or later), then do the following: (1) make a back-up copy of the EHP© diskette, and (2) transfer a copy of the EHP© stack that is now on your back-up copy to the hard disk folder containing your HyperCard software. Put the original EHP© diskette and the back-up copy away for safekeeping and work with the stack that you have transferred to your hard disk. Once the EHP© stack is opened (using HyperCard),

---

* HyperCard and Macintosh are trademarks of Apple Computer, Inc.

the stack's operation is self-explanatory.

The less experienced user of HyperCard should read on. As you do, keep the following points in mind. The expression "EHP© diskette" refers to the original diskette shipped with the textbook (or to the working copy that you are asked to create). The EHP© diskette contains two files: a "Read Me First" text file and the EHP© stack.

HARDWARE AND SOFTWARE REQUIREMENTS

Hardware Requirements

The EHP© HyperCard stack can be used only with Apple Macintosh computers containing at least 2 megabytes of random access memory (i.e., "2 Mb of RAM") and an internal hard disk drive. The presumption is made that you are familiar with the fundamentals of operation of a Macintosh computer, the computer's mouse, and the Macintosh's graphical user interface. If you are not familiar with the Macintosh computer and the meanings of such terms as "desktop," "icon," "window," "click," "double-click," "click and drag," "close box," "message box," etc., you should spend some time looking over your Macintosh computer's *user manual* before proceeding.

Software Requirements

To use the EHP© stack, your Macintosh computer's hard disk must contain either (1) the Apple Computer, Inc. version or the Claris Corporation version of HyperCard (software version 2.0 or later) or (2) the "HyperCard Player" software that is furnished with newly-purchased Macintosh computers. If your computer's hard disk does not have one of these HyperCard programs installed, it will be necessary to obtain and install one of these versions of the Hyper-Card software *before* you attempt to use the EHP© stack.

The Macintosh you are using must be running *System* version 6.0.5 (or a later version of system 6) or *System* 7.1 (or a later version of system 7), and the associated *Finder* file. If you are using System 7.1 (or later), you probably will need at least 3 (rather than 2) megabytes of RAM. To find out what System and Finder versions your computer is running, and to determine whether you have the required amount of RAM, start your computer and choose "About the Finder..." from the Apple *desk accessory menu* in the top left corner of your computer screen (if you are running a version of System 6) or "About This Macintosh..." (if you are running a version of System 7). When you do this, a window will appear on your screen identifying the start-up System and Finder versions as well as the total random access memory that is available. The EHP© stack will not operate properly if you are not using acceptable versions of the Macintosh System and Finder or do not have the required amount of RAM.

Also present on your hard disk must be a special HyperCard stack called "Home." No HyperCard stack (EHP© included) will launch properly unless the Home stack is accessible. The most trouble-free performance of HyperCard is obtained when the HyperCard program, the Home stack, and the EHP© stack are at the same hierarchical level. For example, their icons should appear together on the desktop (i.e., at the "root level") after starting your computer or should

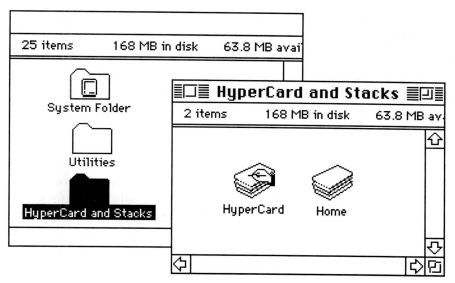

Figure A.1
The HyperCard program and the Home stack should be at the same hierarchical level on your computer's internal hard drive. In the desktop depicted in this figure, HyperCard and Home are enclosed within the same folder called "HyperCard and Stacks."

be inside of the same folder. Figure A.1 shows an arrangement in which the Hyper-Card software and Home stack are within the same folder called "HyperCard and Stacks."

THE EHP© DISKETTE

Backing Up the EHP© Diskette

Once you have determined that your computer satisfies the hardware and software requirements described above, and before you do anything else, you should make a back-up copy of your original EHP© diskette. First, examine the EHP© diskette to make sure that it is *locked*. The locking tab is a small plastic square in the upper left corner of the back of the diskette. In a locked diskette, this tab is pushed upward, so that you

can see through the square hole that is formed. If your EHP© diskette is not locked (i.e., if you can't see through the hole), slide the tab upward. So long as the EHP© diskette is locked, you cannot inadvertently erase the diskette's files or contaminate the diskette with a computer virus. Do not unlock the original diskette. After you have made a back-up copy of the original diskette, put the original diskette away for safekeeping and use the back-up copy for the installation that is described below.

The "Read Me First" File

The back-up copy (and original) of the EHP© diskette contains a file called "Read Me First." The Read Me First file contains important information about the EHP© stack–

431

information that is so recent that it could not be included in this appendix. Therefore, it is important that you read this file before proceeding further. The "Read Me First" file can be opened and read using Apple's "TeachText" utility or any compatible word processing software.

Installing the EHP© Stack on Your Hard Disk

"Boot" (i.e., turn on) your Macintosh computer, and when the hard disk icon appears on the screen, open the hard disk window by double-clicking on the hard disk icon. If the HyperCard software icon and the Home stack icon do not appear on the desktop, open the folder containing these items (both items should be in the same folder). Insert the back-up copy of your EHP© diskette into the disk drive. When the diskette's icon appears on the screen, double-click on the icon to open the diskette window (fig. A.2). Adjust the positions of the windows so that you can see the EHP© stack icon and the HyperCard and Home stack icons. Click on the EHP© icon, and while holding the mouse button down, drag the stack into the window containing the HyperCard and Home icons. The Macintosh computer will now make a copy of the EHP© stack on the computer's hard disk. Once the copy is made, close the EHP© diskette window by clicking inside the close box. The HyperCard icon, the Home icon, and the EHP© icon should now be visible on the desktop (or within the same folder, as is depicted in figure A.3). You can now eject the back-up copy of the EHP© diskette by dragging its icon to the trash can at the bottom-right of the screen. Put the back-up diskette away. If at some future time, your hard disk's copy of the EHP© stack fails to run properly (this would be quite unusual), re-install the stack using your back-up diskette.

Figure A.2
The EHP© diskette contains two files: "Read Me First" and "EHP©." The Read Me First file contains the most recent information about the software and should be read immediately. The file can be opened using Apple's "TeachText" utility or using any compatible word processing program. The EHP© file is the HyperCard stack that should be transferred to the HyperCard folder on your hard disk.

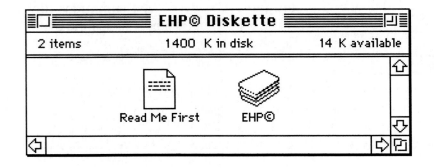

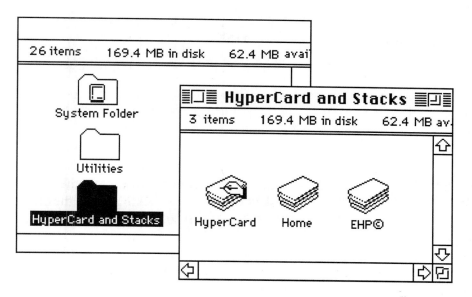

Figure A.3
If the EHP© stack has been correctly transferred to your computer's hard disk, the stack's icon will be at the same hierarchical level as the Hyper-Card program and the Home stack. In the example shown here, the HyperCard program and stacks are in a single folder.

USING THE EHP© STACK

Opening the Stack

To open the EHP© stack, double-click on the stack's icon. Depending upon the fonts that are installed in your System, you may or may not see the *message box* shown in figure A.4. Because a limited number of fonts were used in creating the EHP© stack, there should be no problem using the stack *even if your system does not have the fonts listed in the message box*. If the font message is displayed, just click inside the "OK" box and the EHP© stack will soon open to the *Title Card*. After the title card has been displayed for several seconds, the EHP© stack will proceed to the *Main Index Card* (which is depicted in figure A.5).

The Main Index Card is a home base from which you can navigate to the card sequences corresponding to any of the book's 15 chapters. For example, to move to the first of the series of cards associated with Chapter 4, you simply use the mouse to maneuver the "cursor" (a small hand with its index finger raised) until the tip of the finger is positioned inside the "Chapter 4" button, and then click the mouse. Several other parts of the EHP© stack can also be reached from the Main Index Card. For example, clicking the "Instructions" button takes you to the first of several cards explaining how the stack works and what various card buttons do. Clicking the "Exams" button takes the user to the *Exam Index Card*, from which short objec-

433

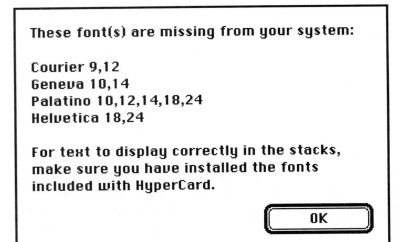

These font(s) are missing from your system:

Courier 9,12
Geneva 10,14
Palatino 10,12,14,18,24
Helvetica 18,24

For text to display correctly in the stacks, make sure you have installed the fonts included with HyperCard.

OK

Figure A.4
The font message box. Depending on the fonts installed in your system, you may or may not get this message box. If you do, simply click the OK button.

tive exams on each of the book's chapters can be reached.

Clicking the "Sound Control" button allows the user to turn off the stack's sound or to adjust the sound to the desired loudness. Clicking the "Quick-Find" button takes the user to a card from which individual cards of the stack may be accessed directly. Clicking

the "Credits" button takes the user to a card containing credits for the *Essentials of Human Physiology* textbook and the EHP© stack, as well as other information.

Finally, at the bottom of the Main Index Card are two buttons that (1) access Hyper-Card's "Home" stack (described below) and (2) allow the user to quit the EHP© stack (i.e.,

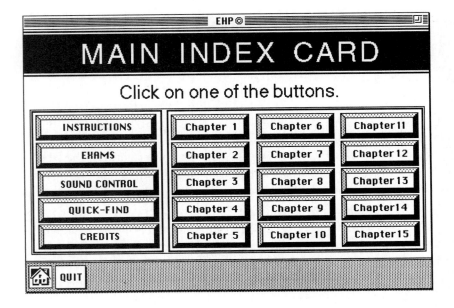

EHP©

MAIN INDEX CARD

Click on one of the buttons.

INSTRUCTIONS	Chapter 1	Chapter 6	Chapter 11
EXAMS	Chapter 2	Chapter 7	Chapter 12
SOUND CONTROL	Chapter 3	Chapter 8	Chapter 13
QUICK-FIND	Chapter 4	Chapter 9	Chapter 14
CREDITS	Chapter 5	Chapter 10	Chapter 15

QUIT

Figure A.5
The Main Index Card. The principal function of this card is to allow the user to navigate to specific chapters. Also accessed from this card are stack instructions, chapter exams, the sound control card, the "quick-find" card, and credits.

end the current session). You can explore the actions of the "Instructions," "Exams," "Sound Control," "Quick-Find," and "Credits" buttons later. For now, click the "Chapter 5" button so that you are taken to the first card in chapter 5. (Notice that when you click on a button, it "flashes;" that is, while the mouse button is depressed, the black pixels that comprise the button temporarily become white, and the white pixels temporarily become black. This is done to reassure you that you have, in fact, activated the button.)

Buttons of the Button Bar

Card 5-1 is shown in figure A.6 and illustrates some rather general features of the chapter cards. The expression "button bar" refers to the row of buttons at the base of each card. Not all buttons appear in the button bar of every card, but a number of the buttons are common to most cards. What follows is an

explanation of the functions of the buttons in the button bar.

When the "Quit" button (figure A.6, also shown below) is clicked, the HyperCard

program is closed and the user is returned to the Macintosh desktop. Click the Quit button when you want to end a session. When the Quit button is used, all cards in the stack are returned to their original status, so that the next time you use the stack, it is as though you are using it for the first time.

When you click near the center of the "Index" button (shown below), you are taken

to the *Main Index Card* (i.e., figure A.5). However, if you click on the right edge of the

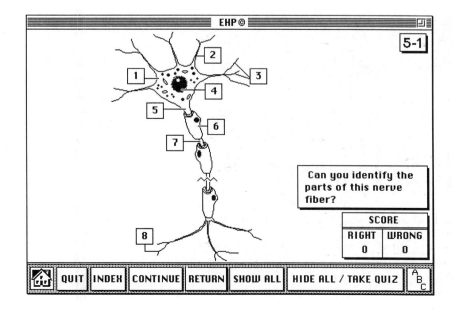

Figure A.6
Card 5-1 in the EHP© stack.

Index button, you are taken instead to the *Quick-Find* card, which gives you immediate access to any individual card in the stack.

The "Continue" button advances you to the next card in the stack. If the card shown

CONTINUE

is the *last* card of the chapter, then clicking the Continue button advances you to the *first* card of the next chapter.

When you click the "Return" button, you are returned to the previous card in the chap-

RETURN

ter. If you already are on the first card of the chapter, then you will be taken to the last card of the previous chapter.

Many of the stack's cards contain anatomical structures whose parts you are challenged to identify. Clicking the "Show All" button simultaneously reveals the names

SHOW ALL

of all of the anatomical parts in question.

The "Hide All/Take Quiz" button conceals the names of anatomical structures depicted on the card (if they already are displayed). In

HIDE ALL/TAKE QUIZ

this way, you can test yourself on the names of the indicated structures. (Clicking "Show All" reveals all of the names simultaneously.) On some cards, testing takes the form of a multiple choice quiz in which you can select

among several alternative choices. Clicking the "Hide All/Take Quiz" button resets the card so that the multiple choice quiz may be taken.

To take a multiple choice quiz (such as the one depicted in figure A.6), you click on the numbered parts of the figure. When you do this, you are presented with a number of choices. Click on the answer of your choice. If you make the correct choice, a chime will be heard and you will receive credit in the "Score Field" (seen in figure A.6 and below).

SCORE	
RIGHT	WRONG
0	0

If you make the wrong choice, you will be told so in no uncertain terms. Once a choice is made, the correct answer flashes five times. When taking a multiple choice quiz, the numbered questions may be answered in any order (i.e., you don't have to begin with item 1 and end with 8; see figure A.6).

Each time that you take a multiple choice quiz on a quiz card, your score is recorded in the the card's Score Field. If, during a session with the stack, you return to a card after having previously taken the card's quiz, then clicking on the "ABC" button will reveal

your earlier score. However, clicking the "Hide All/Take Quiz" button before clicking the ABC button automatically erases any previous score earned on a card. Similarly, an earlier score cannot be recalled if you have since closed and then reopened the EHP© stack.

The button at the far left of the button bar is a "Home" button. This button is provided

for the experienced HyperCard user who may have reason to access HyperCard's "home stack." Under most circumstances, clicking on this button is neither warranted nor needed.

Other buttons will occasionally appear in the button bar. Some of these (e.g., "Animate" [Card 4-4], "Apply Stimulus" [Card 5-2], "Measure Blood Pressure" [Card 7-5], and "Show Stages" [Card 8-5]) activate

simple animations that help explain a particular concept. Occasionally, buttons appear above the button bar (e.g., "Show Isotonic Contraction" [Card 4-6] and "Show Plasma Formation" [Card 8-1]). Clicking these buttons may also initiate a simple animation sequence or may reveal additional card information.

Do not be reluctant to explore the effects of clicking the card buttons. Remember that you cannot do permanent harm to the stack by clicking the mouse. Any changes that are made to the stack during its use are eliminated when you end your session with the stack. By using the EHP© software in conjunction with reading the text, you may have an easier time grasping the concepts and learning the terminology of this science.

USING THE
EHP© FOR WINDOWS
SOFTWARE

This human physiology textbook is accompanied by a high-density 3.5 inch floppy computer diskette containing software that may be used in conjunction with the reading. The software is a Windows* application titled "EHP© " that can be used with an IBM compatible computer ("EHP" stands for *Essentials of Human Physiology*). The EHP© program assists in the illustration of concepts presented in the chapters using interactive demonstrations and simple animations; the program also tests you on your understanding of the reading material.

The "units" of the EHP© program are keyed to the text using small labels similar to the labels used in the program itself. Each time that you encounter one of these labels while reading the text, you are referred to one or more specific units in the EHP© program.

For example, the label illustrated to the right directs you to the fourth unit in the series of units associated with the concepts that are presented in Chapter 7.

| 7-4 |

The remainder of this appendix is devoted to instructions for the installation and use of the EHP© software with Windows. If you are an experienced Windows user, then do the following: (1) make a back-up copy of the EHP© diskette, and (2) run the *setup.exe* program. Once the *setup.exe* program has been run, put the original EHP© diskette and your back-up copy away for safekeeping. You can now work with the *ehp.exe* program that has been created in directory *c:\ehp* on your hard disk. Once the *ehp* program is running, the program's operation is self-explanatory.

The less experienced user of Windows should read on. As you do, keep the following points in mind. The expression "EHP© diskette" refers to the original diskette shipped

---

* Windows is a trademark of Microsoft Corporation.

with the textbook (or to the working copy that you are asked to create). The EHP© diskette contains the *setup.exe* program, together with its associated files.

COMPUTER REQUIREMENTS

The EHP© for Windows software can be used with any IBM-compatible system running Windows 3.1 (or a later version) and equipped with a 3.5 inch high-density disk drive, 4 megabytes of random access memory (i.e., "4 Mb of RAM"), a hard disk drive with at least 2 Mb of free space, a VGA monitor, and a mouse. The presumption is made that you are familiar with the fundamentals of operation of Windows. If you are not, you should spend some time looking over your computer's *user manual* before proceeding.

THE EHP© DISKETTE

Backing Up the EHP© Diskette

Once you have determined that your computer satisfies the hardware and software requirements described above, and before you do anything else, you should make a back-up copy of your original EHP© diskette. First, examine the EHP© diskette to make sure that it is *locked*. The locking tab is a small plastic square in the upper left corner of the back of the diskette. In a locked diskette, this tab is pushed upward, so that you can see through the square hole that is formed. If your EHP© diskette is not locked (i.e., if you can't see through the hole), slide the tab upward. So long as the EHP© diskette is locked, you cannot inadvertently erase the diskette's files or contaminate the diskette

with a computer virus. Do not unlock the original diskette. After you have made a back-up copy of the original diskette, put the original diskette away for safekeeping and use the back-up copy for the installation that is described below.

Installing the EHP© Software on Your Hard Disk

Turn on your computer, and when Windows is running, insert the back-up copy of your EHP© diskette into the disk drive. Under the File menu for the Windows Program Manager, click on "Run...". You will then be prompted for the name of the file to run. Type *a:\setup.exe* (if your 3.5 inch disk drive is designated by a letter other than "*a*," then use that letter in the prefix of your file name).

The *setup* program will create a new directory (*ehp*) on your hard drive and transfer several files into that directory. The *setup* program will also transfer several files to your *c:\Windows\System directory*.

If you already happen to have a directory named *ehp* on your hard disk, the *setup* program will transfer files to that directory *unless* you rename it beforehand. The files for the EHP© software *must* be in a directory named *ehp*.

USING THE EHP© PROGRAM

Starting the Program

To start the EHP© program, double-click on the program's icon. After several seconds, a title page will appear. When you are ready to begin, click on the Start button displayed on this page. The EHP© program will now proceed to the *Main Index* (which is depicted in figure A.7).

The Main Index is a home base from which you can navigate to the unit sequences corresponding to any of the book's 15 chapters. For example, to move to the first of the series of units associated with Chapter 4, you simply use the mouse to maneuver the "cursor" (a bold arrow) until the tip of the arrow is positioned inside the "Chapter 4" button, and then click the mouse. Several other parts of the EHP© software can also be reached from the Main Index. For example, clicking the "Instructions" button takes you to the first of several pages explaining how the program works and what various buttons do. Clicking the "Exams" button takes the user to the *Exam Index*, from which short objective exams on each of the book's chapters can be reached.

Clicking the "About Sound..." button provides the user with information about using sound with the EHP© program. Clicking the "Quick-Find" button takes the user to a page containing a menu from which individual units of the program may be accessed directly. Clicking the "Credits" button takes the user to a page containing credits for the *Essentials of Human Physiology* textbook and the EHP© program, as well as other information.

Finally, at the bottom of the Main Index are two buttons that (1) allow the user to quit the EHP© program (i.e., end the current session) or (2) allow the user to go to an individual program unit directly. (On the Main Index page, this button has the same function as the Quick-Find button, but the latter button will not appear at other times.). You can explore the actions of the "Instructions," "Exams," "About Sound...," "Quick-Find," and "Credits" buttons later. For now, click the "Chapter 5" button so that you are taken to the first unit in chapter 5. (Notice that when you click on a button, it appears to

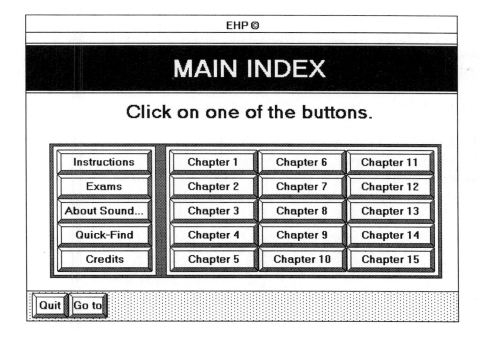

Figure A.7
The Main Index. The principal function of this display is to allow the user to navigate to specific chapters. Also accessed from this page are program instructions, chapter exams, sound control information, the "quick-find" menu, and credits.

be depressed. This is done to reassure you that you have, in fact, activated the button.)

Buttons of the Button Bar

Unit 5-1 is shown in figure A.8 and illustrates some rather general features of the chapter units. The expression "button bar" refers to the row of buttons at the base of each page. Not all buttons appear in the button bar of every unit, but a number of the buttons are common to most units. What follows is an explanation of the functions of the buttons in the button bar.

When the "Quit" button (figure A.8, also shown below) is clicked, the EHP© program

is terminated and the user is returned to Windows. Click the Quit button when you want to end a session. When the Quit button is used, all units in the program are returned

to their original status, so that the next time you use the program, it is as though you are using it for the first time.

When you click the "Index" button (shown below), you are taken to the *Main Index* (i.e.,

figure A.7). If you click the "Go To" button, you are taken instead to the *Quick-Find* menu, which gives you immediate access to any individual unit.

The "Next" button advances you to the next unit in the chapter. If the current unit is the *last* unit of the chapter, then clicking the Next button advances you to the *first* unit of the next chapter.

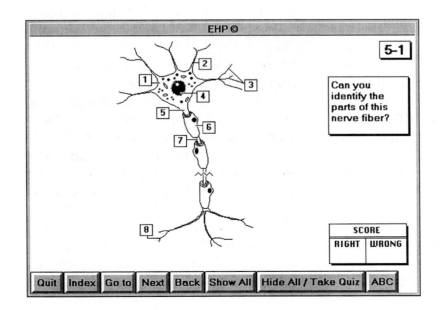

Figure A.8
Unit 5-1 in the EHP©
program.

When you click the "Back" button, you are returned to the previous unit in the chapter. If

you already are on the first unit of the chapter, then you will be taken to the last unit of the previous chapter.

Many of the program's units contain anatomical structures whose parts you are challenged to identify. Clicking the "Show All" button simultaneously reveals the names of all of the anatomical parts in question.

The "Hide All/Take Quiz" button serves two purposes. Clicking this button will con-

ceal the names of the anatomical structures depicted. In this way, you can test yourself on the names of the indicated structures. (Clicking "Show All" reveals all of the names simultaneously.) In some units, testing takes the form of a multiple choice quiz in which you can select among several alternative choices. Clicking the "Hide All/Take Quiz" button also resets the unit so that the multiple choice quiz may be taken.

To take a multiple choice quiz (such as the one depicted in figure A.8), you click on the numbered parts of the figure. When you do this, you are presented with a number of choices. Click on the answer of your choice. If you make the correct choice, you will receive credit in the "Score Field" (fig. A.8).

If you make the wrong choice, you will be so informed. Whatever your choice, the correct answer will be highlighted in bold face type, and the program will not proceed until you click the *OK* button. When taking a multiple choice quiz, the numbered questions may be answered in any order (i.e., you don't have to begin with item 1 and end with 8; see figure A.8).

Each time that you take a quiz, your score

SCORE	
RIGHT	WRONG
0	0

is recorded in the Score Field. If you return to a unit after having previously taken the unit's quiz, then clicking on the "ABC" but-

ton will reveal your earlier score. However, clicking the "Hide All/Take Quiz" button before clicking the ABC button automatically erases any previous score earned on a unit. Similarly, an earlier score cannot be recalled if you have since quit and then restarted the EHP© program .

Other buttons will occasionally appear in the button bar. Some of these (e.g., "Animate" [Unit 4-4], "Apply Stimulus" [Unit 5-2], "Measure Blood Pressure" [Unit 7-5], and "Show Stages" [Unit 8-5]) activate simple animations that help explain a particular concept. Occasionally, buttons appear above the button bar (e.g., "Show Isotonic Contraction" [Unit 4-6] and "Show Plasma Formation" [Unit 8-1]). Clicking these buttons may also initiate a simple animation sequence or may reveal additional information.

Do not be reluctant to explore the effects of clicking the buttons. Remember that you cannot do permanent harm to the program by clicking the mouse. Any changes that are made to the program during its use are eliminated when you end your session. By using the EHP © software in conjunction with reading the text, you may have an easier time grasping the concepts and learning the terminology of this science.

ANSWERS TO SELF TEST QUESTIONS

CHAPTER **1**

True/False Questions

1. True
2. True
3. False
4. False
5. True

Multiple Choice Questions

1. B
2. B
3. C
4. C
5. C

CHAPTER **2**

True/False Questions

1. True
2. False
3. False
4. True
5. True
6. True
7. True

Multiple Choice Questions

1. C
2. D
3. C
4. D

5. D
6. A
7. B
8. C

CHAPTER **3**

True/False Questions

1. True
2. True
3. True
4. False
5. True

Multiple Choice Questions

1. D
2. C
3. C
4. B
5. B
6. A

CHAPTER **4**

True/False Questions

1. False
2. True
3. True
4. True
5. True

Multiple Choice Questions

1. A
2. C
3. A
4. E
5. B
6. B
7. B
8. B
9. C

CHAPTER **5**

True/False Questions

1. False
2. True
3. True
4. True
5. True
6. True

Multiple Choice Questions

1. B
2. A
3. B
4. A
5. C
6. B
7. C

CHAPTER **6**

True/False Questions

1. True
2. False
3. False
4. False
5. False
6. False
7. False
8. True

Multiple Choice Questions

1. C
2. C
3. A
4. B
5. A
6. D
7. C
8. E
9. A
10. C
11. D

CHAPTER **7**

True/False Questions

1. False
2. True
3. True
4. True
5. False
6. True
7. True
8. False
9. True
10. False

Multiple Choice Questions

1. B
2. D
3. C
4. A
5. C
6. C
7. B
8. C
9. A
10. C
11. E

CHAPTER **8**

True/False Questions

1. True
2. False
3. False
4. True
5. False
6. True

Multiple Choice Questions

1. C
2. D
3. C
4. B
5. E
6. D
7. D

CHAPTER **9**

True/False Questions

1. True
2. False
3. False
4. True
5. True
6. False
7. False
8. True
9. False
10. True

Multiple Choice Questions

1. B
2. C
3. B
4. C
5. D
6. C
7. B
8. C
9. D
10. C

CHAPTER **10**

True/False Questions'

1. True
2. False
3. True
4. False

5. False
6. True

Multiple Choice Questions

1. D
2. D
3. D
4. D
5. B

CHAPTER **11**

True/False Questions

1. False
2. True
3. True
4. False
5. True
6. True
7. False
8. False
9. True
10. False
11. False
12. True

Multiple Choice Questions

1. D
2. D
3. A
4. B
5. E
6. B
7. C
8. A
9. D
10. B

CHAPTER **12**

True/False Questions

1. True
2. True
3. False
4. True
5. False
6. True
7. True
8. True
9. True
10. False

Multiple Choice Questions

1   C.
2. A
3. D
4. E
5. A
6. D
7. C
8. D

CHAPTER 13

True/False Questions

1. False
2. True
3. False
4. False
5. True
6. False
7. True
8. False
9. True
10. True

Multiple Choice Questions

1. C
2. D
3. B
4. A

5. E
6. B
7. B
8. D
9. C

CHAPTER 14

True/False Questions

1. True
2. False
3. True
4. True
5. True
6. False
7. True
8. False
9. True
10. False

Multiple Choice Questions

1. E
2. C
3. B
4. D
5. D
6. A
7. C
8. A
9. C

10. A

CHAPTER 15

True/False Questions

1. False
2. False
3. True
4. False
5. True
6. True
7. False
8. False
9. False
10. True
11. True

Multiple Choice Questions

1. D
2. A
3. E
4. D
5. B
6. A
7. A
8. D
9. A
10. D
11. C

INDEX

INDEX

U

ulcers, 280
ulna, 55
umbilical arteries, 414
umbilical cord, 413
umbilical vein, 414
unipolar neurons, 84
universal donor (of blood), 245
universal recipient (of blood), 245
unsaturated fatty acids, 33, 317
uracil, 30ff
urea, 351
ureters, 336
urethra, 336-337, 353, 383, 408
uridine diphosphoglucose, 305
urinary bladder, 336, 408
urine, 345ff
urobilinogen, 285
uterus, 407
utricle, 154
uvula, 253

V

vagal inhibition, 176
vagina, 407
valence, 14-15
valine, 27

valves of Kerckring, 287
vas deferens, 383, 408
vasomotor center, 102
vasopressin, 345, 368
veins, 163, 166
ventral roots, 108ff
ventricles (of the brain), 98
venules, 162
vertebral column, 103
very-low-density lipoproteins, 293
vestibular apparatus, 154ff
vestibular membrane, 150, 152
vestibulo-cochlear nerve, 149, 151ff
villi, 288
viral nucleic acids, 50
viruses, 48-49
vision, 132ff
visual association area, 98
visual cortex, 98, 143
vital capacity, 267
vitamin A, 140, 322
vitamin B1, 309, 321
vitamin B2, 321
vitamin B3, 321
vitamin B5, 321
vitamin B6, 321
vitamin B12, 282, 296, 321
vitamin C, 322
vitamin D, 322
vitamin E, 323

vitamin H, 322
vitamin K, 296, 323
vitamin M, 322
vitamins, 33, 35, 320ff, 328-329
vitreous humor, 136
vocal cords, 253
voluntary muscle, 54
von Willebrand factor, 210

W

water, 14, 17, 18, 20ff, 296, 329ff
wavelength, 133
white fibers (white muscle), 53, 77
white matter, 94, 103

Y

yolk sac, 412, 413

Z

Z-disk, 57
Z-line 57ff
zona pellucida, 390
zonules of Zinn, 137
zygote, 411
zymogens, 281